Operator Theory
Advances and Applications
Vol. 89

Editor
I. Gohberg

On Spectral Theory of Elliptic Operators

Yuri Egorov
Vladimir Kondratiev

Birkhäuser Verlag
Basel · Boston · Berlin

Authors' addresses:

Yuri Egorov
UPS, UFR MIG
118, route de Narbonne
31062 Toulouse
France

Vladimir Kondratiev
Moscow State University
Faculty of Mathematics and Mechanics
119 899 Moscow
Russia

1991 Mathematics Subject Classification 35PXX, 35J10

A CIP catalogue record for this book is available from the Library of Congress, Washington D.C., USA

Deutsche Bibliothek Cataloging-in-Publication Data
Egorov, Yuri:
On spectral theory of elliptic operators / Yuri Egorov ;
Vladimir Kondratiev. – Basel ; Boston ; Berlin : Birkhäuser,
1996
(Operator theory ; Vol. 89)
ISBN-13: 978-3-0348-9875-1 e-ISBN-13: 978-3-0348-9029-8
DOI: 10.1007/978-3-0348-9029-8

NE: Kondratiev, Vladimir:; GT

Softcover reprint of the hardcover 1st edition 1996
Printed on acid-free paper produced from chlorine-free pulp. TCF ∞
Cover design: Heinz Hiltbrunner, Basel

Contents

Preface IX

1 Hilbert Spaces
1.1 Definition and basic properties 1
1.2 Examples 3
1.3 Orthonormal base 5
1.4 Fourier series 6
1.5 Subspaces, orthogonal sums 8
1.6 Linear functionals 9
1.7 Weak convergence 11
1.8 Linear operators 15
1.9 Adjoint operators 16
1.10 The spectrum of an operator 17
1.11 Compact operators 18
1.12 Compact self-adjoint operators 19
1.13 Integral operators 21
1.14 The Lax-Milgram theorem 23

2 Functional Spaces
2.1 Notation and definitions 25
2.2 Lebesgue integral 27
2.3 Level sets of functions of a real variable 29
2.4 Symmetrization 32
2.5 The space $L_1(\Omega)$ 35
2.6 The space $L_2(\Omega)$ 36
2.7 The space $L_p(\Omega)$, $p > 1$ 38
2.8 Density of the set of continuous functions in $L_1(\Omega)$ 40
2.9 Density of the set of continuous functions in $L_p(\Omega), p > 1$ 41
2.10 Separability of $L_p(\Omega)$ 43
2.11 Global continuity of functions of $L_p(\Omega)$ 44
2.12 Averaging 45
2.13 Compactness of a subset in $L_p(\Omega)$ 46
2.14 Fourier transform 47

2.15 The spaces $W_p^m(\Omega)$ . . . 50
2.16 The averaging and generalized derivatives . . . 54
2.17 Continuation of functions of $W_p^m(\Omega)$ . . . 56
2.18 The Sobolev integral representation . . . 58
2.19 The space $W_p^1(\Omega, E)$ . . . 58
2.20 Properties of the space $W_{p,0}^m(\Omega)$ . . . 59
2.21 Sobolev's embedding theorems . . . 60
2.22 Poincaré's inequality . . . 61
2.23 Interpolation inequalities . . . 68
2.24 Compactness of the embedding . . . 70
2.25 Invariance of $W_p^m(\Omega)$ under change of variables . . . 72
2.26 The spaces $W_p^m(\Omega)$ for a smooth domain Ω . . . 73
2.27 The traces of functions of $W_p^1(\Omega)$ . . . 75
2.28 The space H^s . . . 78
2.29 The traces of functions of $W_2^k(\mathbf{R}^n)$ . . . 83
2.30 The Hardy inequalities . . . 89
2.31 The Morrey embedding theorem . . . 105

3 Elliptic Operators
3.1 Strongly elliptic equations . . . 109
3.2 Elliptic equations . . . 115
3.3 Regularity of solutions . . . 120
3.4 Boundary problems for elliptic equations . . . 124
3.5 Smoothness of solutions up to boundary . . . 130

4 Spectral Properties of Elliptic Operators
4.1 Variational principle . . . 133
4.2 The spectrum of a self-adjoint operator . . . 139
4.3 The Friedrichs extension . . . 143
4.4 Examples of linear unbounded operators . . . 144
4.5 Self-adjointness of the Schrödinger operator . . . 148

5 The Sturm-Liouville Problem
5.1 Elementary properties . . . 153
5.2 On the first eigenvalue of a Sturm-Liouville problem . . . 170
5.3 On other estimates of the first eigenvalue . . . 176
5.4 On a more general estimate of the first eigenvalue of the Sturm-Liouville operator . . . 194
5.5 On estimates of all eigenvalues . . . 202

6 Differential Operators of Any Order
6.1 Oscillation of solutions of an equation of any order . . . 207
6.2 On estimates of the first eigenvalue for operators of higher order . . . 213

6.3 Introduction to a Lagrange problem 217
6.4 Preliminary estimates 220
6.5 Precise results 225

7 Eigenfunctions of Elliptic Operators in Bounded Domains
7.1 On the Dirichlet problem for strongly elliptic equations 233
7.2 Estimates of eigenfunctions of strongly elliptic operators 239
7.3 Equations of second order 242
7.4 Estimates of eigenfunctions of operator pencils 246
7.5 The method of stationary phase 248
7.6 Asymptotics of a fundamental solution of an elliptic operator with constant coefficients 253
7.7 Estimates of the eigenfunctions of an elliptic operator with constant coefficients 257
7.8 Estimates of the first eigenvalue of an elliptic operator in a multi-connected domain 265
7.9 Estimates of the first eigenvalue of the Schrödinger operator in a bounded domain 271

8 Negative Spectra of Elliptic Operators
8.1 Introduction 275
8.2 One-dimensional case 278
8.3 Some inequalities and embedding theorems 287
8.4 Estimates of the number of points of the negative spectrum 294
8.5 Some generalizations 298
8.6 Lower estimates for the number N 303
8.7 Other results 305
8.8 On moments of negative eigenvalues of an elliptic operator 309

Bibliography 319

Index 327

Preface

It is well known that a wealth of problems of different nature, applied as well as purely theoretic, can be reduced to the study of elliptic equations and their eigen-values. During the years many books and articles have been published on this topic, considering spectral properties of elliptic differential operators from different points of view. This is one more book on these properties.

This book is devoted to the study of some classical problems of the spectral theory of elliptic differential equations. The reader will find hardly any intersections with the books of Shubin [Sh] or Rempel-Schulze [ReSch] or with the works cited there. This book also has no general information in common with the books by Egorov and Shubin [EgShu], which also deal with spectral properties of elliptic operators. There is nothing here on oblique derivative problems; the reader will meet no pseudodifferential operators.

The main subject of the book is the estimates of eigenvalues, especially of the first one, and of eigenfunctions of elliptic operators. The considered problems have in common the approach consisting of the application of the variational principle and some a priori estimates, usually in Sobolev spaces. In many cases, important for physics and mechanics, as well as for geometry and analysis, this rather elementary approach allows one to obtain sharp results.

The first half of the book, namely Chapters 1 through 4 and Section 5.1, have an educational character. Here the reader will find the basic theory of Hilbert spaces and Sobolev spaces and the classical theory of elliptic equations, serving as a background for the remaining Chapters. We have tried to present the material as suitably as possible for senior university students.

The last four chapters contain primarily the authors' results, most of which were obtained during the past two decades.

In Chapter 5 we consider the dependence of eigenvalues, in particular the first one, of the Sturm-Liouville problem for the equation $(P(x)y')' + \lambda Q(x)y = 0$ on the coefficients P and Q. The estimates of λ are obtained in terms of the integrals $\int P(x)^{\alpha}\,dx$ and $\int Q(x)^{\beta}\,dx$, where α and β are arbitrary reals, and the exact solutions are found.

In Chapter 6 a similar problem is considered for a differential equation of order m. The results on oscillation of solutions are presented in Section 6.1. In Section 6.3 a classical Lagrange problem on the optimal form of a column is studied which is reduced to the estimates of the first eigenvalue λ for the fourth order equation $(P(x)y'')'' - \lambda y'' = 0$ with the restriction $\int P(x)^{\alpha} dx = 1$. We prove the existence of an optimal solution and obtain precise results for this problem. The most important case for applications, case $\alpha = 1/2$, was studied by Keller and Tadjbakhsh [KeTa]. Recently some authors have attempted to refute their answer. We have proved that it is correct, and give a rigorous proof of the existence of optimal solutions in the more general case.

Chapter 7 deals with solvability results for the first boundary problem for strongly elliptic equations depending on the volume of a domain or its inner radius. We also consider here estimates of eigenfunctions for general operator pencils. In Section 7.7 estimates of eigenfunctions are obtained for elliptic equations with constant coefficients satisfying a Vainberg-Grushin condition (see [Va]). These results generalize some theorems of Eidus and Shishmarev. In Section 7.8 the estimates of the first eigenvalue are obtained depending on the inner diameter of the domain and its connectivity. These results generalize some theorems of Osserman and Taylor.

Chapter 8 is devoted to a study of the negative spectrum of an elliptic operator of the Schrödinger type $L = L_0 - V(x)$ depending on V. Here the Rosenblyum theorem and some of its generalizations are proved, and estimates of sums of $\sum(-\lambda_j)^{\gamma}$ are considered for negative eigenvalues.

There are many important problems of the same type that were not considered in this book. We have presented a number of examples; each has its own interest and serves to illustrate some useful methods. We hope that the material is interesting for specialists in pure as well as applied mathematics and can be used in the students' work.

Chapter 1

Hilbert Spaces

1.1 Definition and basic properties

A *linear space* is a nonempty set L further defined by the following data:
A. To any pair of elements $x \in L$, $y \in L$ there corresponds an element z called their *sum* and noted by $x + y$ such that

1. $x + y = y + x$;
2. $(x + y) + z = x + (y + z)$;
3. There is an element 0 such that $x + 0 = x$ for all $x \in L$;
4. For any $x \in L$ there is an element x_1 such that $x + x_1 = 0$.

B. To any element $x \in L$ and each $\lambda \in \mathbf{C}$ there corresponds an element $\lambda \cdot x \in L$ such that for any $x, y \in L$, $\alpha, \beta \in \mathbf{C}$:

1. $\alpha(x + y) = \alpha x + \alpha y$;
2. $(\alpha + \beta)x = \alpha x + \beta x$;
3. $(\alpha\beta)x = \alpha(\beta x)$;
4. $1 \cdot x = x$.

The first four properties mean that L is a commutative group with *addition* as the group operation.

A linear space is *pre-Hilbert* if an *inner product* is defined on it, i.e. to each pair of elements $x \in L$, $y \in L$ there corresponds a complex number (x, y) such that

A. $(x, y) = \overline{(y, x)}$;
B. $(\alpha x_1 + \beta x_2, y) = \alpha(x_1, y) + \beta(x_2, y)$;
C. $(x, x) > 0$ if $x \neq 0$.

Theorem 1. *The inner product satisfies the following conditions:*

(a) $(\lambda x, \lambda x) = |\lambda|^2 (x, x)$;
(b) $|(x, y)| \leq |(x, x)(y, y)|^{1/2}$ *(Cauchy-Bounyakovsky inequality);*
(c) $(x + y, x + y)^{1/2} \leq (x, x)^{1/2} + (y, y)^{1/2}$ *(triangle inequality).*

The inequalities in (b) *and* (c) *become equalities if and only if* $x = \mu y$ *for some complex* μ.

Proof. Indeed, (a) follows from the equalities:

$$(\lambda x, \lambda x) = \lambda(x, \lambda x) = \lambda\overline{(\lambda x, x)} = \lambda\bar{\lambda}(x, x).$$

To prove the Cauchy-Bounyakovsky inequality observe that for any real t

$$0 \leq (x + ty, x + ty) = (x, x) + t(x, y) + t(y, x) + t^2(y, y)$$

and therefore the discriminant of the quadratic polynomial is negative, i.e.

$$|(x, y) + (y, x)|^2 \leq 4(x, x)(y, y)$$

for any x, y. Substituting in the last inequality $x = x'e^{-i\varphi}$, where $\varphi = \arg(x', y)$, we obtain the result. Observe that the inequality becomes an equality if and only if $x = \mu y$ for some complex μ.

In order to prove (c) observe that

$$(x + y, x + y) = (x, x) + (x, y) + (y, x) + (y, y).$$

From (b) we have

$$(x + y, x + y) \leq (x, x) + 2(x, x)^{1/2}(y, y)^{1/2} + (y, y).$$

Therefore

$$(x + y, x + y) \leq ((x, x)^{1/2} + (y, y)^{1/2})^2.$$

Here also the inequality becomes an equality if and only if $x = \mu y$ for some complex μ. □

A *norm* in a linear space is a function $x \mapsto \|x\| \in \mathbf{R}$ such that:

(a) $\|x\| \geq 0$ and $\|x\| > 0$, if $x \neq 0$;
(b) $\|\alpha x\| = |\alpha| \, \|x\|$ for any $\alpha \in \mathbf{C}$;
(c) $\|x_1 + x_2\| \leq \|x_1\| + \|x_2\|$.

A *normed linear space* is a linear space together with a norm.

A sequence $x_1, x_2, \ldots, x_n, \ldots$ of elements of a linear normed space is a *Cauchy sequence* if for any $\varepsilon > 0$ there exists an $N(\varepsilon)$ such that $\|x_p - x_q\| \leq \varepsilon$ if $p \geq N$, $q \geq N$. A normed linear space is called *complete* if any Cauchy sequence has a limit, i.e. there is an element x such that $\lim_{n\to\infty} \|x_n - x\| = 0$. A complete normed linear space is called a *Banach* space.

If a linear normed space B is not complete, one can always extend it up to a complete one. To do this, consider the set of all Cauchy sequences. Two such sequences $\{y_n\}$ and $\{z_n\}$ are called equivalent if $\|y_n - z_n\| \to 0$ as $n \to \infty$. The set $\tilde{B}$ of classes of equivalent sequences is called a *completion* of B. It is a Banach space such that B is dense in it. It is easy to see that $\tilde{B}$ is unique up to isometry.

Theorem 1 implies that in a pre-Hilbert space one can take as a norm $\|x\| = (x,x)^{1/2}$.

If a pre-Hilbert space is complete with respect to convergence in this norm, it is called a *Hilbert space*.

A Hilbert space is *separable* if it contains a denumerable dense subset.

A linear space is called *finite dimensional*, if it has a finite set of elements $x_1, x_2, \dots, x_N$ such that any element $x \in X$ can be represented as $x = \sum_{i=1}^{N} \alpha_i x_i$ with some $\alpha_1, \dots, \alpha_N \in \mathbf{C}$.

1.2 Examples

This section shows some examples of Hilbert spaces.

1. Consider the set l^2 of sequences of complex numbers

$$x = (x_1, x_2, \dots, x_k, \dots)$$

such that the series $\sum_{k=1}^{\infty} |x_k|^2$ converges. This set is a linear space. The inner product in this space is defined as

$$(x,y) = \sum_{k=1}^{\infty} x_k \cdot \overline{y}_k.$$

Thus l^2 is a pre-Hilbert space.

Theorem 2. *The space l^2 is a Hilbert space.*

Proof. Let $\{x^{(n)}\}$ be a Cauchy sequence in l^2. This means that for any $\varepsilon > 0$ there is a natural N such that

$$\|x^{(n)} - x^{(m)}\|^2 = \sum_{j=1}^{\infty} |x_j^{(n)} - x_j^{(m)}|^2 < \varepsilon \text{ for } m, n > N,$$

if $x^{(n)} = (x_1^{(n)}, x_2^{(n)}, \dots, x_j^{(n)}, \dots)$. This implies that for $m, n > N$ the estimate

$$|x_j^{(n)} - x_j^{(m)}|^2 < \varepsilon$$

is true for all j, so that for any fixed j the sequence $\{x_j^{(n)}\}$ is a Cauchy sequence and thus is converging. Let $x_j = \lim_{n\to\infty} x_j^{(n)}$ and $x = (x_1, \dots, x_k, \dots)$.

Let us prove that $x \in l^2$ and $\|x^{(n)} - x\| \to 0$. As we have seen

$$\sum_{j=1}^{p} |x_j^{(n)} - x_j^{(m)}|^2 < \varepsilon \quad \text{for} \quad m, n > N$$

for any fixed p. Since the latter sum is finite, one can pass to the limit as $m \to \infty$. We get

$$\sum_{j=1}^{p} |x_j^{(n)} - x_j|^2 < \varepsilon \quad \text{for} \quad n > N(\varepsilon).$$

Since this inequality is true for each p, the series of positive numbers

$$\sum_{j=1}^{\infty} |x_j^{(n)} - x_j|^2$$

converges and its sum does not exceed ε.

Taking $\varepsilon = 1$ one can find an n_0 such that $\|x^{(n)} - x^{(n_0)}\| < 1$ for all $n > n_0$. Therefore, $\|x^{(n)}\| < \|x^{(n_0)}\| + 1$ for $n > n_0$ and $\|x^{(n)}\| \leq C$ for all n.

By the triangle inequality for any $n > N$ we have

$$\sum_{j=1}^{\infty} |x_j|^2 \leq [(\sum_{j=1}^{\infty} |x_j^{(n)} - x_j|^2)^{1/2} + (\sum_{j=1}^{\infty} |x_j^{(n)}|^2)^{1/2}]^2.$$

Therefore, the series $\sum_{j=1}^{\infty} |x_j|^2$ converges, i.e. $x \in l^2$. Besides, the inequality

$$\sum_{j=1}^{\infty} |x_j^{(n)} - x_j|^2 < \varepsilon \quad \text{for} \quad n > N(\varepsilon)$$

being valid for any $\varepsilon > 0$ means that

$$\|x^{(n)} - x\| \to 0. \qquad \square$$

The space l^2 is separable since the set of finite sequences with rational real and imaginary parts is a countable dense subset.

2. Let Ω be a domain in $\mathbf{R}^n$. Consider the set $L^2(\Omega)$ of measurable functions f, defined in Ω such that the integrals $\int_\Omega |f(x)|^2\, dx$ converge. This set is a linear space. The inner product in this space is defined as

$$(f, g) = \int_\Omega f(x) \cdot \overline{g}(x)\, dx.$$

Thus $L^2(\Omega)$ is a pre-Hilbert space. Later we shall see that the space $L^2(\Omega)$ is a Hilbert space.

3. An example of a nonseparable Hilbert space. Consider the set of functions defined on $[0, 1]$ taking real nonzero values on a finite subset of $[0, 1]$ and vanishing outside of it. Let us define a scalar product as

$$(f, g) = \sum_i f(x_i) g(x_i),$$

where $\{x_i\}$ is the set of points where $f \neq 0$. It is easy to see that the completion of the space of such functions is nonseparable. We will omit the elementary proof.

4. The space l_p is a Banach space for $1 \leq p < \infty$. The elements of this space are the sequences of complex numbers $(a_1, a_2, \ldots, a_n, \ldots)$ such that

$$\|a\| = (\sum_{i=1}^{\infty} |a_i|^p)^{1/p} < \infty.$$

One can verify that this space is complete.

1.3 Orthonormal base

Elements x, y of a Hilbert space H are called *orthogonal* if $(x, y) = 0$. A set $B = (x_1, \ldots, x_n, \ldots)$ of elements of H is called an *orthonormal base* if any two elements x_j, x_k are orthogonal, $\|x_k\| = 1$ for all k, and the set of linear combinations of elements of B is dense in H, i.e. for any element $x \in H$ and any $\varepsilon > 0$ there is a linear combination y of elements of B such that $\|y - x\| < \varepsilon$.

Example 3. In the space l^2 the elements

$$e_1 = (1, 0, 0, \ldots),\ e_2 = (0, 1, 0, \ldots),\ e_3 = (0, 0, 1, \ldots), \ldots$$

form an orthonormal base.

Theorem 4. *Any separable Hilbert space H has an orthonormal base.*

Proof. Let x_1, x_2, ..., x_k ... be a countable dense set. It is easy to find a dense subset of linearly independent elements. To do this one has to eliminate recurrently those x_j that can be represented as a linear combination of elements x_1, x_2, ..., x_{j-1}.

Let now $y_1 = x_1/\|x_1\|$. Set

$$y_2 = C(x_2 - (x_2, y_1)y_1),$$

where the constant C is such that $\|y_2\| = 1$. Then $(y_1, y_2) = 0$.

Using induction, if $y_1, \ldots, y_{k-1}$ are found, one can define

$$y_k = C_k(x_k - (x_k, y_1)y_1 - \cdots - (x_k, y_{k-1})y_{k-1})$$

and choose the constant C_k so that $\|y_k\| = 1$. Then $(y_k, y_j) = 0$ for $j = 1, \ldots, k-1$. □

1.4 Fourier series

Let $x_1, \dots, x_n, \dots$ be an orthonormal base in a Hilbert space H. The numbers $c_k = (f, x_k)$ are called *Fourier coefficients* of an element $f \in H$, and the series $\sum_{k=1}^{\infty} c_k x_k$ is its *Fourier series.*

Let $S_N = \sum_{k=1}^{N} c_k x_k$. Then

$$\begin{aligned}\|f - S_N\|^2 &= (f - \sum_{k=1}^{N} c_k x_k, f - \sum_{k=1}^{N} c_k x_k) \\ &= (f,f) - (f, \sum_{k=1}^{N} c_k x_k) - (\sum_{k=1}^{N} c_k x_k, f) + (\sum_{k=1}^{N} c_k x_k, \sum_{k=1}^{N} c_k x_k) \\ &= (f,f) - \sum_{k=1}^{N} \bar{c}_k c_k - \sum_{k=1}^{N} c_k \bar{c}_k + \sum_{k=1}^{N} |c_k|^2 = (f,f) - \sum_{k=1}^{N} |c_k|^2.\end{aligned}$$

Since $\|f - S_N\|^2 \geq 0$, it follows that

$$\sum_{k=1}^{N} |c_k|^2 \leq (f,f).$$

Therefore, the series $\sum_{k=1}^{\infty} |c_k|^2$ converges and its sum does not exceed (f,f). The inequality

$$\sum_{k=1}^{\infty} |c_k|^2 \leq (f,f)$$

is called the *Bessel inequality.*

An orthonormal system of elements $x_1, \dots, x_n, \dots$ is called *closed* if for any $f \in H$ the equality

$$\sum_{k=1}^{\infty} |c_k|^2 = (f,f) \quad \text{holds.}$$

The above obtained equality

$$\|f - S_N\|^2 = (f,f) - \sum_{k=1}^{N} |c_k|^2$$

implies that the system $x_1, \dots, x_n, \dots$ is closed if and only if the Fourier series of any element $f \in H$ converges to this element.

Theorem 5 (Riesz-Fischer). *Let $x_1, \dots, x_n, \dots$ be an orthonormal base in a Hilbert space H, $c_1, \dots, c_n \dots$ be complex numbers, and the series $\sum_{k=1}^{\infty} |c_k|^2$ converge. Then there exists an element $f \in H$ such that $c_k = (f, x_k)$ and $\sum_{k=1}^{\infty} |c_k|^2 = (f,f)$.*

Proof. Let $f_k = \sum_{j=1}^{k} c_j x_j$. Then

$$\|f_{N+p} - f_N\|^2 = \|\sum_{j=N+1}^{N+p} c_j x_j\|^2 = \sum_{j=N+1}^{N+p} |c_j|^2.$$

Therefore, the sequence $\{f_k\}$ is a Cauchy sequence and since the space H is complete there exists $\lim f_k = f$. Since $|(f_k - f, x_j)| \le \|f_k - f\|$ for all $j = 1, 2, \ldots$, we see that

$$(f, x_j) = \lim_{k\to\infty} (f_k, x_j) = c_j.$$

At last, the equality

$$\|f - f_N\|^2 = (f - \sum_{j=1}^{N} c_j x_j, f - \sum_{j=1}^{N} c_j x_j) = (f, f) - \sum_{j=1}^{N} |c_j|^2$$

implies that the series $\sum_{j=1}^{\infty} |c_j|^2$ converges and its sum is equal to (f, f). □

Corollary 6. *An orthonormal system is closed if and only if there is no $z \in H \setminus \{0\}$ orthogonal to all its elements.*

Proof. If such an element z exists, then

$$\|z - \sum_{k=1}^{N} a_k x_k\|^2 = \|z\|^2 + \|\sum_{k=1}^{N} a_k x_k\|^2 \ge \|z\|^2 > 0$$

for any $N, a_1, \ldots, a_N$. Therefore, the system $(x_1, \ldots, x_k, \ldots)$ is not closed.

Inversely, if the system $(x_1, \ldots, x_k, \ldots)$ is not closed, then there exists an element $g \in H$ such that the series $\sum_{j=1}^{\infty} |c_j|^2$ with $c_j = (g, x_j)$ does not converge to (g, g), i.e.

$$(g, g) - \sum_{k=1}^{N} |c_k|^2 \ge \delta > 0.$$

By the Riesz-Fischer theorem there exists an element $f \in H$ such that

$$(f, f) = \sum_{k=1}^{\infty} |c_k|^2, \quad (f, x_k) = c_k.$$

The element $z = g - f$ is orthogonal to all x_k and is not zero since $(f, f) \le (g, g) - \delta$. □

Theorem 7. *Any separable Hilbert space H is isomorphic to l^2.*

Proof. Let us find in H an orthonormal base $x_1, \ldots, x_k, \ldots$. For any $f \in H$ the sequence $c = (c_1, \ldots, c_k, \ldots)$, where $c_j = (f, x_j)$, satisfies the inequality

$$\sum_{k=1}^{\infty} |c_k|^2 \leq (f, f)$$

and hence belongs to l_2.

Inversely, by the Riesz-Fischer theorem any $c = (c_1, \ldots, c_k, \ldots)$ of l_2 defines an element $f = \sum_{j=1}^{\infty} c_j x_j$ of H. The definition implies immediately that the sum of elements defines the sum of their images, the product zc by a complex number z corresponds to the product zf. Since the inner product (f_1, f_2) in H is equal to the inner product $(c^{(1)}, c^{(2)})$ in l_2, where $c^{(i)}$ is a sequence of the Fourier coefficients of f_i, this correspondence is an isomorphism. □

1.5 Subspaces, orthogonal sums

A subspace of a Hilbert space H is a closed linear space $M \subset H$.

Theorem 8. *A subspace M of a separable Hilbert space H has an orthonormal base.*

Proof. Let $x_1, x_2, \ldots, x_k \ldots$ be a countable dense in H set. Set $r_k = \inf_{y \in M} \|x_k - y\|$. For any k and l there exists a point $z_{kl} \in M$ such that $\|z_{kl} - x_k\| < r_k + 1/l$. For any $\varepsilon > 0$ and any point $y \in M$ there are k and l such that $1/l < \varepsilon/3$, $\|x_k - y\| < \varepsilon/3$. Therefore, $r_k < \varepsilon/3$ and hence

$$\|y - z_{kl}\| \leq \|y - x_k\| + \|x_k - z_{kl}\| < \varepsilon.$$

This means that the set of elements z_{kl} is dense in M. Now we can apply to this set the procedure of orthogonalization described in Theorem 4. Since M is a linear manifold, the obtained base belongs to M. □

Let $M^{\perp}$ be the set of elements in H that are orthogonal to all elements of M. It is evident that $M^{\perp}$ is a linear manifold. If a sequence $\{f_k\}$ of elements of $M^{\perp}$ converges to f, then for any $g \in M$

$$(g, f) = \lim_{k \to \infty} (g, f_k) = 0,$$

so that $f \in M^{\perp}$. Therefore, $M^{\perp}$ is a subspace.

Theorem 9. *If M is a subspace of a separable Hilbert space H, then H is a direct sum $M \oplus M^{\perp}$, i.e. each element f of H is represented in a unique way as a sum $f = x + y$, where $x \in M$, $y \in M^{\perp}$.*

Proof. Let $x_1, \ldots, x_k, \ldots$ be an orthonormal base in M and $x = \sum_{j=1}^{\infty} c_j x_j$, where $c_j = (f, x_j)$. Let $y = f - x$. It is evident that $(y, x_j) = 0$ for all j so that $y \in M^{\perp}$.

If $f = x' + y'$ is another representation of f with $x' \in M$, $y' \in M^{\perp}$, then $(x', x_j) = (f, x_j) = c_j$ and hence $x' = x$, $y' = y$. □

The notion of direct sum is easily generalized to any finite or even countable set of subspaces, so that

$$H = M_1 \oplus M_2 \oplus \cdots \oplus M_k \oplus \dots,$$

if each element of M_j is orthogonal to every element of M_k for $j \neq k$.

If $f \in H$, then there exists a unique representation of f as a sum

$$f = f_1 + f_2 + \cdots + f_k + \dots,$$

where $f_k \in M_k$ and $\|f\|^2 = \sum_j \|f_j\|^2$.

On the other hand, if H_1 and H_2 are Hilbert spaces, then one can define their *direct sum* $H = H_1 \oplus H_2$ as a Hilbert space, whose elements are the pairs (x_1, x_2), where $x_1 \in H_1$, $x_2 \in H_2$, and the inner product for such pairs is equal to

$$((x_1, x_2),\ (y_1, y_2)) = (x_1, y_1)_{H_1} + (x_2, y_2)_{H_2}.$$

It is clear that H_1 is isomorphic to the subspace in H containing all pairs of the form $(h_1, 0)$, and H_2 is isomorphic to the subspace containing all pairs of the form $(0, h_2)$. Analogously, one can define a direct sum of any finite or countable set of Hilbert spaces.

1.6 Linear functionals

A linear functional is a map $f : D \to \mathbf{C}$ defined on a linear subspace D of a Hilbert space H such that

$$f(x + y) = f(x) + f(y), \quad x, y \in D;$$

$$f(\lambda x) = \lambda f(x), \quad x \in D,\ \lambda \in \mathbf{C}.$$

A linear functional f is *continuous* if $D = H$ and $f(x) \to 0$ as $x \to 0$ in H.

Theorem 10. *A linear functional f is continuous on H if and only if there exist positive constants A and B such that $|f(x)| \leq A$ for all x with $\|x\| < B$.*

Proof. Let f be continuous at a point $x_0 \in H$. Then for $\varepsilon = 1$ there exists $\delta > 0$ such that $|f(x)| = |f(x_0 + x) - f(x_0)| < 1$ if $\|x\| < \delta$.

Inversely, let there exist numbers A and B with this property. Let $x_0 \in H$ and ε be a positive number. Then for any $y \in H$ such that $\|y - x_0\| < \varepsilon B/A$ the element $x = A(y - x_0)/\varepsilon$ satisfies the condition $\|x\| < B$ and therefore $|f(x)| < A$. But then

$$|f(y) - f(x_0)| = |\varepsilon f(x)/A| < \varepsilon$$

and therefore the functional f is continuous at the point x_0. □

Example 11. If $y \in H$, then the linear functional $f(x) = (x, y)$ is continuous.

Theorem 12 (Riesz). *If a linear functional f is continuous on H, then there exists a unique element $y \in H$ such that $f(x) = (x, y)$.*

Proof. Let N be the set of all elements x of H such that $f(x) = 0$. Since f is a linear continuous functional, the set N is a subspace in H. If $N = H$, then $f = 0$ and $y = 0$. Otherwise, let $N^\perp$ be the orthogonal supplement of N in H.

It is easy to see that the space $N^\perp$ is one dimensional. Indeed, if $y_1 \in N^\perp$, $y_1 \neq 0$, $y_2 \in N^\perp$, then $f(y_1) \neq 0$ and there exists a constant C such that

$$f(Cy_1 + y_2) = 0,$$

i.e. $Cy_1 + y_2 \in N$. But $Cy_1 + y_2 \in N^\perp$, since $N^\perp$ is linear. Therefore, $Cy_1 + y_2 = 0$, which means that y_1 and y_2 are linearly dependent.

Let $z \in N^\perp$, $\|z\| = 1$. Each element $x \in H$ is represented in a unique way as $x = x_1 + Cz$, where $x_1 \in N$, $C \in \mathbf{C}$. Therefore

$$f(x) = f(x_1) + f(Cz) = Cf(z).$$

On the other hand, $(x, z) = C(z, z) = C$. If $f(z) = \alpha$ and $y = \bar{\alpha} z$, then

$$f(x) = C\alpha = \alpha(x, z) = (x, y)$$

for all $x \in H$. □

The set of all linear continuous functionals in H is a linear space and is denoted by H^*. The Riesz theorem allows us to identify the spaces H and H^*.

Definition 13. A *linear functional* T in a Banach space B is a map of B in $\mathbf{C}$ such that

$$T(\alpha_1 x_1 + \alpha_2 x_2) = \alpha_1 T(x_1) + \alpha_2 T(x_2)$$

for any $\alpha_1, \alpha_2 \in \mathbf{C}$ and any $x_1, x_2 \in B$. A linear functional is called *bounded* if there is a C such that $\|Tx\| \leq C\|x\|$ for all $x \in B$.

The least upper bound of the numbers C is called the *norm* of T and is denoted by $\|T\|$. The topology in B^* induced by this norm is called *strong.*

A linear functional T in a Banach space is bounded iff it is continuous.

The space of linear bounded functionals in B is called an *adjoint* space and is denoted as B^*.

In the same way one can define the space B^{**}. Obviously, $B^{**} \supset B$. If $B = B^{**}$, then the space B is called *reflexive.*

Example 14. Let $X = l_p$, $1 < p < \infty$. Any element of the space $l_{p'}$, where p' is such that $1/p + 1/p' = 1$, defines a linear continuous functional. If $a = (a_1, \dots, a_n, \dots) \in l_p$, $b = (b_1, \dots, b_n, \dots) \in l_{p'}$, then $T(a) = \sum a_i \bar{b}_i$ is a bounded linear functional. It follows from the Hölder inequality

$$\left(\sum_1^\infty a_i b_i\right) \leq \left(\sum_1^\infty |a_i|^p\right)^{1/p} \left(\sum_1^\infty |b_i|^{p'}\right)^{1/p'}.$$

The space of linear functionals is linear if we set

$$(T_1 + T_2)x = T_1x + T_2x; \; \lambda T_1 x = T_1(\lambda x).$$

Theorem 15. *The space of linear functionals in a Banach space B is complete.*

Proof. Let $x \in B$ and the sequence $T_1, T_2, \dots, T_k, \dots$ of linear functionals be a Cauchy sequence. The sequence $T_1x, T_2x, \dots, T_kx, \dots$ is also a Cauchy sequence in B. Let $T_\infty x$ be its limit. It is evident that T_∞ is a linear functional. Besides, $\|T_k\| \leq C$. Therefore, $|T_\infty x| = \lim_{k\to\infty} |T_k x| \leq C\|x\|$, i.e. T_∞ is a bounded functional. For any $\varepsilon > 0$ there exists N such that $\|T_n - T_m\| < \varepsilon$, i.e. $|T_nx - T_mx| < \varepsilon\|x\|$ for $n > N, m > N$. For m tending to ∞ we obtain $|T_nx - T_\infty x| < \varepsilon\|x\|$, i.e. $\|T_n - T_\infty\| < \varepsilon$ for $n > N$. Therefore, $T_n \to T_\infty$. □

1.7 Weak convergence

Definition 16. A sequence $\{x_k\}$ in a Hilbert space H is called *weakly convergent* if for any linear continuous functional $f \in H^*$ the sequence $\{f(x_k)\}$ has a finite limit.

Theorem 17. *If a sequence $\{x_k\}$ weakly converges in H, then there exists a constant C such that $\|x_k\| \leq C$ for all k.*

Proof. Let $f \in H^*$. The sequence $\{f(x_k)\}$ converges and therefore is bounded. Let M_n be a set of functionals f of H^* such that

$$|f(x_k)| \leq n, \quad k = 1, 2, \dots.$$

The set M_n is closed, $M_n \subset M_{n+1}$, and the union of all M_n is equal to H^*.

Let us show that then there exists a set M_{n_0} containing a ball

$$B = \{f : \; \|f - f_0\| \leq r\}$$

with some f_0 and $r > 0$. Indeed, otherwise the set M_1 does not contain at least one point x_1 such that

$$x_1 \in B_0 = \{x : \|x\| < 1\}.$$

Since M_1 is closed, there is a ball

$$B_1 = \{x : \|x - x_1\| < r_1\}$$

such that $B_1 \subset B_0$ and $M_1 \cap B_1 = \emptyset$. The set M_2 does not contain at least one point x_2 of B_1 and thus there is a ball

$$B_2 = \{x : \|x - x_2\| < r_2\}$$

such that $B_2 \subset B_1$ and $M_2 \cap B_2 = \emptyset$. Continuing the argument we obtain a sequence of balls

$$B_0 \supset B_1 \supset B_2 \supset \dots$$

such that

$$M_k \cap B_j = \emptyset \quad \text{for all} \quad k = 1, \dots, j.$$

The intersection B_∞ of all B_j is not empty and $B_\infty \cap M_k = \emptyset$ for all k. But then the union of all M_k does not contain the whole space H, which contradicts the assumption.

If $g \in H^*$, then the functional $h = f_0 + rg\|g\|^{-1}$ belongs to B and hence $|h(x_k)| \le n_0$ for all k. Therefore

$$|g(x_k)| = \|g\| r^{-1} |(h - f_0)(x_k)| \le \|g\| r^{-1}(n_0 + c),$$

for $k = 1, 2, \dots$, where c is a constant such that $|f_0(x_k)| \le c$.

Let $g(x) = (x, x_k)$. Then $|g(x_k)| = \|x_k\|^2$ and $\|g\| = \|x_k\|$. The obtained inequality implies that

$$\|x_k\| \le r^{-1}(n_0 + c), \quad k = 1, 2, \dots,$$

i.e. the sequence $\{x_k\}$ is bounded. □

Theorem 18. *A sequence $\{x_k\}$ weakly converges to x in H if and only if the two following conditions are fulfilled:*

(i) There exists a constant C such that $\|x_k\| \le C$ for all k;

(ii) $f(x_k) \to f(x)$ for each $f \in G$, where G is a subset in H^* such that the linear combinations of the elements of G are dense in H^*.

Proof. The "only if" part follows from Theorem 17.

Let conditions (i) and (ii) be satisfied for a sequence $\{x_k\}$. Let f be an arbitrary functional from H^* and $\{f_n\}$ be a sequence of linear combinations of elements of G, converging to f. One has to prove that $f(x_k) \to f(x)$. To this end observe that

$$|f(x_k) - f(x)| \le |f(x_k) - f_n(x_k)| + |f_n(x_k) - f_n(x)|.$$

Let ε be an arbitrary positive number. Then there exists N such that $\|f - f_n\| < \varepsilon/2C$ for $n \ge N$. On the other hand, there is a number K such that

$$|f_N(x_k) - f_N(x)| < \varepsilon/2$$

for $k \ge K$. Therefore

$$|f(x_k) - f(x)| \le \|f - f_N\| \|x_k\| + |f_N(x_k) - f_N(x)| < \varepsilon/2C \cdot C + \varepsilon/2 = \varepsilon$$

for all $k \ge K$. However, this means that $f(x_k)$ converges to $f(x)$. The inverse statement follows from Theorem 17. □

Example 19. The sequence $x_k = (0, \ldots, 0, 1, 0, \ldots)$, where 1 is on the kth place, converges weakly to 0 in l_2. However, $\|x_k\| = 1$ so this sequence is not convergent.

Theorem 20. *Let H be a separable Hilbert space. Then a bounded sequence $\{x_k\}$ of elements of H contains a weakly convergent subsequence.*

Proof. By Theorem 12, the space H^* is isomorphic to H and therefore is separable. Let us find a countable dense in H^* set $(f_1, \ldots, f_n, \ldots)$. The sequence $\{f_1(x_k)\}$ is bounded. Therefore, there exists a subsequence $\{x_k^{(1)}\}$ such that the sequence $\{f_1(x_k^{(1)})\}$ converges. Furthermore, from the sequence $\{x_k^{(1)}\}$ one can choose a subsequence $\{x_k^{(2)}\}$ such that the sequence $\{f_2(x_k^{(2)})\}$ converges. Continuing this process, we get a set of sequences

$$b_j = (x_1^{(j)}, \ldots, x_k^{(j)}, \ldots),$$

such that $b_j \subset b_{j-1}$ and the sequences $\{f_i(x_k^{(j)})\}$ are converging as $k \to \infty$, if $1 \leq i \leq j$.

This implies that all sequences $\{f_j(x_k^{(k)})\}$ are converging as $k \to \infty$. From Theorem 18 it follows then that the sequence $\{x_k^{(k)}\}$ is weakly converging in H. □

Theorem 21. *Let $\{x_k\}$ be a sequence of elements of a Hilbert space H weakly converging to x such that $\|x_k\| \to \|x\|$. Then $\{x_k\}$ converges to x in H.*

Proof. The identity

$$\|x_k - x\|^2 = (x_k - x, x_k - x) = \|x_k\|^2 - (x, x_k) - (x_k, x) + \|x\|^2$$

implies that $\|x_k - x\| \to 0$ because

$$(x, x_k) + (x_k, x) \to 2\|x\|^2$$

in virtue of the weak convergence. □

Theorem 22 (Banach-Saks). *If a sequence $\{x_n\}$ of elements of a Hilbert space converges weakly to x, then there exists a subsequence $\{x_{n_k}\}$ such that the sequence of the arithmetic means*

$$v_k = \frac{1}{k}(x_{n_1} + \cdots + x_{n_k})$$

converges to x.

Proof. Substituting x_k by $x_k - x$ one can reduce the proof to the case when $x = 0$. Since $(x_k, x_n) \to 0$ as $n \to \infty$, there are $n_1, n_2, \ldots$ such that

$$n_1 = 1;\ |(x_{n_2}, x_{n_1})| \leq 1;\ |(x_{n_3}, x_{n_1})| \leq 1/2,\ |(x_{n_3}, x_{n_2})| \leq 1/2; \ldots$$

$$|(x_{n_k}, x_{n_1})| \leq 1/(k-1), \ldots |(x_{n_k}, x_{n_{k-1}})| \leq 1/(k-1) \text{ for all } k.$$

The weak convergence of the sequence $\{x_k\}$ implies its boundedness, so that $\|x_k\| \le C$. Therefore

$$\begin{aligned}\|v_k\|^2 &= \|\frac{1}{k}\sum_{j=1}^{k} x_{n_j}\|^2 = \frac{1}{k^2}\sum_{j,l=1}^{k}(x_{n_j}, x_{n_l})\\ &\le \frac{1}{k^2}(kC^2 + 2\sum_{j=1}^{k}\sum_{l=1}^{j-1}|(x_{n_j}, x_{n_l})|)\\ &\le \frac{1}{k^2}(kC^2 + 2\cdot 1 + 4\cdot\frac{1}{2} + \cdots + 2(k-1)\cdot\frac{1}{k-1})\\ &< (kC^2 + 2k)/k^2 \to 0 \text{ as } k \to \infty.\end{aligned}$$

Therefore, $v_k \to 0$ in H. □

Definition 23. A sequence of elements $\{x_n\}$ in a Banach space is said to be *weakly converging* to an element x, if it is bounded, i.e. $\|x_n\| \le C$, and for any bounded linear functional T we have $Tx_n \to Tx$.

Theorem 24. *If a space B^* is separable, then any bounded subset in B contains a weakly convergent subsequence.*

Proof. Let $T_1, T_2, \ldots, T_m, \ldots$ be a sequence of elements of the space B^* whose linear combinations are dense in B^*, i.e. for any $T \in B^*$ and any $\varepsilon > 0$ there are numbers α_i such that $\|T - \sum_1^N \alpha_i T_i\| \le \varepsilon$. Let $x_1, x_2, \ldots, x_m, \ldots$ be a bounded in B sequence. The number sequence $T_1(x_1), \ldots, T_1(x_m), \ldots$ is bounded. Let us find a convergent subsequence $T_1(x_{11}), \ldots, T_1(x_{m1}), \ldots$. The sequence x_{i1}, $i = 1, 2, \ldots$, contains a subsequence x_{i2}, $i = 1, 2, \ldots$, such that the corresponding sequence $T_2(x_{12}), \ldots, T_2(x_{m2}) \ldots$ is converging. Continuing the process we find a countable set of sequences, each of which is contained in the previous one. The sequence $x_{pp} = y_p$ is such that the sequence $T_i(y_p)$ is convergent as $p \to \infty$ for all i.

Now let T be an arbitrary element of B^*. Show that the sequence $T(y_n)$ is convergent. Let a ε be fixed and the numbers α_i be such that $\|T - \sum_1^l \alpha_i T_i\| \le \varepsilon$. There exists $N > 0$ such that

$$|\alpha_i||T_i(y_n - y_m)| \le \varepsilon/l$$

for $i = 1, \ldots, l$ and $m > N$, $n > N$. Then

$$\begin{aligned}|T(y_n) - T(y_m)| &\le |T(y_n) - \sum_1^l \alpha_i T_i(y_n)| + |T(y_m) - \sum_1^l \alpha_i T_i(y_n)|\\ &+ |\sum_1^l \alpha_i T_i(y_n - y_m)| \le \varepsilon\|y_n\| + \varepsilon\|y_m\| + \varepsilon \le C\varepsilon\end{aligned}$$

for n, m sufficiently large. Therefore, the sequence $T(y_n)$ is a Cauchy sequence and hence is convergent. □

1.8 Linear operators

A *linear operator* in H is a linear map $A : H_1 \to H_2$, where H_1 and H_2 are linear manifolds in H. The set H_1 is called a *domain* of A and is denoted by $D(A)$. The set of elements x of $D(A)$, such that $Ax = 0$, is called a *kernel* of the operator A and is denoted by $\ker A$. The set of elements y of H_2, such that there exists an x in $D(A)$ such that $y = Ax$, is called an *image* of the operator A and is denoted by $\operatorname{im} A$ or $R(A)$.

An operator A is called *continuous at a point* $x_0 \in D(A)$ if for any $\varepsilon > 0$ there exists a $\delta > 0$ such that the condition $x \in D(A)$, $\|x - x_0\| < \delta$, implies the inequality $\|Ax - Ax_0\| < \varepsilon$. An operator A is called *continuous* if it is continuous at each point $x_0 \in D(A)$. A linear operator A is called *bounded* if $D(A) = H$ and $\sup_{\|x\|=1} \|Ax\| < \infty$.

Theorem 25. *Let A be a linear operator in H. If $D(A) = H$, then A is continuous if and only if it is bounded.*

Proof. If the operator A is not bounded, then there exists a sequence $\{x_k\}$ of elements of $D(A)$ such that $\|x_k\| = 1$ and $\|Ax_k\| \geq k$. These conditions imply that the sequence $\{y_k\} = \{x_k/k\}$ tends to 0 in H as $k \to \infty$, and $\|Ay_k\| \geq 1$, so that the operator A is not continuous at the point $x = 0$.

On the other hand, if an operator A is not continuous, then there exists a sequence $\{x_k\}$ converging to x_0 in H as $k \to \infty$ such that $\|A(x_k - x_0)\| \geq 1$. Then the sequence $\{y_k\}$, where $y_k = (x_k - x_0)/\|x_k - x_0\|$, is bounded in H, but $\|Ay_k\| \to \infty$ as $k \to \infty$. Therefore, the operator A is not bounded. □

If an operator A is bounded in H, then

$$\|Ax\| = \|x\| \cdot \|A(\frac{x}{\|x\|})\| \leq C\|x\|, \quad \text{where} \quad C = \sup_{\|x\|=1} \|Ax\|.$$

A norm of an operator A is defined as

$$\|A\| = \sup_{\|x\| \leq 1} \|Ax\|.$$

It is evident that

$$\|A\| = \sup_{x \neq 0} \frac{\|Ax\|}{\|x\|}.$$

Since the set of all linear operators, defined on H, is a linear space, the subset $\mathcal{L}(H)$ of bounded linear operators is a linear normed space.

Theorem 26. *The space $\mathcal{L}(H)$ is complete.*

Proof. Let $\{A_k\}$ be a Cauchy sequence of linear bounded operators. Then the sequence $\{A_k x\}$ is a Cauchy sequence for each $x \in H$. Therefore, there exists

$$\lim_{k \to \infty} A_k x = y.$$

Let A be the linear operator such that $Ax = y$. For any $\varepsilon > 0$ there exists N such that $\|A_n - A_m\| < \varepsilon$, i.e. $\|A_n x - A_m x\| < \varepsilon\|x\|$ for $n > N$, $m > N$. For m tending to ∞ we obtain $\|A_n x - Ax| < \varepsilon\|x\|$, i.e. $\|A_n - A\| < \varepsilon$ for $n > N$. Therefore, $A_n \to A$. □

A linear operator A is called *invertible*, if to each element $y \in \operatorname{im} A$ there corresponds a unique element $x \in D(A)$, for which $Ax = y$. The operator realizing this correspondence, is called *inverse* to A and is denoted by A^{-1}. It is easy to see that it is a linear operator.

Theorem 27. *Let a linear operator A be bounded and invertible and let the inverse operator also be bounded. If $B \in \mathcal{L}(H)$ is such that $\|B\| < 1/\|A^{-1}\|$, then the operator $A + B$ has a bounded inverse operator.*

Proof. The definition of the norm implies that $\|BA^{-1}\| \leq \|B\| \cdot \|A^{-1}\|$ so that $\|BA^{-1}\| < 1$. Therefore, the series

$$I - BA^{-1} + (BA^{-1})^2 - \cdots + (-1)^{j-1}(BA^{-1})^j + \ldots$$

converges in $\mathcal{L}(H)$ and defines a bounded operator C. Here I is as usual the unit operator. It is easy to check that

$$(I + BA^{-1})C = I.$$

Therefore, $(A+B)A^{-1}C = I$, i.e. the operator $A^{-1}C$ is inverse to $A+B$. Furthermore,

$$\|C\| \leq 1 + \|B\|\|A^{-1}\| + \|B\|^2\|A^{-1}\|^2 + \ldots = (1 - \|B\|\|A^{-1}\|)^{-1},$$

$$\|A^{-1}C\| \leq \|A^{-1}\|(1 - \|B\|\|A^{-1}\|)^{-1}.$$ □

1.9 Adjoint operators

If A is a bounded linear operator in a Hilbert space H, then for any $y \in H$ one can define a linear functional

$$f(x) = (Ax, y).$$

By the Riesz theorem of Section 1.6 there is an element $z \in H$ such that $f(x) = (x, z)$. Thus to each element $y \in H$ there corresponds an element $z \in H$, i.e. an operator is defined. It is easy to see that this operator is linear and bounded. It is called an *adjoint* operator and is denoted by A^*. By the definition

$$(Ax, y) = (x, A^*y)$$

for any elements $x, y \in H$.

Theorem 28. $\|A^*\| = \|A\|$.

Proof. By the definition of the norm we have

$$|(x, A^*y)| = |(Ax, y)| \le \|A\| \cdot \|x\| \cdot \|y\|.$$

So $\|A^*y\| \le \|A\| \cdot \|y\|$ and therefore $\|A^*\| \le \|A\|$. Analogously

$$|(Ax, y)| = |(x, A^*y)| \le \|x\| \cdot \|A^*\| \cdot \|y\|,$$

so that $\|Ax\| \le \|A^*\| \cdot \|x\|$ and therefore $\|A\| \le \|A^*\|$. Thus $\|A^*\| = \|A\|$. □

A bounded linear operator A in a Hilbert space H is called *self-adjoint* if $A^* = A$, i.e. if $(Ax, y) = (x, Ay)$ for any $x, y \in H$.

Let A be an unbounded linear operator with its domain $D(A)$. We say that $y \in D(A^*)$, if there is an element $z \in H$ such that

$$(Ax, y) = (x, z)$$

for all $x \in D(A)$. If $y \in D(A^*)$, then we set $z = A^*y$. The operator A^* is called *adjoint* to A. An unbounded operator A is called *symmetric*, if

$$(Ax, y) = (x, Ay)$$

for all $x, y \in D(A)$. A symmetric operator A is called *self-adjoint* if $D(A^*) = D(A)$.

1.10 The spectrum of an operator

Let A be a linear operator in H. If a complex number λ is such that the operator $(A - \lambda I)^{-1}$ is defined on the whole H and is bounded, then λ is called a *regular* number for A. The set of all nonregular numbers is called the *spectrum* of the operator A.

A complex number λ is called an *eigenvalue* of an operator A if there exists an element $x \in H$, $x \neq 0$ such that $Ax = \lambda x$. This vector x is called an *eigenvector* of the operator A. If λ is an eigenvalue, then the operator $A - \lambda I$ is not invertible and λ belongs to the spectrum of A. The set of all eigenvalues of A is its *point spectrum*. A number λ is called a point of the *continuous spectrum* of A, if the operator $(A - \lambda I)^{-1}$ exists but is not defined on the whole space H.

If A is a bounded linear operator and λ is a regular point of A, then by Theorem 27 of Section 1.8 the operator $(A - \mu I)^{-1}$ is defined and bounded if $|\mu - \lambda| < 1/\|(A - \lambda I)^{-1}\|$. Therefore, all these values μ are regular. Thus the set of regular points is open and the spectrum is a closed set.

The operator $(A - \lambda I)^{-1}$ is called the *resolvent* of the operator A.

Theorem 29. *If A is a bounded linear operator in a Hilbert space H, then its spectrum is lying in the disk $|\lambda| \le r$, where*

$$r = \limsup_{n\to\infty} \sqrt[n]{\|A\|^n}.$$

Proof. Let $|\lambda| > r$. Then the series

$$\sum_{j=0}^{\infty} A^j \lambda^{-j-1}$$

is converging and defines an operator. It is easy to check that

$$(A - \lambda I) \sum_{j=0}^{\infty} A^j \lambda^{-j-1} = \sum_{j=0}^{\infty} A^j \lambda^{-j-1} (A - \lambda I) = -I.$$

Therefore

$$R_\lambda = -\sum_{j=0}^{\infty} A^j \lambda^{-j-1} \quad \text{and} \quad \|R_\lambda\| \leq \sum_{j=0}^{\infty} \|A\|^j |\lambda|^{-j-1} \quad \text{for} \quad |\lambda| > r. \qquad \square$$

1.11 Compact operators

A bounded linear operator A in a Hilbert space H is called *compact* if for any bounded sequence $\{x_k\}$ of elements of H the sequence $\{Ax_k\}$ contains a Cauchy sequence.

For example, any linear continuous operator A for which that the set $\operatorname{im} A = \{Ax,\, x \in H\}$ is finite dimensional, is compact.

The definition implies that the operators AB and BA are compact if A is compact and B is bounded. The unit operator in an infinite-dimensional Hilbert space is not compact since the bounded sequence of elements $x_1, \ldots, x_k, \ldots$ of an orthonormal base is such that $\|x_i - x_j\|^2 = 2$ for $i \neq j$ and cannot contain a Cauchy subsequence. It follows, in particular, that a compact operator in a infinite-dimensional Hilbert space cannot have a bounded inverse operator.

Theorem 30. *If A is a compact linear operator in a Hilbert space H, then the adjoint operator A^* is also compact.*

Proof. Let B be the unit ball in H. The set $M = A(B)$ is compact in H. Let us verify that the set F of the functions $\{f : \; f(x) = (x, y),\; y \in B\}$, defined in M, is compact in $C(M)$, the space of continuous functions defined on M. Indeed, this set is uniformly bounded since $|f(x)| \leq \|x\| \cdot \|y\| \leq \|x\|$ and the set M is bounded, and is equicontinuous, since $|f(x) - f(x')| \leq \|x - x'\|$. By the Arzela theorem (obviously valid in this situation) F is compact.

This means that for any $\varepsilon > 0$ there exists a finite ε-net, i.e. a set of functions $f_1(x), \ldots, f_N(x)$ such that for any $f \in F$ we have $|f(x) - f_j(x)| < \varepsilon$ for some j and all $x \in M$. Moreover, $f_j(x) = (x, y_j)$.

Therefore, for any $y \in B$ there is an element y_j such that

$$|(Ax, y - y_j)| < \varepsilon \quad \text{for all} \quad x \in B.$$

This means that

$$|(x, A^*y - A^*y_j)| < \varepsilon \quad \text{for all} \quad x \in B.$$

But then

$$\|A^*y - A^*y_j\| < \varepsilon,$$

i.e. the set of the elements A^*y_j for $j = 1, \dots, N$ is an ε-net in the set $A^*(B)$. Since ε is arbitrary, the closure of the set $A^*(B)$ is compact and hence the operator A is compact. □

1.12 Compact self-adjoint operators

Theorem 31. *Let A be a bounded self-adjoint operator in a Hilbert space H. Then*

(i) *The number (Ax, x) is real for any $x \in H$.*
(ii) *The eigenvalues of the operator A are real.*
(iii) *The eigenvectors of the operator A corresponding to different eigenvalues are mutually orthogonal.*
(iv) $\|A\| = \sup_{\|x\|=1} |(Ax, x)|$.

Proof. Since $(Ax, x) = (x, Ax) = \overline{(Ax, x)}$, (i) follows immediately. If $Ax = \lambda x$, then $(Ax, x) = \lambda(x, x) = \lambda\|x\|^2$ and (ii) follows from (i).

Let $Ax = \lambda x,\ Ay = \mu y$ and $\lambda \neq \mu$. Then

$$(Ax, y) = \lambda(x, y), \quad (x, Ay) = \mu(x, y).$$

Since $(Ax, y) = (x, Ay)$ and $\lambda \neq \mu$, it follows that $(x, y) = 0$.

Set

$$M = \sup_{\|x\|=1} |(Ax, x)|.$$

It is clear that $|(Ax, x)| \leq \|Ax\|\|x\| \leq \|A\|\|x\|^2$. Therefore, $M \leq \|A\|$. Obviously, for any $x, y \in H$

$$(A(x + y), x + y) - (A(x - y), x - y) = 4\,\mathrm{Re}(Ax, y).$$

Thus

$$\mathrm{Re}(Ax, y) \leq \frac{1}{4}M(\|x + y\|^2 + \|x - y\|^2) = \frac{1}{2}M(\|x\|^2 + \|y\|^2).$$

If $\|x\| = 1$ and $y = Ax\|Ax\|^{-1}$, it implies that $\|Ax\| = \mathrm{Re}(Ax, y) \leq M$ and therefore $\|A\| \leq M$. □

Theorem 32. *A compact self-adjoint operator A in a Hilbert space H has at least one eigenvalue.*

Proof. Let $A \neq 0$, so that

$$M = \sup_{\|x\|=1} |(Ax, x)| \neq 0.$$

For definiteness, let $M = \sup_{\|x\|=1}(Ax, x)$. There exists a sequence $\{x_k\}$ such that

$$\|x_k\| = 1, \quad (Ax_k, x_k) \to M \text{ as } k \to \infty.$$

Since A is compact, the sequence $\{Ax_k\}$ contains a convergent subsequence $\{Ax'_k\}$. Let y_0 be the limit of this subsequence. Observe that $\|A\| = M$ by Theorem 31 and

$$\begin{aligned}\|Ax'_k - Mx'_k\|^2 &= \|Ax'_k\|^2 - 2M(Ax'_k, x'_k) + M^2 \\ &\leq \|A\|^2 - 2M(Ax'_k, x'_k) + M^2 = 2(M^2 - M(Ax'_k, x'_k)).\end{aligned}$$

However, $M(Ax'_k, x'_k) \to M^2$, so that the right-hand side tends to 0 and

$$\|Ax'_k - Mx'_k\| \to 0.$$

Therefore, $x'_k \to y_0/M = x_0$ and $\|x_0\| = 1$. Since $Ax_0 = Mx_0$, M is an eigenvalue. If

$$M = -\inf_{\|x\|=1}(Ax, x) \neq 0,$$

the proof is similar. □

Theorem 33. *A compact self-adjoint operator A in a Hilbert space H has at most a finite number of eigenvalues whose absolute values are greater than r for any $r > 0$. In particular, each eigenspace with nonzero eigenvalue is finite dimensional.*

Proof. Let $\lambda_1, \ldots, \lambda_k, \ldots$ be the eigenvalues of A whose absolute values are greater than r. Let $x_1, \ldots, x_k, \ldots$ be the corresponding sequence of the eigenvectors and $\|x_j\| = 1$. Since the eigenvectors with different eigenvalues are mutually orthogonal, we can suppose without loss of generality that $(x_j, x_k) = 0$ if $j \neq k$. The sequence $y_k = x_k/\lambda_k$ is bounded in H and therefore the sequence $\{Ay_k\}$ contains a convergent subsequence. However, this is impossible, since $Ay_k = x_k$ and $\|x_k - x_j\| = \sqrt{2}$ for any j and $k \neq j$. □

Theorem 34 (Hilbert-Schmidt). *Let A be a compact self-adjoint operator in a Hilbert space H. Then there exists an orthonormal system $\{x_k\}$ of eigenvectors of the operator A with nonzero eigenvalues $\{\lambda_k\}$ such that each element $x \in H$ is represented in a unique way as*

$$x = \sum_k c_k x_k + x_0,$$

where $x_0 \in \ker A$, c_k *are constants. Moreover,* $Ax = \sum_k c_k \lambda_k x_k$ *and if the system* $\{x_k\}$ *is infinite, then* $\lim_{k\to\infty} \lambda_k = 0$.

Proof. The last statement follows at once from Theorem 33.

As was shown in Theorem 32, there is an element x_1 such that $Ax_1 = \pm M_1 x_1$, where $M_1 = \sup_{\|x\|=1} |(Ax, x)|$. Let X_1 be the one-dimensional subspace generated by x_1 and $H = X_1 \oplus H_1$. If the restriction of A on H_1 is the zero operator, the theorem is proved. Otherwise, there exists an element $x_2 \in H_1$ such that $\|x_2\| = 1, Ax_2 = \pm M_2 x_2$, where

$$M_2 = \sup_{\|x\|=1,\, (x,x_1)=0} |(Ax, x)|.$$

Repeating these arguments, we obtain the sequence $\{x_k\}$ such that $\|x_k\| = 1$ and $Ax_k = \pm M_k x_k$, where

$$M_k = \sup_{\|x\|=1,\, (x,x_j)=0,\, j=1,\dots,k-1} |(Ax, x)|.$$

There are two possibilities: either after a finite number of steps the restriction of A on the subspace H_k orthogonal to $x_1, \dots, x_k$ is the zero operator or the sequence $\{x_k\}$ is infinite. In both cases the statement is true. □

Remark 35. Theorem 34 means that for each compact self-adjoint operator there is an orthonormal base in H of its eigenvectors. Indeed, in order to obtain such a base it is sufficient to add to the constructed system $\{x_k\}$ an orthonormal base of the subspace $\ker A$. This theorem generalizes in a natural way the theorem on reduction of a symmetric matrix to diagonal form.

Remark 36. If a compact operator A is not self-adjoint, it can have no eigenvectors. For example, the operator A in l_2 such that for the element x with coordinates $\{x_k\}$ the element $Ax = y$ has the coordinates

$$y_1 = 0,\ y_k = 2^{-k} x_{k-1}, \quad k = 2, 3, \dots,$$

is compact, but from $Ax = \lambda x$ it follows readily that $x = 0$.

1.13 Integral operators

Consider an example of a compact operator which plays an important role in what follows.

Let Ω be a bounded domain in $\mathbf{R}^n$, $K(x, y)$ be a measurable function defined in $\Omega \times \Omega$ and

$$Af(x) = \int_\Omega K(x, y) f(y)\, dy, \quad x \in \Omega.$$

Theorem 37. *If the integral*

$$\int_\Omega \int_\Omega |K(x, y)|^2\, dx\, dy$$

is finite and equal to N^2, then the linear operator A is compact in $L_2(\Omega)$ and $\|A\| \leq N$.

The proof is based on the following lemma, which is of interest on its own.

Lemma 38. *Let $\{A_k\}$ be a sequence of linear compact operators in a Hilbert space H and A be a linear operator in H such that*

$$\|A_k - A\| \to 0 \text{ as } k \to \infty.$$

Then A is compact.

Proof. Let $\{x_j\}$ be an arbitrary bounded sequence of elements of H such that $\|x_j\| \leq C$ for all j. Since A_1 is compact, the sequence $\{A_1 x_j\}$ contains a convergent subsequence $\{A_1 x_j^{(1)}\}$. Moreover, the sequence $\{A_2 x_j^{(1)}\}$ contains a convergent subsequence $\{A_2 x_j^{(2)}\}$. Of course, the sequence $\{A_1 x_j^{(2)}\}$ is convergent, too. Repeating the argument, one can find for any m a subsequence $\{x_j^{(m)}\}$ such that all the sequences $\{A_1 x_j^{(m)}\}, \dots, \{A_m x_j^{(m)}\}$ are convergent. Consider the diagonal sequence $y_j = x_j^{(j)}$, $j = 1, 2, \dots$. Each operator A_j transforms it in a convergent sequence.

Let ε be an arbitrary positive number. Let k be so big that $\|A_k - A\| < \varepsilon/3C$. Choose N so big that $\|A_k x_j^{(j)} - A_k x_l^{(l)}\| < \varepsilon/3$ for $j > N$, $l > N$. We have

$$\|Ay_j - Ay_l\| \leq \|Ay_j - A_k y_j\| + \|A_k y_j - A_k y_l\| + \|A_k y_l - Ay_l\|$$

$$\leq \|A - A_k\|(\|y_j\| + \|y_l\|) + \|A_k x_j^{(j)} - A_k x_l^{(l)}\| < \varepsilon.$$

This shows that the sequence $Ax_j^{(j)}$ is a Cauchy sequence. Therefore, the operator A is compact. □

Proof of Theorem 37. In virtue of the Fubini theorem the integral

$$\int_\Omega |K(x,y)|^2 \, dy$$

is defined for almost all $x \in \Omega$, so that for $f \in L_2(\Omega)$ the function $g(x) = Af(x)$ is defined almost everywhere in Ω. We have

$$|g(x)|^2 \leq \int_\Omega |K(x,y)|^2 \, dy \int_\Omega |f(y)|^2 \, dy,$$

so that

$$\int_\Omega |g(x)|^2 \, dx \leq \int_\Omega \int_\Omega |K(x,y)|^2 \, dy \, dx \int_\Omega |f(y)|^2 \, dy,$$

and hence $\|Af\| \leq N\|f\|$, i.e. $\|A\| \leq N$.

Let $\{\varphi_j(x)\}$ be a complete orthonormal system in $L_2(\Omega)$. Then

$$K(x,y) = \sum_{i,j=1}^{\infty} a_{ij}\varphi_i(x)\overline{\varphi_j(y)},$$

where

$$a_{ij} = \int_\Omega \int_\Omega K(x,y)\varphi_i(x)\overline{\varphi_j(y)}\,dx\,dy,$$

and the series converges in the space $L_2(\Omega) \times L_2(\Omega)$. Let

$$K_N(x,y) = \sum_{i,j=1}^{N} a_{ij}\varphi_i(x)\overline{\varphi_j(y)}$$

and

$$A_N f(x) = \int_\Omega K_N(x,y)f(y)\,dy, \quad x \in \Omega.$$

This operator is compact, since its image has a finite dimension N.

On the other hand, as we have seen,

$$\|A - A_N\|^2 \le \int_\Omega \int_\Omega |K(x,y) - K_N(x,y)|^2\,dx\,dy,$$

so that $\|A - A_N\| \to 0$ as $N \to \infty$. Applying Lemma 38 we obtain the result. □

1.14 The Lax-Milgram theorem

The following theorem, generalizing the Riesz theorem, is very useful in the theory of elliptic differential equations.

Theorem 39. *Let $B(x,y)$ be a functional defined on the Hilbert space $H \times H$ having the following properties:*

(i) $B(\alpha_1 x_1 + \alpha_2 x_2, y) = \alpha_1 B(x_1,y) + \alpha_2 B(x_2,y)$;
$B(x, \alpha_1 y_1 + \alpha_2 y_2) = \overline{\alpha_1} B(x,y_1) + \overline{\alpha_2} B(x,y_2)$;
(ii) $|B(x,y)| \le C\|x\|\|y\|$;
(iii) $|B(x,x)| \ge c_0\|x\|^2$, $c_0 = \text{const} > 0$ for all $x, y \in H$, $\alpha_1, \alpha_2 \in \mathbf{C}$.

Then there exists one and only one bounded linear operator T such that the inverse operator T^{-1} exists, is bounded, and

$$B(x,y) = (x, Ty), \quad x \in H,\ y \in H.$$

Besides, $\|T\| \le C$, $\|T^{-1}\| \le c_0^{-1}$.

Proof. For a fixed y the functional $B(x,y)$ is a linear bounded functional of x. Therefore, by the Riesz theorem there is an element $z \in H$ such that

$$B(x,y) = (x,z).$$

It is clear that z depends on y linearly, so that $z = Ty$, where T is a linear operator, defined on H.

The operator T is bounded since by (ii)

$$\|T\| = \sup_{\|x\|=1,\ \|y\|=1} (x, Ty) = \sup_{\|x\|=1,\ \|y\|=1} B(x,y) \leq C.$$

On the other hand, by (iii) we have

$$c_0\|x\|^2 \leq |B(x,x)| = |(x,Tx)| \leq \|x\| \cdot \|Tx\|,$$

i.e. $\|Tx\| \geq c_0\|x\|$. Therefore, the operator T has a bounded inverse and $\|T^{-1}\| \leq c_0^{-1}$. □

Chapter 2

Functional Spaces

2.1 Notation and definitions

We state here some standard definitions and notation that will be used in what follows. $\mathbf{R}^n$ is an n-dimensional real space. The *points* of $\mathbf{R}^n$ are denoted by x (or $y, z, \ldots$) so that $x = (x_1, \ldots, x_n)$, $|x|^2 = \sum_{j=1}^n x_j^2$. If Ω is a set in $\mathbf{R}^n$, i.e. $\Omega \subset \mathbf{R}^n$, then $\overline{\Omega}$ is its *closure* and $\mathbf{R}^n \setminus \Omega$ is its *complement*, $\partial\Omega$ is its *boundary*. A *domain* in $\mathbf{R}^n$ is a connected open set.

A domain Ω is *star-like* with respect to a point x_0 if for any $x \in \Omega$ the segment connecting x and x_0 belongs to Ω. In particular, a *convex* domain is star-like with respect to its every point. A domain Ω_1 is *compactly embedded* in Ω $(\Omega_1 \subset\subset \Omega)$, if $\overline{\Omega}_1 \subset \Omega$ and Ω_1 is bounded.

The set of continuous functions defined in Ω is denoted by $C(\Omega)$. $C^k(\Omega)$ for $k > 0$ is the set of functions continuous in Ω whose derivatives up to order k are also continuous. The set of infinitely differentiable functions is denoted by $C^\infty(\Omega)$.

The *support* of a function f is the closure of the set of points x, in which $f(x) \neq 0$. It is denoted by $\operatorname{supp} f$. $C_0^k(\Omega)$ is the set of functions of $C^k(\Omega)$ having their supports compactly embedded in Ω, $0 \le k \le \infty$.

If $\Omega_1 \subset \Omega$, f_1 is defined on Ω_1, f_2 is defined on Ω, and $f_1(x) = f_2(x)$ for $x \in \Omega_1$, then f_2 is called a *continuation* of f_1 from Ω_1 on Ω and f_1 is called a *restriction* of f_2 from Ω to Ω_1.

A function f satisfies the *Lipschitz condition* in Ω if there is a constant L such that

$$|f(x) - f(y)| \le L|x - y|$$

for all $x \in \Omega, y \in \Omega$. A function f satisfies the *Hölder condition in Ω with an exponent* γ, $0 < \gamma \le 1$, if there is a constant L such that

$$|f(x) - f(y)| \le L|x - y|^\gamma$$

for all $x \in \Omega$, $y \in \Omega$.

A *multi-index* $\alpha = (\alpha_1, \ldots, \alpha_n)$ is a set of nonnegative integers,

$$|\alpha| = \alpha_1 + \ldots + \alpha_n, \ \alpha! = \alpha_1! \ldots \alpha_n!,$$

$$x^\alpha = x_1^{\alpha_1} \ldots x_n^{\alpha_n}, \ D^\alpha = \frac{\partial^{|\alpha|}}{\partial x_1^{\alpha_1} \ldots \partial x_n^{\alpha_n}}.$$

The inequality $\alpha \le \beta$ means that $\alpha_j \le \beta_j$ for all j, $1 \le j \le n$. The set of all derivatives $D^\alpha u$ such that $|\alpha| = m$ is denoted by $\nabla_m u$. By definition $\nabla u = \nabla_1 u$.

The *characteristic function* of a set $X \subset \mathbf{R}^n$ is the function f such that $f(x) = 1$ if $x \in X$ and $f(x) = 0$ if $x \notin X$.

By $C_1, C_2, \ldots$ we denote positive constants, whose values are not essential for us.

The terms *measure, Lebesgue measure*, and *measurable function* should be familiar.

The *Hausdorff measure of dimension d* is defined as follows. Let M be a set in $\mathbf{R}^n$ and U_ε be a covering of M by balls of radii $r_j < \varepsilon$. Then the Hausdorff d-measure of M is $H_d(M) = \lim_{\varepsilon \to 0} \inf_{U_\varepsilon} \sum_j r_j^d$. The limit exists since the inf depends on ε monotonically.

The volume of the unit ball in $\mathbf{R}^n$ is equal to $V_n = \frac{2\pi^{n/2}}{n\Gamma(n/2)}$, where Γ is the Euler function $\Gamma(t) = \int_0^\infty e^{-x} x^{t-1}\, dx, \quad t > 0$. The measure of the surface of the unit sphere in $\mathbf{R}^n$ is equal to $\omega_n = nV_n$.

A surface $S \subset \mathbf{R}^n$ is a set

$$\{x : \ x_j = f_j(t_1, \ldots, t_{n-1}), \ j = 1, \ldots, n\},$$

where $(t_1, \ldots, t_{n-1}) \in \Omega$, Ω is a domain in $\mathbf{R}^{n-1}$, $f_j \in C^1(\Omega)$, and the rank of the matrix $(\partial f_j / \partial t_k)$, $j = 1, \ldots, n$; $k = 1, \ldots, n-1$, is equal to $n-1$ at each point of Ω. If besides $f_j \in C^m(\Omega)$, then the surface S is said to be a surface of class C^m.

Analogously, S is a *k-dimensional surface in* $\mathbf{R}^n$ *of the class* C^m, where $m \ge 1$, if

$$S = \{x : \ x_j = f_j(t_1, \ldots, t_k), \ j = 1, \ldots, n\},$$

where $(t_1, \ldots, t_k) \in \Omega$, Ω is a domain in $\mathbf{R}^k$, $f_j \in C^m(\Omega)$ and the rank of the matrix $(\partial f_j / \partial t_l)$, $j = 1, \ldots, n$; $l = 1, \ldots, k$ is equal to k at each point of Ω.

A domain $\Omega \subset \mathbf{R}^n$ is a domain of the class C^k, if for any point $P \in \partial\Omega$ there is a neighborhood U_P of this point and a function $f \in C^k(U_P)$, such that $\partial f / \partial x_j \ne 0$ for some j and the intersection of Ω with U_P coincides with the intersection of U_P with the set $\{x : \ f(x) > 0\}$. If the function f satisfies the Lipschitz condition, then Ω is called a *Lipschitz domain.*

The *distance between two sets* M_1, M_2 in $\mathbf{R}^n$ is defined as

$$\rho(M_1, M_2) = \inf_{x \in M_1, y \in M_2} |x - y|.$$

$S(x_0, R)$ is the sphere $\{x : \ |x - x_0| = R\}$. The ball $\{x : \ |x - x_0| \le R\}$ is denoted by $B(x_0, R)$. The cube $\{x : \ |x_j - x_{0j}| \le R/2, \ j = 1, \ldots, n\}$ is denoted by $Q(x_0, R)$.

Let Ω be a bounded domain in $\mathbf{R}^n$. The classical *isoperimetric inequality* says that the Lebesgue measure $\mu_n(\Omega)$ and the $(n-1)$- Hausdorff measure of its boundary satisfy the relation

$$H_{n-1}(\partial\Omega) \geq nV_n^{1/n}\mu_n(\Omega)^{(n-1)/n}.$$

The equality in this relation is valid if and only if Ω is a ball. It means that among all domains of a given volume the ball has the minimal surface of the boundary. There are many ways to prove this inequality. Remark that if $n = 2$ and Ω is a polygon with p sides, then $\mu_1^2(\partial\Omega) \geq 4p\mu_2(\Omega)\tan\frac{\pi}{p}$. The equality is true here if and only if the polygon is regular. (See [Ba].)

2.2 Lebesgue integral

Recall some principal facts of theory of the Lebesgue integral. A function f, defined on a set $X \subset \mathbf{R}^n$, is called *simple* if it takes a finite or a countable set of values. Let f be a simple function such that $f(x) = a_j$ if $x \in A_j \subset \mathbf{R}^n$ and $a_j \neq a_k$ if $j \neq k$. By definition,

$$\int_X f\,d\mu = \sum_{j=1}^{\infty} a_j\mu(A_j)$$

if the series absolutely converges. Here $\mu(A_j)$ is a measure of the set A_j. If this is the n-dimensional Lebesgue measure, then $\int_X f\,d\mu$ is denoted by $\int_X f\,dx$.

Assume that a set X has a finite measure. Say that a function f is *summable over* X or $f \in L_1(X)$ if there is a sequence $\{f_k(x)\}$ of simple functions defined in X, converging uniformly to f and such that there exists $\lim_{k\to\infty}\int_X f_k\,d\mu$. This limit is denoted by $\int_X f\,d\mu$. It is easy to see that this definition is independent of the choice of the sequence $\{f_k(x)\}$.

If the measure $\mu(X)$ is infinite, the integral is defined as follows. Let $X = \lim_{j\to\infty} X_j$, where each X_j has a finite measure. If there exists $\lim_{j\to\infty}\int_{X_j} f\,d\mu = I$ and this limit does not depend on the choice of X_j, then $f \in L_1(X)$ and $\int_X f\,d\mu = I$.

One can easily demonstrate the following properties of the Lebesgue integral:

(a) If $f_1, f_2 \in L_1(X)$, then $a_1f_1 + a_2f_2 \in L_1(X)$ for any complex numbers a_1, a_2.
(b) If $f \in L_1(X)$, then $|f| \in L_1(X)$ and inversely, if $|f| \in L_1(X)$, then $f \in L_1(X)$.
(c) If f is bounded and measurable on a set X of a finite measure, then $f \in L_1(X)$.
(d) If $0 \leq f_1(x) \leq f_2(x)$ almost everywhere in X (i.e. for all $x \in X$ except of a set of zero measure) and $f_2 \in L_1(X)$, then $f_1 \in L_1(X)$ and $\int_X f_1(x)\,d\mu \leq \int_X f_2(x)\,d\mu$.

The proofs of the following theorems can be found in textbooks on function theory.

Theorem 1 (Lebesgue). *Let $f_k \in L_1(\Omega)$, $k = 1, 2, \ldots$, where $\Omega \subset \mathbf{R}^n$, and $|f_k(x)| \leq F(x)$, where $F \in L_1(\Omega)$. Let f_k converge to f almost everywhere in Ω as $k \to \infty$. Then $f \in L_1(\Omega)$ and*

$$\lim_{k\to\infty} \int_\Omega f_k(x)\,dx = \int_\Omega f(x)\,dx, \quad \lim_{k\to\infty} \int_\Omega |f(x) - f_k(x)|\,dx = 0.$$

Theorem 2 (Fatou). *If f_k, $k = 1, 2, \ldots$, are nonnegative summable in $\Omega \subset \mathbf{R}^n$ functions converging to f almost everywhere in Ω as $k \to \infty$, and such that*

$$\int_\Omega f_k(x)\,dx \leq C,$$

where the constant C is independent of k, then $f \in L_1(\Omega)$ and

$$\int_\Omega f(x)\,dx \leq C.$$

Theorem 3 (Levi). *If $f_k \in L_1(\Omega)$, $k = 1, 2, \ldots$, $\Omega \subset \mathbf{R}^n$, and*

$$0 \leq f_k(x) \leq f_{k+1}(x), \quad \int_\Omega f_k(x)\,dx \leq C,$$

where the constant C is independent of k, then there exists the limit

$$\lim_{k\to\infty} f_k(x) = f(x)$$

almost everywhere, the function $f \in L_1(\Omega)$, and

$$\lim_{k\to\infty} \int_\Omega f_k(x)\,dx = \int_\Omega f(x)\,dx \leq C.$$

Theorem 4 (Fubini). *If a function $f(x, y)$, $x \in \mathbf{R}^n$, $y \in \mathbf{R}^m$, is summable in $\mathbf{R}^{n+m}$, then the integral $\int_{\mathbf{R}^n} f(x, y)\,dx$ exists there for almost all $y \in \mathbf{R}^m$ and the integral $\int_{\mathbf{R}^m} f(x, y)\,dy$ exists there for almost all $x \in \mathbf{R}^n$, and also*

$$\int_{\mathbf{R}^{n+m}} f(x, y)\,dx\,dy = \int_{\mathbf{R}^m} \left(\int_{\mathbf{R}^n} f(x, y)\,dx \right) dy$$
$$= \int_{\mathbf{R}^n} \left(\int_{\mathbf{R}^m} f(x, y)\,dy \right) dx.$$

Corollary 5. *If $u \in L_1(\mathbf{R}^n)$, $u(x) \geq 0$, then*

$$\int_{\mathbf{R}^n} u(x)\,dx = \int_0^\infty \mu(t)\,dt,$$

where $\mu(t) = \operatorname{meas}_n\{x : u(x) > t\}$ and meas_n is the n-dimensional Lebesgue measure.

Moreover, if $u(x) \geq 0, v(x) \geq 0$*, and the functions* u, v, uv *are summable, then*

$$\int_{\mathbf{R}^n} u(x)v(x)\,dx = \int_0^\infty [\int_{u(x)>t} v(x)\,dx]\,dt.$$

Proof. Let $t > 0$ and $E_t = \{(t,x) \in \mathbf{R}^{n+1} : \ x \in \mathbf{R}^n, \ u(x) > t\}$, $\theta(t,x)$ be the characteristic function of E_t. By the Fubini theorem,

$$\int_{\mathbf{R}^{n+1}} \theta(t,x)\,dt\,dx = \int_0^\infty (\int_{\mathbf{R}^n} \theta(t,x)\,dx)\,dt = \int_{\mathbf{R}^n} (\int_0^\infty \theta(t,x)\,dt)\,dx.$$

Since $\int_{\mathbf{R}^n} \theta(t,x)\,dx = \mu(t)$, $\int_0^\infty \theta(t,x)\,dt = u(x)$, the proof is complete.

The second statement follows in the same way by consideration of the integral $\int_{\mathbf{R}^{n+1}} \theta(t,x)v(x)\,dt\,dx$. □

2.3 Level sets of functions of a real variable

Let a function u be defined in a domain $\Omega \subset \mathbf{R}^n$. Its *level set* $K = \{x \in \Omega : \ u(x) = \text{const}\}$ is called *critical* if $\nabla u(x) = 0$ for some $x \in K$. As $u(M)$ we will denote the image of a set M by the map $x \mapsto u(x)$.

Theorem 6 (Sarde). *If* $u \in C^\infty(\Omega)$ *and* K *is its critical set, then* $\text{meas}_1\, u(K) = 0$.

Corollary 7. *If* $u \in C^\infty(\Omega)$*, then for almost all* t *the set* E_t *is a surface of the class* C^∞.

This follows from Theorem 6 and the implicit function theorem.

The proof of Theorem 6 is based on the following lemma.

Lemma 8 (Vitali). *Let* E *be a bounded subset of the line* $\mathbf{R}^1$ *covered by a system of closed segments such that for any* $\varepsilon > 0$ *and any* $x \in E$ *there exists a segment of this system containing* x *and having length less than* ε*. Then there exists a countable set of nonintersecting segments* $I_1, \dots, I_N, \dots$ *such that*

$$\text{meas}_1(E \setminus \cup_{j=1}^\infty I_j) = 0.$$

Proof. A covering of E satisfying the conditions of Lemma 8 is called a *Vitali covering.* Let $E \subset [a,b]$, where a and b are finite. We can choose from the given covering the segments contained in $[a,b]$. Let the upper bound of lengths of these segments be c_1. Let I_1 be a segment of the covering having length $> c_1/2$. Let $E_1 = [a,b] \setminus I_1$. Let us take those segments of the covering that are contained in E_1. They give a Vitali covering of the set $E \setminus I_1$. Let the upper bound of lengths of these segments be c_2. As I_2 take a segment from this covering of length $\geq c_2/2$. If the segments $I_1, \dots, I_k$ are already defined, set $E_k = [a,b] \setminus \cup_{j=1}^k I_j$. Let us consider the segments of the covering contained in E_k. They give a Vitali covering of the set $E \setminus \cup_{j=1}^k I_j$. Let the upper bound of their lengths be c_k. As I_{k+1} take a segment of this covering of length $\geq c_k/2$. In this way we obtain a set of nonintersecting segments $I_1, \ldots, I_k, \ldots$. Let us show that

$$\text{meas}_1(E \setminus \cup_{j=1}^{\infty} I_j) = 0.$$

Let the length of I_j be equal to d_j. Since I_j are mutually nonintersecting and they all are lying in the finite segment $[a, b]$, we have $\sum_{j=1}^{\infty} d_j < \infty$. Let J_j be the segment concentric with I_j and having length $5d_j$. We show that for any $N > 1$

$$\cup_{j=N}^{\infty} J_j \supset E \setminus \cup_{j=1}^{\infty} I_j.$$

Since $\sum_{j=1}^{\infty} \text{meas}_1 J_j < \infty$, this implies that $\text{meas}_1(E \setminus \cup_{j=1}^{\infty} I_j) = 0$. Fix N and let $x \in E \setminus \cup_{j=1}^{\infty} I_j$. It is clear that $x \notin I_1 \cup I_2 \cup \ldots \cup I_N$. Let I be a segment of the original covering, whose length δ is less than $\min(d_1, \ldots, d_N)$, containing x and not intersecting with $I_1, \ldots, I_N$. There exists a number $L > N$ such that $d_L < \delta/2$. Assume that L is the minimal number satisfying this condition. Since the least upper bound of lengths of the segments covering $E \setminus \cup_{j=1}^{L-1} I_j$ and nonintersecting $I_1, \ldots, I_{L-1}$ is less than δ, one of the segments $I_1, \ldots, I_{L-1}$ intersects I. Let it be a segment I_s. Then $J_s \ni x$ and $s \geq N$. Therefore $\cup_{j=N}^{\infty} J_j \supset E \setminus \cup_{j=1}^{\infty} I_j$ and thus the system of segments $I_1, \ldots, I_j, \ldots$ satisfies all the conditions. □

Proof of Theorem 6. It is sufficient to consider the case of a bounded domain Ω. Let

$$K_n = \{x : \ x \in \Omega, \ D^{\alpha} u(x) = 0 \quad \text{if} \quad 1 \leq |\alpha| \leq n\}.$$

If G is a domain in $\mathbf{R}^n$ and $u \in C(\overline{G})$, we set $\text{osc}(u, G) = \max_{\overline{G}} u - \min_{\overline{G}} u$. Let $\varepsilon > 0$ and $x \in K_n$. There exists a ball $B = B(x, r) \subset \Omega$, such that $\text{osc}(u, B) \leq \varepsilon r^n$. Let $t \in u(K_n)$ and $x(t)$ be a point in K_n such that $u(x(t)) = t$. Consider the segment $[t - \delta, t + \delta]$, where $\delta = \varepsilon r^n/2$. The set of all such segments with arbitrary $\varepsilon > 0$ forms a Vitali covering of $u(K_n)$. In virtue of Lemma 8, there is a sequence of such segments $I_1, \ldots, I_j \ldots$ covering the set $u(K_n)$ excluding a set of zero measure. Let $I_m = [t_m - \delta_m, t_m + \delta_m]$. Let $x_m \in K_n$, $u(x_m) = t_m$. By construction, the ball $B(x_m, \gamma_m)$, where $\gamma_m = (\delta_m/\varepsilon)^{1/n}$, belongs to the preimage of the segment I_m under the map $x \mapsto u(x)$. Therefore, $\text{meas}_n u^{-1}(I_m) \geq V_n \delta_m/\varepsilon$. Since the segments I_m are mutually nonintersecting, their preimages are nonintersecting, too. Thus

$$\sum_{m=1}^{\infty} \delta_m \leq \varepsilon/V_n \sum_{m=1}^{\infty} \text{meas}_n u^{-1}(I_m) \leq \text{meas}_n \Omega \cdot \varepsilon/V_n,$$

i.e. $\text{meas}_1 u(K_n) = 0$.

Now we prove Theorem 6 using induction in n. The case $n = 1$ has already been considered, since then $K = K_1$. Assume that Theorem 6 is true for some n. If $x \in K \setminus K_{n+1}$, then for some α with $|\alpha| < n + 1$ and for some i we have

$$D^{\alpha} u(x) = 0, \ D_i D^{\alpha} u(x) \neq 0.$$

Let $X_{\alpha,i}$ be a set of points x, at which Theorem 6 is true. Let us show that $\text{meas}_1 u(X_{\alpha,i}) = 0$. Let, for example, $i = n+1$. Set $\hat{x} = (x_1, \ldots, x_n)$, if $x = (x_1, \ldots, x_{n+1})$, and $f = D^\alpha u$. Then we have $f(x) = 0$, $\partial f(x)/\partial x_{n+1} \neq 0$, if $x \in X_{\alpha,n+1}$. By the implicit function theorem there is a neighborhood U_{x_0} of any point $x_0 \in X_{\alpha,n+1}$, such that the part of the surface $f(x) = 0$ lying in U_{x_0} is described by the equation $x_{n+1} = \psi(\hat{x})$ and $\psi \in C^1$. Since from any covering of $X_{\alpha,n+1}$ one can choose a countable subcovering, it suffices to prove that $\text{meas}_1 u(X_{\alpha,n+1} \cap U) = 0$. If $x \in X_{\alpha,n+1} \cap U$, then $u(x) = u(\hat{x}, \psi(\hat{x}))$. Let P be the projection of the set $X_{\alpha,n+1} \cap U$ on the hyperplane $x_{n+1} = 0$. Since $D_i u(\hat{x}, \psi(\hat{x})) = 0$ for all $i = 1, \ldots, n$, $\hat{x} \in P$, by the inductive assumption it follows that $\text{meas}_1 u(P, \psi(P)) = 0$. Since $u(X_{\alpha,n+1} \cap U) = u(P, \psi(P))$ and

$$\text{meas}_1 u(K) \leq \text{meas}_1 u(K_{n+1} + \sum_{|\alpha| \leq n} \sum_{i=1}^{n+1} u(X_{\alpha,i}) = 0,$$

the proof is complete. □

Remark 9 (Sarde). The condition $u \in C^\infty$ can be weakened. The sharp condition is $u \in C^n$.

Lemma 10. *Let Ω be a C^1-domain in the space $\mathbf{R}^n$, $u(x) \in C_0^\infty(\Omega)$, $\Phi(x)$ satisfy the Lipschitz condition in $\overline{\Omega}$, $u(x) \geq 0$ in Ω, $\Phi(x) \geq 0$, and ds be the $(n-1)$-dimensional Lebesgue measure. Then*

$$\int_\Omega \Phi(x) |\nabla u(x)| \, dx = \int_0^\infty dt \int_{E_t} \Phi(x) \, ds,$$

where $E_t = \{x : \ u(x) = t\}$.

Proof. Observe that for almost all t the set E_t is a surface of the class C^1. Set $\vec{w}(x) = (\varepsilon^2 + |\nabla u|^2)^{-1/2} \Phi(x) \nabla u$. From the Green formula and Corollary 5 it follows that

$$\sum_1^n \int_\Omega w_i \frac{\partial u}{\partial x_i} \, dx = -\int_\Omega u \div \vec{w} \, dx = -\int_0^\infty dt \int_{u(x) \geq t} \div \vec{w} \, dx.$$

Applying once again the Green formula we obtain that for almost all $t > 0$

$$\int_{u(x) > t} \div \vec{w} \, dx = \sum_1^n \int_{u(x) = t} w_i n_i \, ds,$$

where $\vec{n} = (n_1, \ldots, n_n)$ is the unit normal vector, directed outside the domain $u(x) > t$. Since

$$\sum_1^n \int_\Omega w_i \frac{\partial u}{\partial x_i} \, dx = -\int_0^\infty dt \int_{E_t} \vec{w} \vec{n} \, ds,$$

we have

$$\int_\Omega \Phi(x) \frac{|\nabla u|^2}{(\varepsilon^2 + |\nabla u|^2)^{1/2}}\, dx = -\int_0^\infty dt \int_{E_t} \frac{\Phi(x)(\nabla u, \vec{n})}{(\varepsilon^2 + |\nabla u|^2)^{1/2}}\, ds.$$

Note that $\vec{n} = -\nabla u/|\nabla u|$. For ε tending to 0, and using the Levi theorem, we obtain that

$$\int_\Omega \Phi(x)|\nabla u|\, dx = \int_0^\infty dt \int_{E_t} \Phi(x)\, ds. \qquad \square$$

2.4 Symmetrization

Let u be a summable function. Consider the function

$$\mu(t) = \text{meas}_n\{x :\ u(x) > t\}.$$

This is a nonincreasing function, and for any t we have $\lim_{\varepsilon\to 0} \mu(t+\varepsilon) = \mu(t)$; this follows immediately from countable additivity of the Lebesgue measure. The function μ can have a countable set of points of discontinuity. If the function u is continuous, then $\mu(t)$ is strictly monotone, since the set $\{x \in \mathbf{R}^n :\ t_1 < u(x) < t_2\}$ for any $t_1 < t_2$ contains a ball. The function μ is discontinuous at a point t if the set $\{x \in \mathbf{R}^n :\ u(x) = t\}$ has a positive Lebesgue measure. A set

$$E_t = \{x \in \mathbf{R}^n :\ u(x) = t\}$$

is called *a level set* of the function u.

Let u be a measurable nonnegative function in $\mathbf{R}^n$. Then the function

$$u^*(x) = \sup\{t :\ \mu(t) > V_n|x|^n\},$$

where $V_n = \pi^{n/2}/\Gamma(1+n/2)$ is the volume of the unit ball in $\mathbf{R}^n$ and Γ is the Euler gamma function, is called the *symmetrization* of the function u. This function depends on $|x|$ only and has the same function $\mu(t)$ as u. Corollary 5 implies that

$$\int_{\mathbf{R}^n} u(x)^p\, dx = \int_{\mathbf{R}^n} u^*(x)^p\, dx$$

for any p.

We shall list some elementary properties of $u^*(x)$.

This function has nonnegative values. If $|x| \geq |y|$, then $u^*(y) \leq u^*(x)$. If $|x| \geq [V_n^{-1}\,\text{meas}\{x :\ u(x) > 0\}]^{1/n}$, then $u^*(x) = 0$ and $u^*(x) > 0$, if $|x| < [V_n^{-1}\,\text{meas}\{x :\ u(x) > 0\}]^{1/n}$. In general, the function $u^*(x)$ can be unbounded. However, if u is bounded, then u^* is bounded, too. Namely, $u^*(x) \leq u^*(0) = ess\ \sup\ u(x)$, where $\text{ess sup}\ u(x) = M$ and M is a number such that meas $\{x :\ u(x) > M\} = 0$ and for any $\varepsilon > 0$ we have $meas\{x :\ u(x) > M - \varepsilon\} > 0$.

The definition of $u^*(x)$ implies that $u^*(x) \geq t$, if $\mu(t) > V_n|x|^n$.

Observe that
$$\mu(u^*(x)) \le V_n|x|^n.$$
Indeed, let $\mu(u^*(x)) > V_n|x|^n$. Since $\lim_{\varepsilon\to 0+}\mu(t+\varepsilon) = \mu(t)$, we obtain that $\mu(u^*(x)+\varepsilon) > V_n|x|^n$ for $\varepsilon > 0$, contrary to the definition of $u^*(x)$. If $\mu(t)$ is continuous at $t = u^*(x)$, then $\mu(u^*(x)) = V_n|x|^n$.

Let us show that the function $u_1(|x|) = u^*(x)$ is continuous from the left. If a sequence of points $\{x_m\}$ is such that $x_m \to x$ as $m \to \infty$, $|x_m| \le |x|$, then $u^*(x_m) \to u^*(x)$. Assume the contrary. In view of the monotonicity of $u^*(x)$ in $|x|$ there are $\alpha > 0$ and a sequence $y_1, \dots, y_m, \dots$ such that $|y_m| < |y_{m+1}| < |x|$, and $u^*(y_m) > u^*(x) + \alpha$. By the definition for any m there is a t_m such that $u^*(x) + \alpha < t_m \le u^*(y_m)$ and $\mu(t_m) > V_n|y_m|^n$. Let t' be a limit point of the sequence t_m. Then $\mu(t') \ge V_n|x|^n$. Therefore, $u^*(x) \ge t'$, but $t' \ge u^*(x) + \alpha$ and we come to a contradiction.

The function $u^*(x)$ can be discontinuous. Let us prove that continuity of u implies continuity of $u^*(x)$. Since the function $u_1(|x|) = u^*(x)$ is nondecreasing and $0 \le u_1(|x|) \le M$, where $M = \operatorname{ess\,sup}\ u(x)$, it is continuous if it takes all values from the segment $[0, M]$. Let $0 < \tau < M$. Set $|x'| = V_n^{-1}\mu(\tau)$. By definition $u^*(x') = \sup\{t,\ \mu(t) > \mu(\tau)\}$. Therefore, $u^*(x') \le \tau$ and $\mu(u^*(x')) \le \mu(\tau)$. If $u^*(x') < \tau$ then $\mu(u^*(x')) \ge \mu(\tau)$ in virtue of the monotonicity of the function μ. Therefore, $\mu(t)$ is constant on the segment $[u^*(x'), \tau]$, which is impossible for a continuous function $u^*(x)$.

Lemma 11. *If a nonnegative function u satisfies the Lipschitz condition, then u^* satisfies the Lipschitz condition, too.*

Proof. By Lemma 10 we have
$$\int_{t-h<u(x)\le t} |\nabla u(x)|\, dx = \int_{t-h}^{t} \mu_{n-1}(E(\tau))\, d\tau.$$
Hence and from the isoperimetric inequality it follows that
$$\int_{t-h<u(x)\le t} |\nabla u(x)|\, dx \ge V_n \mu(t)^{1-1/n} h.$$
If $L = \operatorname{ess\,sup}|\nabla u|$, then
$$V_n\mu(t)^{1-1/n}h \le L(\mu(t-h) - \mu(t)).$$
Observe that for any $x, y \in \mathbf{R}$ there exists $z \in]x, y[$ such that
$$x^n - y^n = nz^{n-1}(x-y).$$
Therefore for any a and b there is a $\tau \in (a, b)$ such that
$$\mu^{1/n}(a) - \mu^{1/n}(b) = 1/n\mu(\tau)^{1/n-1}(\mu(a) - \mu(b)) \ge 2V_n/(nL).$$

This means that

$$\frac{|u^*(x)-u^*(y)|}{|(|x|-|y|)|}=\frac{|u_1(|x|)-u_1(|y|)|}{V_n^{-1/n}|\mu^{1/n}(u^*(x))-\mu^{1/n}(u^*(y))|}\le 2LV_n^{-1+1/n}n=L_1$$

and therefore $|u^*(x)-u^*(y)|\le L_1|x-y|$. Here, as above, $u_1(|x|)=u^*(x)$. □

Theorem 12. *If* $u(x)\in C_0^\infty(\mathbf{R}^n),\ u(x)\ge 0,$ *then*

$$\int_{\mathbf{R}^n}|\nabla u^*|^p\,dx\le\int_{\mathbf{R}^n}|\nabla u|^p\,dx,\ p\ge 1.$$

Proof. Let us prove the inequality

$$\int_{E_t}|\nabla u|^{p-1}\,d\mu_{n-1}\ge[-\mu'(t)]^{1-p}[\mu_{n-1}(E_t)]^p, \tag{1}$$

where $E_t=\{x:\ u(x)=t\}$. Note that this inequality holds for any function u satisfying the Lipschitz condition, if its level sets E_t are smooth for almost all t. For $p=1$ this inequality is obvious. Let $p>1$. The Hölder inequality yields

$$\frac{1}{h}\int_{t<u(x)\le t+h}|\nabla u(x)|\,dx\le[\frac{1}{h}\int_{t<u(x)\le t+h}|\nabla u|^p\,dx]^{1/p}[-\frac{\mu(t+h)-\mu(t)}{h}]^{1-1/p}.$$

Therefore

$$-\frac{d}{dt}\int_{u(x)>t}|\nabla u|\,dx\le[-\mu'(t)]^{1-1/p}(-\frac{d}{dt}\int_{u(x)>t}|\nabla u|^p\,dx)^{1/p}$$

for almost all t. Here we have used the fact that a monotone function is differentiable almost everywhere. Moreover, by Lemma 10

$$\int_{u(x)>t}|\nabla u(x)|^p\,dx=\int_t^\infty d\tau\int_{E(\tau)}|\nabla u(x)|^{p-1}\,d\mu_{n-1}.$$

Therefore

$$-\frac{d}{dt}\int_{u(x)>t}|\nabla u(x)|^p\,dx=\int_{E(t)}|\nabla u(x)|^{p-1}\,d\mu_{n-1}.$$

This is true for all $p\ge 1$, in particular,

$$-\frac{d}{dt}\int_{u(x)>t}|\nabla u|\,dx=\mu_{n-1}(E_t).$$

These relations imply (1). Furthermore, we obtain that

$$\begin{aligned}\int_{\mathbf{R}^n}|\nabla u(x)|^p\,dx&=\int_0^\infty d\tau\int_{E(\tau)}|\nabla u(x)|^{p-1}\,d\mu_{n-1}\\&\ge\int_0^\infty[\mu_{n-1}(E_t)]^p(-\mu'(t))^{1-p}\,dt.\end{aligned}$$

Using the isoperimetric inequality: $\mu_{n-1}(E_t) \geq \mu(t)^{(n-1)/n} V_n^{1/n} n$, we see that

$$\int_{\mathbf{R}^n} |\nabla u|^p \, dx \geq (\sqrt{\pi} n)^p \Gamma(1+\frac{n}{2})^{-p/n} \int_0^\infty \mu(t)^{p(1-1/n)} \cdot (-\mu'(t))^{1-p} \, dt.$$

Let a function u depends on $|x|$ only. Then we have

$$\int_{\mathbf{R}^n} |\nabla u|^p \, dx = \int_0^\infty d\tau \int_{E(\tau)} |\nabla u|^{p-1} \, d\mu_{n-1} = \int_0^\infty dt [\mu_{n-1}(E_t)]^p (-\mu'(t))^{1-p},$$

since the sets E_t are spheres and we have equality in the isoperimetric inequality. In particular, if a function $u(x)$ satisfies the Lipschitz condition, then it is true for u^* and

$$\int_{\mathbf{R}^n} |\nabla u^*|^p \, dx = \int_t^\infty d\tau \int_{E(\tau)} |\nabla u^*|^{p-1} \, d\mu_{n-1} = \int_0^\infty [\mu_{n-1}(E_t)]^p (-\mu'(t))^{1-p} \, dt,$$

and we obtain the result. □

2.5 The space $L_1(\Omega)$

Let Ω be a domain in $\mathbf{R}^n$. Consider the set of complex-valued summable functions f defined in Ω. Denote it by $L_1(\Omega)$. It is clear that it is a linear space. Let us identify two functions f_1, f_2 of $L_1(\Omega)$, if their difference vanishes almost everywhere in Ω, i.e. everywhere except a set of zero Lebesgue measure.

Define a norm in $L_1(\Omega)$, setting

$$\|f\|_{L_1(\Omega)} = \int_\Omega |f(x)| \, dx.$$

It is evident that for all complex α and β and all functions $f_1, f_2 \in L_1(\Omega)$ we have

$$\|\alpha f_1 + \beta f_2\|_{L_1(\Omega)} \leq |\alpha| \|f_1\|_{L_1(\Omega)} + |\beta| \|f_2\|_{L_1(\Omega)}.$$

Theorem 13. *The space $L_1(\Omega)$ is complete.*

Proof. Let the sequence $\{f_j(x)\}$ be a Cauchy sequence in $L_1(\Omega)$. Then for any integer $p \geq 1$ there exists an integer $N(p)$ such that

$$\|f_k - f_l\|_{L_1(\Omega)} \leq 2^{-p}$$

for $k > N(p)$, $l > N(p)$. Let $\{f_{N_j}\}$ be a subsequence such that

$$\|f_{N_j} - f_l\|_{L_1(\Omega)} \leq 2^{-j}, \text{ if } l > N_j.$$

Set $g_j = f_{N_j}$ and

$$\varphi_j(x) = |g_1(x)| + |g_2(x) - g_1(x)| + \cdots + |g_{j+1}(x) - g_j(x)|.$$

The sequence of nonnegative functions φ_j is increasing and

$$\int_\Omega |\varphi_j(x)|\,dx \le \int_\Omega |g_1(x)|\,dx + 2^{-1} + \cdots + 2^{-j} \le 1 + \int_\Omega |g_1(x)|\,dx.$$

By Theorem 3 there exists a function $\varphi(x) \in L_1(\Omega)$ such that the sequence $\{\varphi_j(x)\}$ converges to it almost everywhere, $\varphi_j(x) \le \varphi(x)$, and

$$\int_\Omega |\varphi_j(x) - \varphi(x)|\,dx \to 0$$

as $j \to \infty$.

Since the series

$$g_1 + (g_2 - g_1) + \cdots + (g_{j+1} - g_j) + \cdots$$

converges absolutely almost everywhere in Ω, it follows that the sequence $\{g_j\}$ converges almost everywhere to a measurable function $f(x)$. Since $|g_j(x)| \le \varphi(x)$, by Theorem 1 the function f is summable and

$$\lim_{j\to\infty} \int_\Omega |f_{N_j}(x) - f(x)|\,dx = 0$$

as $j \to \infty$. Moreover

$$\int_\Omega |f_k(x) - f(x)|\,dx \le \int_\Omega |f_{N_j}(x) - f_k(x)|\,dx + \int_\Omega |f_{N_j}(x) - f(x)|\,dx$$

$$\le 2^{-j} + \int_\Omega |f_{N_j}(x) - f(x)|\,dx \le 2^{1-j},$$

if $k \ge N_j$. Therefore, $\int_\Omega |f_j(x) - f(x)|\,dx \to 0$ as $j \to \infty$. □

2.6 The space $L_2(\Omega)$

The space $L_2(\Omega)$ has been previously considered in Chapter 1. It is a Hilbert space with the inner product

$$(f, g) = \int_\Omega f(x)\overline{g(x)}\,dx.$$

Now we will study it in more detail. This space contains functions f such that $|f(x)|^2 \in L_1(\Omega)$. Recall some elementary inequalities.

The Cauchy-Bounyakovsky inequality:

$$|\int_\Omega f(x)g(x)\,dx| \le (\int_\Omega |f(x)|^2\,dx)^{1/2}(\int_\Omega |g(x)|^2\,dx)^{1/2}.$$

The Minkovsky inequality:

$$\left(\int_\Omega |f(x)+g(x)|^2\,dx\right)^{1/2} \le \left(\int_\Omega |f(x)|^2\,dx\right)^{1/2} + \left(\int_\Omega |g(x)|^2\,dx\right)^{1/2}.$$

The latter inequality implies that $\alpha f_1 + \beta f_2 \in L_2(\Omega)$ if $f_1 \in L_2(\Omega)$, $f_2 \in L_2(\Omega)$, and α and β are constants. As usual, we identify two functions if they coincide almost everywhere. The norm in the space $L_2(\Omega)$ is defined as

$$\|f\| = \left(\int_\Omega |f(x)|^2\,dx\right)^{1/2}.$$

Theorem 14. *The space $L_2(\Omega)$ is complete.*

Proof. Let the sequence of functions $\{f_j(x)\}$ be a Cauchy sequence in $L_2(\Omega)$. Then for any integer $p \ge 1$ there exists an integer $N(p)$ such that

$$\|f_k - f_l\|_{L_2(\Omega)} \le 2^{-p}$$

for $k > N(p)$, $l > N(p)$. Let $\{f_{N_j}\}$ be a subsequence such that

$$\|f_{N_j} - f_l\|_{L_2(\Omega)} \le 2^{-j}, \text{ if } l > N_j.$$

Set $g_j = f_{N_j}$ and

$$\varphi_j(x) = |g_1(x)| + |g_2(x) - g_1(x)| + \cdots + |g_{j+1}(x) - g_j(x)|.$$

Since the sequence of nonnegative functions $\{|\varphi_j(x)|^2\}$ is increasing, we have by the Minkovsky inequality

$$\left(\int_\Omega |\varphi_j(x)|^2\,dx\right)^{1/2} \le \left(\int_\Omega |g_1(x)|^2\,dx\right)^{1/2} + 2^{-1} + \cdots + 2^{-j}$$
$$\le 1 + \left(\int_\Omega |g_1(x)|^2\,dx\right)^{1/2}.$$

By Theorem 3 there exists a function $\varphi(x)$ such that $\varphi(x)^2$ is summable and the sequence $\varphi_j(x)^2$ converges to $\varphi(x)^2$ almost everywhere.

Since the series $g_1+(g_2-g_1)+\cdots+(g_{j+1}-g_j)+\cdots$ converges absolutely almost everywhere in Ω, the sequence g_j converges almost everywhere to a measurable function $f(x)$. The sequence $g_{j+k} - g_j$ converges as $k \to \infty$ almost everywhere to a measurable function $f(x) - g_j(x)$. Besides,

$$|g_{j+k} - g_j|^2 \le (|g_{j+k} - g_{j+k-1}| + \cdots + |g_{j+1} - g_j|)^2 \le |\varphi_{j+k}(x)|^2 \le |\varphi(x)|^2.$$

By Theorem 1 we have

$$\lim_{k\to\infty} \int_\Omega |g_{j+k} - g_j|^2\,dx = \int_\Omega |f(x) - g_j(x)|^2\,dx.$$

But then

$$\int_\Omega |f(x) - g_j(x)|^2\, dx \le 2^{-j},$$

i.e. f is the limit of g_j in $L_2(\Omega)$.

Moreover

$$(\int_\Omega |f_k(x) - f(x)|^2\, dx)^{1/2}$$

$$\le (\int_\Omega |f_{N_j}(x) - f_k(x)|^2\, dx)^{1/2} + (\int_\Omega |f_{N_j}(x) - f(x)|^2\, dx)^{1/2}$$

$$\le 2^{-j} + (\int_\Omega |f_{N_j}(x) - f(x)|^2\, dx)^{1/2} \le 2^{1-j},$$

if $k \ge N_j$. Therefore

$$\int_\Omega |f_k(x) - f(x)|^2\, dx \to 0$$

as $k \to \infty$. □

2.7 The space $L_p(\Omega)$, $p > 1$

The space $L_p(\Omega)$ contains functions f such that $|f(x)|^p \in L_1(\Omega)$. Recall some elementary inequalities.

Theorem 15 (Hölder). *If $f \in L_p(\Omega)$, $g \in L_q(\Omega)$, where p and q are such that $p > 1$, $q > 1$, $1/p + 1/q = 1$, then*

$$|\int_\Omega f(x)g(x)\, dx| \le (\int_\Omega |f(x)|^p\, dx)^{1/p} (\int_\Omega |g(x)|^q\, dx)^{1/q}.$$

Proof. The proof of this inequality is based on the inequality:

$$XY \le X^p/p + Y^q/q,\ X \ge 0,\ Y \ge 0.$$

To prove the latter, note that the function $XY - X^p/p$ of X has on $[0, \infty)$ the only point of maximum at the point X such that $Y = X^{p-1}$ and the maximal value is equal to Y^q/q; therefore, $XY \le X^p/p + Y^q/q$. Now setting

$$X = f(\int_\Omega f^p(x)\, dx)^{-1/p};\ Y = g(\int_\Omega g^q(x)\, dx)^{-1/q}$$

and integrating the proved inequality over Ω, we obtain the Hölder inequality. □

Theorem 16 (Minkovsky). *If $f \in L_p(\Omega)$, $g \in L_p(\Omega), p > 1$, then*

$$(\int_\Omega |f(x) + g(x)|^p \, dx)^{1/p} \leq (\int_\Omega |f(x)|^p \, dx)^{1/p} + (\int_\Omega |g(x)|^p \, dx)^{1/p}.$$

Proof. This inequality is obvious if $p = 1$. Let $p > 1$. Then using the Hölder inequality we obtain that

$$\int_\Omega |f(x) + g(x)|^p \, dx$$

$$\leq \int_\Omega |f(x) + g(x)|^{p-1}|f(x)| \, dx + \int_\Omega |f(x) + g(x)|^{p-1}|g(x)| \, dx$$

$$\leq (\int_\Omega |f(x) + g(x)|^p \, dx)^{(p-1)/p}(\int_\Omega |f(x)|^p \, dx)^{1/p}$$

$$+(\int_\Omega |f(x) + g(x)|^p \, dx)^{(p-1)/p}(\int_\Omega |g(x)|^p \, dx)^{1/p}.$$

Therefore

$$(\int_\Omega |f(x) + g(x)|^p \, dx)^{1/p} \leq (\int_\Omega |f(x)|^p \, dx)^{1/p} + (\int_\Omega |g(x)|^p \, dx)^{1/p}. \qquad \square$$

The Minkovsky inequality implies that $\alpha f_1 + \beta f_2 \in L_p(\Omega)$, if $f_1 \in L_p(\Omega)$, $f_2 \in L_p(\Omega)$, and α and β are constants. As usual, we identify two functions, if they coincide almost everywhere. The norm in the space $L_p(\Omega)$ is defined as

$$\|f\|_p = (\int_\Omega |f(x)|^p \, dx)^{1/p}.$$

Remark 17 (Clarkson).

$$\|\frac{u+v}{2}\|_p^p + \|\frac{u-v}{2}\|_p^p \leq \frac{1}{2}(\|u\|_p^p + \|v\|_p^p), \; 2 \leq p < \infty;$$

$$\|\frac{u+v}{2}\|_p^p + \|\frac{u-v}{2}\|_p^p \geq \frac{1}{2}(\|u\|_p^p + \|v\|_p^p), \; 1 \leq p \leq 2.$$

We do not state the proof. (See [Ad], [So].)

Theorem 18. *The space $L_p(\Omega)$ is complete.*

The proof of this theorem is the same as in the case $p = 2$.

Theorem 19. *The space $L_p(\Omega)^*$ of linear continuous functionals in $L_p(\Omega)$ is isomorphic to the space $L_q(\Omega)$, where $1/p + 1/q = 1$. This space is reflexive, i.e. $L_p(\Omega)^{**} = L_p(\Omega)$.*

Proof. If $p > 1$ and $g \in L_q(\Omega)$, then the linear functional

$$l(f) = \int_\Omega f(x)g(x) \, dx$$

is bounded by Theorem 15. By the Hölder inequality $\|l\| \le \|g\|_q$. On the other hand, taking $f(x) = g(x)|g(x)|^{q-2}$ we see that

$$f \in L_p(\Omega),\ \|f\|_p = \|g\|_q^{q-1} \text{ and } l(f) = \|g\|_q^q.$$

Therefore $\|l\| = \|g\|_q$.

The converse statement is also true: if l is a linear bounded functional in $L_p(\Omega)$, then there exists a function $g \in L_q(\Omega)$ such that $l(f) = \int_\Omega f(x)g(x)\,dx$ for all $f \in L_p(\Omega)$. We omit the proof, which can be found, for example, in [KaAk]. □

Theorem 20. *Each bounded set in* $L_p(\Omega)$, $p > 1$ *is weakly compact.*

We do not state the proof here. See, for example, [Ad], [So], [KaAk].

Remark that these spaces can be considered with an abstract measure $d\mu$ in the place of the Lebesgue measure dx.

2.8 Density of the set of continuous functions in $L_1(\Omega)$

Now we show that any function from $L_1(\Omega)$ can be approximated by continuous functions.

Theorem 21. *If* $f \in L_1(\Omega)$, *then for any* $\varepsilon > 0$ *there exists a continuous function* $\varphi(x) \in C_0(\Omega)$ *(see* 2.2*) with compact support such that*

$$\int_\Omega |f(x) - \varphi(x)|\,dx < \varepsilon.$$

Proof. By definition, f is a limit in $L_1(\Omega)$ of a sequence of simple summable compactly supported functions. Any simple summable in Ω function is a limit in $L_1(\Omega)$ of a sequence of simple functions taking a finite number of values. Indeed, if $\varphi(x) = a_j$ on the set A_j, then by the definition

$$\int_\Omega \varphi(x)\,dx = \sum_{j=1}^{\infty} a_j \mu(A_j)$$

and therefore

$$\lim_{N\to\infty} \int_\Omega |\varphi(x) - \varphi_N(x)|\,dx = 0,$$

where $\varphi_N(x) = a_j$ on A_j for $j \le N$ and $\varphi_N(x) = 0$ on A_j if $j > N$. Thus the set of compactly supported functions, taking a finite number of values, is dense in $L_1(\Omega)$. We show now that any function of this set is a limit of a sequence of continuous compactly supported functions. It is sufficient to prove this for the functions taking two values only: 0 and 1, because any simple function is a linear combination of such functions.

Now let $\varphi(x) = 1$ for $x \in M$ and $\varphi(x) = 0$ for $x \notin M$, where M is a bounded measurable set. By definition of the Lebesgue measure, for any $\varepsilon > 0$ there are a closed set F and an open set G such that $F \subset M \subset G$, $|\mu(F) - \mu(G)| < \varepsilon$. Fix $\varepsilon > 0$ and the sets F, G and set

$$\varphi_\varepsilon(x) = \frac{\rho(x, \mathbf{R}^n \setminus G)}{\rho(x, \mathbf{R}^n \setminus G) + \rho(x, F)},$$

where $\rho(x, A)$ is the distance of x from a set A. It is obvious that $\varphi_\varepsilon(x) = 0$, if $x \in \mathbf{R}^n \setminus G$ and $\varphi_\varepsilon(x) = 1$, if $x \in F$. The function $\varphi_\varepsilon(x)$ is continuous since the function $\rho(x, A)$ is continuous, when A is a closed set. Moreover, $0 \leq \varphi_\varepsilon(x) \leq 1$. Therefore,

$$\int_\Omega |\varphi_\varepsilon(x) - \varphi(x)|\, dx = \int_{G \setminus F} |\varphi_\varepsilon(x) - \varphi(x)|\, dx \leq \mu(G \setminus F) < \varepsilon.$$

Thus a summable function, taking two values, can be approximated in the metrics of the space $L_1(\Omega)$ by a continuous compactly supported function as well as we wish. The proof is complete. □

It is obvious that if a sequence $\{f_j(x)\}$ converges to $f(x)$ in $L_1(\Omega)$, then it *converges to $f(x)$ in measure*, i.e. for any $\varepsilon > 0$ and $\delta > 0$ there exists N such that

$$\text{meas}\{x : \ |f(x) - f_j(x)| > \delta\} < \varepsilon,$$

if $j > N$. In the measure theory it is proved that a sequence, convergent in measure, contains a subsequence, convergent almost everywhere. (See [KolFo].) Therefore, if a sequence $\{f_j(x)\}$ converges to $f(x)$ in $L_1(\Omega)$, then it contains a subsequence, convergent almost everywhere. But this sequence is not necessarily convergent almost everywhere. Indeed, let $\Omega = [0,1]$, $n = 1$, $f_{jk} = 1$, if $x \in [k2^{-j}, (k+1)2^{-j}] \cap [0,1]$ and $f_{jk} = 0$, if $x \notin [k2^{-j}, (k+1)2^{-j}] \cap [0,1]$ for $j = 1, 2, \ldots$, $k = 0, 1, \ldots, 2^j - 1$. If we numerate this set by natural numbers, we shall get the sequence, converging to 0 in $L_1(\Omega)$. However, this sequence has values 0 and 1 infinitely many times at each point of $[0,1]$ and therefore is convergent at no point of Ω.

2.9 Density of the set of continuous functions in $L_p(\Omega), p > 1$

Functions from $L_p(\Omega)$, $p > 1$ are also approximated by continuous functions.

Theorem 22. *Let Ω be a domain in $\mathbf{R}^n$. The set of continuous in Ω functions with compact supports is dense in $L_p(\Omega), p > 1$.*

Proof. Let at first Ω be a bounded domain in $\mathbf{R}^n$, $f \in L_p(\Omega)$. Set $f_k(x) = f(x)$, if $|f(x)| < k$, and $f_k(x) = 0$, if $|f(x)| \geq k$. Then

$$\int_\Omega |f_k(x) - f(x)|^p\, dx \to 0$$

as $k \to \infty$, since

$$\int_\Omega |f(x)|^p\,dx = \sum_{j=0}^{\infty} \int_{\Omega_j} |f(x)|^p\,dx,$$

where

$$\Omega_j = \{x :\ x \in \Omega,\ j < |f(x)| < j+1\}$$

and

$$\int_\Omega |f_k(x) - f(x)|^p\,dx = \sum_{j=k+1}^{\infty} \int_{\Omega_j} |f(x)|^p\,dx.$$

Therefore, for any $\varepsilon > 0$ there exists k such that

$$\int_\Omega |f_k(x) - f(x)|^p\,dx < \varepsilon^p.$$

Since the function f_k is bounded, it is summable and thus, as it has been proved in Section 2.8, there exists a continuous in Ω function g_k such that

$$\int_\Omega |f_k(x) - g_k(x)|\,dx < \varepsilon^p 2^{1-p} k^{1-p}.$$

Since $|f_k(x)| < k$, we can assume that $|g_k(x)| \le k$, because it is always possible to replace it at points where $|g_k(x)| > k$ by $k \cdot \operatorname{sign} g_k(x)$. Therefore, we have

$$\int_\Omega |f_k(x) - g_k(x)|^p\,dx \le \int_\Omega |f_k(x) - g_k(x)||f_k(x) - g_k(x)|^{p-1}\,dx$$

$$\le 2^{p-1}k^{p-1} \int_\Omega |f_k(x) - g_k(x)|\,dx < \varepsilon^p.$$

This gives the inequalities

$$(\int_\Omega |f(x) - g_k(x)|^p\,dx)^{1/p}$$

$$\le (\int_\Omega |f_k(x) - g_k(x)|^p\,dx)^{1/p} + (\int_\Omega |f(x) - f_k(x)|^p\,dx)^{1/p} \le 2\varepsilon.$$

It remains to remove the condition of boundedness of Ω. Set $f(x) = 0$ outside Ω. Then $f \in L_p(\mathbf{R}^n)$. Now let $f_R(x) = f(x)$ for $|x| < R$ and $f_R(x) = 0$ for $|x| \ge R$. Then

$$\int_{\mathbf{R}^n} |f(x) - f_R(x)|^p\,dx = \int_{|x|>R} |f(x)|^p\,dx \to 0$$

as $R \to \infty$. The function f_R belongs to $L_p(B(R))$ and vanishes for $|x| > R$. As we have seen, for any $\varepsilon > 0$ there exists a continuous function $u_R(x)$ such that $u_R(x) = 0$ for $|x| > R+1$ and $\|f_R(x) - u_R(x)\|_{L_p(\mathbf{R}^n)} < \varepsilon$. Therefore, for R large enough we have $\|f(x) - u_R(x)\|_{L_p(\mathbf{R}^n)} < 2\varepsilon$. □

2.10 Separability of $L_p(\Omega)$

The following theorem shows that $L_p(\Omega)$ is a separable space.

Theorem 23. *Let $p \geq 1$ and Ω be a domain in $\mathbf{R}^n$. There is a countable subset S of elements of $L_p(\Omega)$ whose linear combinations are dense in $L_p(\Omega)$. This set S is the set of the characteristic functions $h_j(x)$ of cubes $Q(x,R)$ with rational coordinates of their centers x and rational R.*

Proof. Let $u \in L_p(\Omega)$ and $\varepsilon > 0$. Set $u(x) = 0$ for $x \notin \Omega$ and consider u as a function of $L_p(\mathbf{R}^n)$. Let Q be a cube $Q(0,R)$, where R is an integer and so large that $\int_{\mathbf{R}^n \setminus Q} |u(x)|^p \, dx < \varepsilon^p$. By Theorems 21 and 22 there exists a function v_R continuous in Q and such that

$$\int_Q |u(x) - v_R(x)|^p \, dx < \varepsilon^p.$$

Since the function v_R is uniformly continuous in Q, there exists a δ such that

$$|v_R(x) - v_R(y)| < \varepsilon R^{-n/p}, \text{ if } x \in Q,\ y \in Q,\ |x-y| < \delta.$$

Diminishing δ, one can take $\delta = R\sqrt{n}2^{-N}$ for some integer N. Bisecting Q, represent it as a sum of nonintersecting cubes having sides $R \cdot 2^{-N}$. The set S consists of the characteristic functions of such cubes for all N. Now set

$$w(x) = \sum_j v_R(x_j) h_j(x),$$

where x_j is the center of a small cube. Then

$$|v_R(x) - w(x)| = |v_R(x) - v_R(x_j)| < \varepsilon R^{-n/p},$$

if x belongs to the cube with the center at x_j. We have $\int_Q |v_R(x) - w(x)|^p \, dx \leq \varepsilon^p$. Set $v_R(x) = 0$, $w(x) = 0$ for $x \in \mathbf{R}^n \setminus Q$, so that

$$\Big(\int_{\mathbf{R}^n} |u(x) - w(x)|^p \, dx\Big)^{1/p} \leq \Big(\int_Q |u(x) - w(x)|^p \, dx\Big)^{1/p} + \Big(\int_{\mathbf{R}^n \setminus Q} |u(x)|^p \, dx\Big)^{1/p}$$

$$\leq \Big(\int_Q |u - v_R|^p \, dx\Big)^{1/p} + \Big(\int_Q |v_R - w|^p \, dx\Big)^{1/p} + \Big(\int_{\mathbf{R}^n \setminus Q} |u(x)|^p \, dx\Big)^{1/p} \leq 3\varepsilon.$$

Therefore, the set of linear combinations of the functions h_j is dense in $L_p(\Omega)$. Thus the space $L_p(\Omega)$ is separable. □

2.11 Global continuity of functions of $L_p(\Omega)$

It is important for applications that functions in $L_p(\Omega)$ are globally continuous.

Theorem 24. *Let Ω be a domain in $\mathbf{R}^n$, $u \in L_p(\Omega)$, $p \geq 1$, $u(x) = 0$ outside of Ω. Then for any $\varepsilon > 0$ there is an $h > 0$ such that*

$$\int_\Omega |u(x) - u(x+y)|^p \, dx < \varepsilon,$$

if $|y| < h$.

Proof. Let at first Ω be bounded. By Theorems 21 and 22, for any $\varepsilon_1 > 0$ there is a function $v \in C(\mathbf{R}^n)$ with compact support such that $\int_{\mathbf{R}^n} |u(x) - v(x)|^p \, dx < \varepsilon_1^p$. Since v is uniformly continuous, there is $h > 0$ such that $|v(x) - v(x+y)| \leq \varepsilon_1$, if $|y| < h$. Hence

$$(\int_{\mathbf{R}^n} |u(x) - u(x+y)|^p \, dx)^{1/p} \leq (\int_{\mathbf{R}^n} |u(x) - v(x)|^p \, dx)^{1/p}$$

$$+(\int_{\mathbf{R}^n} |v(x) - v(x+y)|^p \, dx)^{1/p} + (\int_{\mathbf{R}^n} |v(x+y) - u(x+y)|^p \, dx)^{1/p}$$

$$\leq 2\varepsilon_1 + \varepsilon_1 (2 \operatorname{meas} \Omega)^{1/p}.$$

If we now choose ε_1 so that $\varepsilon = \varepsilon_1^p (2 + (2 \operatorname{meas} \Omega)^{1/p})^p$, the proof is complete. If the domain Ω is unbounded, one can choose at first a bounded domain Ω_1 such that

$$\int_{\Omega \setminus \Omega_1} |u(x)|^p \, dx < \varepsilon^p, \quad \int_{\Omega \setminus \Omega_1} |u(x+y)|^p \, dx < \varepsilon^p,$$

for $|y| \leq 1$, and thereafter use the proved statement for Ω_1. □

Theorem 25. *Let Ω be a domain in $\mathbf{R}^n$, star-like with respect to the origin, and $u \in L_p(\Omega), u(x)$ vanishes outside of Ω. For any $\varepsilon > 0$ there is $h > 0$ such that*

$$\int_\Omega |u(x) - u(\lambda x)|^p \, dx < \varepsilon,$$

if $|\lambda - 1| < h$.

Proof. It suffices to consider a bounded domain. However the proof then follows immediately from Theorem 24, since $u(\lambda x) = u(x + (\lambda - 1)x)$. □

2.12 Averaging

Let ω be a real function of $C_0^\infty(\mathbf{R}^n)$, such that $\omega(x) = \omega(-x)$, $\omega(x) \geq 0$, $\omega(x) = 0$ for $|x| > 1$ and $\int_{\mathbf{R}^n} \omega(x)\, dx = 1$. For instance, one may take

$$\omega(x) = Cexp(-\frac{1}{1-|x|^2}) \text{ if } |x| \leq 1, \text{ and } \omega(x) = 0 \text{ if } |x| \geq 1$$

with the corresponding value of the constant C. The function ω is called an *averaging kernel.*

If $u \in L_1(\Omega)$, then the function

$$u_h(x) = h^{-n} \int_\Omega \omega(\frac{x-y}{h}) u(y)\, dy$$

is defined in $\mathbf{R}^n$ and is infinitely smooth. It is called *an averaging* of u.

Theorem 26. *If $u \in L_p(\Omega)$, $p \geq 1$, then* $\lim_{h\to 0} \|u_h - u\|_{L_p(\Omega)} = 0$.

Proof. Set $u(x) = 0$ for $x \in \mathbf{R}^n \setminus \Omega$. Then

$$u_h(x) = h^{-n} \int_\Omega \omega(\frac{x-y}{h}) u(y)\, dy = \int_{\mathbf{R}^n} \omega(z) u(x+hz)\, dz.$$

Therefore

$$u_h(x) - u(x) = \int_{\mathbf{R}^n} \omega(z)[u(x+hz) - u(x)]\, dz,$$

and

$$|u_h(x) - u(x)|^p \leq C \int_{|z|<1} [u(x+hz) - u(x)]^p\, dz.$$

After integrating with respect to x and changing the order of integration we get by the Fubini theorem

$$\int_\Omega |u_h(x) - u(x)|^p \leq C \int_{|z|<1} dz \int_\Omega [u(x+hz) - u(x)]^p\, dx.$$

By Theorem 24 the latter integral tends to 0 as $h \to 0$. □

Proposition 27. *If $f, g \in L_1(\Omega)$, then $\int_\Omega f_h(x) g(x)\, dx = \int_\Omega f(x) g_h(x)\, dx$.*

Proof. Indeed,

$$\int_\Omega f_h(x) g(x)\, dx = h^{-n} \int_\Omega \int_\Omega \omega\left(\frac{x-y}{h}\right) f(y)\, dy\; g(x)\, dx$$

$$= h^{-n} \int_\Omega f(y)\, dy \int_\Omega \omega\left(\frac{y-x}{h}\right) g(x)\, dx = \int_\Omega f(x) g_h(x)\, dx.$$ □

Corollary 28. *If $f \in L_1(\Omega)$ and $\int_\Omega f(x)\varphi(x)\, dx = 0$ for all $\varphi \in C_0^\infty(\Omega)$, then $f = 0$.*

Proof. Let $\psi \in C_0(\Omega)$ be such a function that

$$\int_\Omega |\psi(x) - \operatorname{sign} f_h(x)|\, dx \le 1/2 \int_\Omega |f_h|\, dx / \max_{x\in\Omega} |f_h|.$$

Then

$$0 = \int_\Omega f\psi_h\, dx = \int_\Omega f_h \psi\, dx = \int_\Omega f_h(\psi - \operatorname{sign} f_h + \operatorname{sign} f_h)\, dx$$

$$= \int_\Omega |f_h|\, dx + \int_\Omega f_h(\psi - \operatorname{sign} f_h)\, dx \ge \int_\Omega |f_h|\, dx - 1/2 \int_\Omega |f_h|\, dx.$$

Therefore, $f_h = 0$ and since $\|f - f_h\|_{L_p(\Omega)} \to 0$, we see that $f = 0$ almost everywhere. □

2.13 Compactness of a subset in $L_p(\Omega)$

The following criterion of compactness of a set in $L_p(\Omega)$, $p \ge 1$, is often used.

Theorem 29. *Let Ω be a bounded domain in $\mathbf{R}^n$. A set M of functions $f \in L_p(\Omega)$ is compact if and only if it is bounded and equicontinuous, i.e. for any $\varepsilon > 0$ there is a δ such that $\int_\Omega |f(x+y) - f(x)|^p\, dx < \varepsilon$, when $|y| < \delta$, $f \in M$.*

Proof. Let the set M be bounded and equicontinuous. Set $f(x) = 0$ for $x \in \mathbf{R}^n \setminus \Omega$. Consider the averaging f_h of f after continuing it by zero values outside of Ω. Given $h > 0$ the set of continuous functions f_h is uniformly bounded, since

$$|f_h(x)| = h^{-n} \left| \int_\Omega \omega\left(\frac{|x-y|}{h}\right) f(y)\, dy \right| \le \int_{\mathbf{R}^n} \omega(z) |f(x+hz)|\, dz$$

$$\le C \left(\int_{\mathbf{R}^n} |f(x+hz)|^p\, dz \right)^{1/p} = C h^{-n/p} \left(\int_{\mathbf{R}^n} |f(z)|^p\, dz \right)^{1/p}$$

and equicontinuous since we have as above

$$|f_h(x_1) - f_h(x_2)| \le C \left[\int_{\mathbf{R}^n} |f(x_1 + hz) - f(x_2 + hz)|^p\, dz \right]^{1/p}$$

$$\le C_1 h^{-n/p} \left[\int_{\mathbf{R}^n} |f(y) - f(y + (x_2 - x_1))|^p\, dy \right]^{1/p} < \varepsilon,$$

if $|x_2 - x_1| < \delta$. By the Arzela theorem, the set of the functions f_h is compact in $C(\Omega)$ and there exists a finite ε-net in this set. Obviously, this net is an $\varepsilon(1 + \operatorname{meas}\Omega)^{1/p}$-net in the $L_p(\Omega)$ metrics. Since

$$\|f_h(x) - f(x)\|^p_{L_p(\Omega)} \le \max_{|t|\le 1} \int_{\mathbf{R}^n} |f(x+ht) - f(x)|^p\, dx < \varepsilon^p,$$

if h is small enough, for any $\varepsilon > 0$ there exists a finite ε-net in M, i.e. M is compact.

Inversely, let M be compact. For $\varepsilon = 1$ there is a finite set of elements $f_1, \dots, f_k$ such that for any $f \in M$ we have $\|f - f_j\|_{L_p(\Omega)} < 1$ with certain j and therefore, $\|f\|_{L_p(\Omega)} \le 1 + \max\{\|f_1\|_{L_p(\Omega)}, \dots, \|f_k\|_{L_p(\Omega)}\}$, i.e. M is bounded.

Moreover, given $\varepsilon > 0$, there exists a finite set of elements $f_1, \dots, f_k$ such that for any element $f \in M$ we have $\|f - f_j\|_{L_p(\Omega)} < \varepsilon/3$ with certain j. Furthermore, there exists a positive δ such that $\|f_j(x+y) - f_j(x)\|_{L_p(\Omega)} < \varepsilon/3$ for $|y| < \delta, j = 1, \dots, k$. Therefore we have for some j the inequality

$$\|f(x+y) - f(x)\|_{L_p(\Omega)} < \|f(x) - f_j(x)\|_{L_p(\Omega)} + \|f_j(x+y) - f_j(x)\|_{L_p(\Omega)}$$

$$+\|f_j(x+y) - f(x+y)\|_{L_p(\Omega)} < \varepsilon.$$

Therefore, the set M is equicontinuous. □

2.14 Fourier transform

The Fourier transform is an important tool of modern and classical analysis.

Consider the *Schwartz space* S, that is the set of functions u of $C^\infty(\mathbf{R}^n)$ such that $(1 + |x|^2)^k |D^\alpha u(x)| \le C_{k,\alpha}$ for all k, α. The topology in this space can be defined as follows: the sequence $\{u_m(x)\} \in \mathcal{S}$ converges to u if each of the sequences $\{(1 + |x|^2)^k D^\alpha u_m(x)\}$ is uniformly converging to $(1 + |x|^2)^k D^\alpha u(x)$ as $m \to \infty$. It is obvious that $C_0^\infty(\mathbf{R}^n) \subset \mathcal{S}$, $e^{-|x|^2} P(x) \in \mathcal{S}$ for each polynomial $P(x)$. If $u \in \mathcal{S}$ set

$$\tilde{u}(\xi) = \int_{R^n} u(x) e^{-ix\xi}\, dx, \quad \text{where } x\xi = \sum_{j=1}^n x_j \xi_j.$$

The function $\tilde{u}(\xi)$ is called the *Fourier transform* of the function u. It is denoted often as $\mathcal{F}_{x \to \xi} u$.

Lemma 30. *If $u \in \mathcal{S}$, then $\tilde{u}(\xi) \in \mathcal{S}$.*

Proof. For any α we have

$$D^\alpha \tilde{u}(\xi) = \int_{R^n} (-ix)^\alpha u(x) e^{-ix\xi}\, dx.$$

Thus

$$|\xi|^{2k} D^\alpha \tilde{u}(\xi) = \int_{R^n} (-ix)^\alpha u(x) (-\Delta)^k e^{-ix\xi}\, dx = \int_{R^n} (-\Delta)^k [(-ix)^\alpha u(x)] e^{-ix\xi}\, dx,$$

where Δ is the Laplace operator: $\Delta = \sum_{j=1}^n \frac{\partial^2}{\partial x_j^2}$. Therefore, $|\xi|^{2k} |D^\alpha \tilde{u}| \le C'_{k,\alpha}$, i.e. $\tilde{u} \in \mathcal{S}$. □

Theorem 31. *For any $u \in S$ the formula*

$$u(x) = (2\pi)^{-n} \int_{R^n} \tilde{u}(\xi) e^{ix\xi}\, d\xi,$$

is valid.

Proof. Let at first $n = 1$. We have

$$\int_{-\infty}^{+\infty} \tilde{u}(\xi) e^{ix\xi}\, d\xi = \int_{-\infty}^{+\infty} [\int_{-\infty}^{+\infty} u(t) e^{-it\xi}\, dt] e^{ix\xi}\, d\xi$$

$$= \int_{-\infty}^{+\infty} \lim_{N\to\infty} \int_{-N}^{N} e^{i(x-t)\xi} u(t)\, dt\, d\xi = \lim_{N\to\infty} \int_{-\infty}^{+\infty} \frac{e^{iN(x-t)} - e^{-iN(x-t)}}{i(x-t)} u(t)\, dt$$

$$= 2 \lim_{N\to\infty} \int_{-\infty}^{+\infty} \frac{\sin N(x-t)}{x-t} u(t)\, dt.$$

Set $t = x + \tau/N$. Then

$$\int_{-\infty}^{+\infty} \tilde{u}(\xi) e^{ix\xi}\, d\xi = 2 \lim_{N\to\infty} \int_{-\infty}^{+\infty} \frac{\sin\ \tau}{\tau} u(x + \frac{\tau}{N})\, d\tau.$$

Let us show that for all x

$$\lim_{N\to\infty} \int_{-\infty}^{+\infty} \frac{\sin\ \tau}{\tau} [u(x + \frac{\tau}{N}) - u(x)]\, d\tau = 0.$$

Set $J(R, N) = \int_R^\infty \frac{\sin \tau}{\tau} u(x + \frac{\tau}{N})\, d\tau$. With the help of integration by parts we get

$$J(R, N) = -\int_R^\infty u(x + \frac{\tau}{N})\, d\,(\int_\tau^\infty \frac{\sin t}{t}\, dt)$$

$$= u(x + R/N) \int_R^\infty \frac{\sin t}{t}\, dt + \frac{1}{N} \int_R^\infty (\int_\tau^\infty \frac{\sin t}{t}\, dt) u'(x + \frac{\tau}{N})\, d\tau = J_1 + J_2.$$

The first summand J_1 tends to 0 as $R \to \infty$ since u is bounded. If R is large enough, we have

$$|\int_\tau^\infty \frac{\sin t}{t}\, dt| \le \varepsilon \text{ for } \tau > R$$

and therefore

$$|J_2| \le \varepsilon/N \int_R^\infty |u'(x + \frac{\tau}{N})|\, d\tau \le \varepsilon \int_{-\infty}^\infty |u'(x + z)|\, dz \to 0$$

as $\varepsilon \to 0$, since the function u' is summable over $\mathbf{R}$.

Thus $J(R,N) \to 0$ as $R \to \infty$ uniformly in N. If the value of R is big enough, we have

$$|\int_R^\infty \frac{\sin\tau}{\tau}[u(x+\frac{\tau}{N}) - u(x)]\,d\tau| + |\int_{-\infty}^{-R} \frac{\sin\tau}{\tau}[u(x+\frac{\tau}{N}) - u(x)]\,d\tau| < \varepsilon.$$

Since u is continuous, it follows for fixed R that

$$\lim_{N\to\infty} \int_{-R}^{R} \frac{\sin\tau}{\tau}[u(x+\frac{\tau}{N}) - u(x)]\,d\tau = 0.$$

On the other hand, we get

$$\frac{1}{2\pi}\int \tilde{u}(\xi)e^{ix\xi}\,dx$$

$$= \frac{1}{\pi}\lim_{N\to\infty} \int_{-\infty}^{\infty} \frac{\sin\tau}{\tau}[u(x+\frac{\tau}{N}) - u(x)]\,d\tau + \frac{u(x)}{\pi}\int_{-\infty}^{\infty}\frac{\sin\tau}{\tau} = u(x).$$

The proof in the case $n > 1$ can be obtained with the help of induction on the dimension n. □

Theorem 32. *For any $f, g \in \mathcal{S}$ the Parseval equality*

$$\int_{R^n} f(x)\overline{g}(x)\,dx = (2\pi)^{-n}\int_{R^n} \tilde{f}(\xi)\overline{\tilde{g}(\xi)}\,d\xi$$

is valid.

Proof. In virtue of Theorem 31,

$$f(x) = (2\pi)^{-n}\int_{R^n} \tilde{f}(\xi)e^{ix\xi}\,d\xi.$$

Since the integrands are rapidly decreasing, we obtain

$$\int_{R^n} f(x)\overline{g}(x)\,dx = (2\pi)^{-n}\int_{R^n} \overline{g}(x)\,dx \int_{R^n} \tilde{f}(\xi)e^{ix\xi}\,d\xi$$

$$= (2\pi)^{-n}\int_{R^n} \tilde{f}(\xi)\,d\xi \int_{R^n} \overline{g}(x)e^{ix\xi}\,dx = (2\pi)^{-n}\int_{R^n} \tilde{f}(\xi)\overline{\tilde{g}(\xi)}\,d\xi.$$ □

Corollary 33. *If $f \in \mathcal{S}$, then*

$$\int_{R^n} |f(x)|^2\,dx = (2\pi)^{-n}\int_{R^n} |\tilde{f}(\xi)|^2\,d\xi.$$

Theorem 32 allows us to define the Fourier transform of functions of $L_2(\mathbf{R}^n)$, since $\mathcal{S}(\mathbf{R}^n)$ is dense in $L_2(\mathbf{R}^n)$. Indeed, let $f \in L_2(\mathbf{R}^n)$. Then there is a sequence $\{f_j\}$ in $C_0^\infty(\mathbf{R}^n)$ such that $\|f(x) - f_j(x)\|_{L_2(R^n)} \to 0$ as $j \to \infty$. Moreover, then for any $l \in \mathbf{N}$ $\|f_j(x) - f_{j+l}(x)\|_{L_2(R^n)} \to 0$ as $j \to \infty$. Corollary 33 implies that then $\|\tilde{f}_j(\xi) - \tilde{f}_{j+l}(\xi)\|_{L_2(R^n)} \to 0$ as $j \to \infty$, i.e. the sequence $\{\tilde{f}_j\}$ is a Cauchy sequence. Its limit in $L_2(\mathbf{R}^n)$ is the Fourier transform of the function f. As a result we obtain the Parseval equality for functions $f, g \in L_2(\mathbf{R}^n)$. Therefore, the operator $(2\pi)^{-n/2}\mathcal{F}_{x\to\xi}$ is an isometry of the space $L_2(\mathbf{R}^n)$.

Now we state another, equivalent way of defining the Fourier transform in $L_2(\mathbf{R}^n)$. Let $u \in L_2(\mathbf{R}^n)$ and the support of u be compact. Then its Fourier transform can be defined by the same formula as for functions of $\mathcal{S}(\mathbf{R}^n)$ used in the beginning of the section, i.e. $\tilde{u}(\xi) = \int u(x)e^{-ix\xi}\,dx$.

Let us consider the sequence of the averaging functions $u_h(x)$ for $h = 2^{-l}$, $l = 1, 2, \ldots$. It converges to u in $L_2(\mathbf{R}^n)$ as $l \to \infty$. The functions u_h have compact supports and hence belong to $\mathcal{S}(\mathbf{R}^n)$. The supports of u_h and that of u are in a domain Ω, having compact closure. Since

$$\tilde{u}_h(\xi) - \tilde{u}(\xi) = \int_\Omega [u_h(x) - u(x)]e^{-ix\xi}\,dx,$$

we see that $\tilde{u}_h(\xi) - \tilde{u}(\xi) \to 0$ as $h \to 0$, uniformly in $\xi \in \mathbf{R}^n$. Therefore, the limit of $\tilde{u}_h(\xi)$ in $L_2(\mathbf{R}^n)$ coincides with $\tilde{u}(\xi)$.

Now let $f \in L_2(\mathbf{R}^n)$. Set $f_N(x) = f(x)$ for $|x| < N$ and $f_N(x) = 0$ for $|x| \geq N$. The sequence $\{f_N(x)\}$ converges to $f(x)$ in $L_2(\mathbf{R}^n)$. The Parseval equality implies that the sequence $\{\tilde{f}_N(\xi)\}$ has a limit in $L_2(\mathbf{R}^n)$ that is the Fourier transform of f. It allows us to say that the integral, defining the Fourier transform of a function $f \in L_2(\mathbf{R}^n)$, is "converging in mean", i.e. is defined as the limit in $L_2(\mathbf{R}^n)$ of the sequence $v_N(\xi) = \int_{|x|<N} f(x)e^{-ix\xi}\,dx$.

The following properties of the Fourier transform are easy to check:

1. $D^\alpha\tilde{u}(\xi) = (-i\xi)^\alpha\tilde{u}(\xi)$;
2. $\tilde{u}(x + h) = e^{ixh}\tilde{u}(\xi)$;
3. $\tilde{u}(kx) = k^n\tilde{u}(\xi/k)$, $k \in \mathbf{R}$, $k \neq 0$;
4. $\mathcal{F}_{x\to\xi}(u * v) = \tilde{u}(\xi) \cdot \tilde{v}(\xi)$,
   where
   $$f * g(x) = \int_{R^n} f(x - y)g(y)\,dy$$
   is the *convolution* of f and g.

2.15 The spaces $W_p^m(\Omega)$

Definition 34. A function $u(x) \in L_1(\Omega)$ is a *generalized derivative* of order α of a function $v \in L_1(\Omega)$ if

$$\int_\Omega u(x)\psi(x)\,dx = (-1)^{|\alpha|}\int_\Omega v(x)D^\alpha\psi(x)\,dx$$

for any $\psi \in C_0^\infty(\Omega)$.

The classical Green formula implies that in the case when the function v has a continuous derivative of order α, the latter is also a generalized derivative of order α. On the other hand, the function v can have only one generalized derivative of order α, since if u_1 and u_2 are such derivatives, we have

$$\int_\Omega (u_1(x) - u_2(x))\psi(x)\, dx = 0$$

for any $\psi \in C_0^\infty(\Omega)$. Corollary 28 implies that $u_1 = u_2$ almost everywhere in Ω.

It is easily seen that a function v can have a generalized derivative without having the classical one. For example, if $v(x) = f(x_1) + g(x_2)$, then $\frac{\partial^2 v}{\partial x_1 \partial x_2} = 0$ for any $f, g \in L_1(\mathbf{R})$, but it is possible that the classical derivatives of first and of second orders do not exist everywhere in $\mathbf{R}^2$.

It is evident that the derivative of order α exists in a domain Ω' if it exists in a domain Ω and $\Omega' \subset \Omega$. In this case the derivative in Ω' is the restriction of the derivative defined in Ω to Ω'. It is obvious that

$$D^{\alpha+\beta} v = D^\alpha(D^\beta v),\ aD^\alpha v_1 + bD^\alpha v_2 = D^\alpha(av_1 + bv_2)$$

for any constants a and b.

Consider the case $n = 1$. If a function f is absolutely continuous on a finite interval (a, b), then it has a classical derivative f' almost everywhere in (a, b). (See [KolFo], p. 396.) This derivative is summable and $f(x) = f(a) + \int_a^x f'(t)\, dt$. Therefore, for any function $\psi \in C_0^\infty(a, b)$ we have

$$\int_a^b f'(t)\psi(t)\, dt = \int_a^b (f(t)\psi(t))'\, dt - \int_a^b f(t)\psi'(t)\, dt = -\int_a^b f(t)\psi'(t)\, dt,$$

i.e. f' is a generalized derivative.

Inversely, let f have a generalized derivative $f_1 \in L_1(a, b)$. Let g be a function of $C_0^\infty(a, b)$ and $\int_a^b g(t)\, dt = 1$. Set

$$\varphi(x) = f(x) - \int_a^x f_1(t)\, dt + C,$$

where C is a constant such that $\int_a^b \varphi(t) g(t)\, dt = 0$. Then for any $\psi \in C_0^\infty(a, b)$ we have

$$\int_a^b \varphi(t)\psi'(t)\, dt = 0,$$

since by definition of the generalized derivative

$$\int_a^b \varphi(t)\psi'(t)\, dt = \int_a^b f(t)\psi'(t)\, dt - \int_a^b \int_a^x f_1(t)\, dt\ \psi'(x)\, dx$$

$$= \int_a^b f(t)\psi'(t)\, dt + \int_a^b f_1(t)\psi(t)\, dt = 0.$$

Set $\theta(x) = \psi(x) - g(x)\int_a^b \psi(t)\,dt$. It is clear that $\theta \in C_0^\infty(a,b)$, $\int_a^b \theta(t)\,dt = 0$. Therefore

$$\theta_1(x) = \int_a^x \theta(t)\,dt \in C_0^\infty(a,b)$$

and

$$\int_a^b \varphi(x)\psi(x)\,dx = \int_a^b \varphi(x)\theta(x)\,dx = \int_a^b \varphi(x)\theta_1'(x)\,dx = 0$$

for all $\psi \in C_0^\infty(a,b)$. Therefore, $\varphi(x) \equiv 0$ and $f(x) = \int_a^x f_1(t)\,dt - C$, i.e. the function f is absolutely continuous and $f'(x) = f_1(x)$ almost everywhere.

Definition 35. A function $u(x)$ belongs to the *Sobolev space* $W_p^m(\Omega)$ if it has generalized derivatives of all orders α with $|\alpha| \le m$ and

$$\|u\|_{W_p^m(\Omega)} = (\int_\Omega \sum_{|\alpha|\le m} |D^\alpha u(x)|^p\,dx)^{1/p} < \infty.$$

Theorem 36. *If* $p \ge 1$, $m \ge 0$, *then the space* $W_p^m(\Omega)$ *is complete.*

Proof. Let $\{u_N\}$ be a Cauchy sequence, i.e. for any natural k

$$\int_\Omega \sum_{|\alpha|\le m} |D^\alpha(u_N - u_{N+k})|^p\,dx \to 0$$

as $N \to \infty$. For each α the sequence $\{D^\alpha u_N\}$ is a Cauchy sequence in $L_p(\Omega)$. Hence there is a function $u_\alpha \in L_p(\Omega)$ such that

$$\int_\Omega |D^\alpha u_N - u_\alpha|^p\,dx \to 0$$

as $N \to \infty$. In particular, $u_0 \in L_p(\Omega)$. Let us show that u_α is the generalized derivative of the function u_0 of the order α. The definition of the generalized derivative gives the relation

$$\int_\Omega u_N D^\alpha \psi\,dx = (-1)^{|\alpha|}\int_\Omega D^\alpha u_N \cdot \psi\,dx,$$

valid for all $\psi \in C_0^\infty(\Omega)$. Passing to the limit we get the equality

$$\int_\Omega u_0 D^\alpha \psi\,dx = (-1)^{|\alpha|}\int_\Omega u_\alpha \psi\,dx,$$

showing that u_α is the generalized derivative of u_0 of order α. Therefore, $\|u_N - u_0\|_{W_p^m(\Omega)} \to 0$ as $N \to \infty$. □

Theorem 37. *If $n > 1$ and $u \in W_p^1(\Omega)$, then the function $x_1 \mapsto u(x_1, x_2, \dots, x_n)$ is absolutely continuous and belongs to the space $W_p^1(I_1)$ for almost all $(x_2, \dots, x_n)$, where I_j is the intersection of Ω with the line parallel to the x_j-axis.*

Inversely, if a function $u(x)$ is absolutely continuous in x_j for almost all $(x_1, \dots, x_{j-1}, x_{j+1}, \dots, x_n)$ and has a generalized derivative $\frac{\partial u(x)}{\partial x_j} \in L_p(I_j)$ such that $\partial u(x)/\partial x_j \in L_p(\Omega)$, then $u \in W_p^1(\Omega)$.

Proof. It is sufficient to consider the case when Ω is a cube $\Omega = \{x \in \mathbf{R}^n : |x_j - x_{0j}| < a,\ j = 1, \dots, n\}$, because one can cover Ω by a countable system of such cubes. Let $\psi(x_1) \in C_0^\infty(x_{01} - a, x_{01} + a)$. We have

$$\int_\Omega \frac{\partial \psi(x_1)}{\partial x_1} u(x) \varphi(x_2, \dots, x_n)\, dx = -\int_\Omega \frac{\partial u(x)}{\partial x_1} \psi(x_1) \varphi(x_2, \dots, x_n)\, dx,$$

where $\psi(x_1)\varphi(x_2, \dots, x_n) \in C_0^\infty(\Omega)$. Since $\varphi(x_2, \dots, x_n)$ is arbitrary, we see that

$$\int_{x_{01}-a}^{x_{01}+a} \frac{\partial \psi(x_1)}{\partial x_1} u(x)\, dx_1 = -\int_{x_{01}-a}^{x_{01}+a} \frac{\partial u(x)}{\partial x_1} \psi(x_1)\, dx_1,$$

for $x' = (x_2, \dots, x_n) \in M \subset Q' = \{x \in \mathbf{R}^{n-1} : |x_j - x_{0j}| < a,\ j = 1, \dots, n-1\}$ and $\operatorname{meas} M = (2a)^{n-1}$. The set M can depend on ψ. Let M_j correspond on $\psi_j, j = 1, 2, \dots$, and the sequence $\{\psi_j\}$ be dense in $W_p^1(Q')$. Then this equality holds for $x \in M_0 = \cap M_j$ and $\operatorname{meas} M_0 = (2a)^{n-1}$. Therefore, $u(x_1, x') \in W_p^1(I_1)$ for $x' \in M_0$.

In order to prove the second statement of Theorem 37, it also suffices to consider the case when Ω is a cube. We have

$$\int_\Omega u(x) \frac{\partial \psi(x)}{\partial x_1}\, dx = \int dx_2 \dots dx_n \int_{x_{01}-a}^{x_{01}+a} u(x) \frac{\partial \psi(x)}{\partial x_1}\, dx_1 = -\int_\Omega \frac{\partial u(x)}{\partial x_1} \psi(x)\, dx.$$

□

Corollary 38. *If $u \in W_p^1(\Omega)$ and a function $f(t)$ satisfies the Lipschitz condition, then $f(u(x)) \in W_p^1(\Omega)$.*

The corollary is obvious for functions of one variable and Theorem 37 implies that it is true in the general case. In particular, if $u \in W_p^1(\Omega)$, then $|u| \in W_p^1(\Omega)$ and for any real k the function $\max(u(x), k)$ belongs to $W_p^1(\Omega)$. For instance, the function $u_+(x) = \max(u(x), 0)$ is in $W_p^1(\Omega)$, if $u \in W_p^1(\Omega)$.

Corollary 39. *If $u \in W_p^1(\Omega_1)$, $u \in W_p^1(\Omega_2)$, $u \in C(\Omega_1 \cup \Omega_2)$, where $\Omega_1 = (a, 0) \times Q$, $\Omega_2 = (0, b) \times Q$, $a < 0$, $b > 0$, $Q \subset \mathbf{R}^{n-1}$, then $u \in W_p^1(\Omega_1 \cup \Omega_2)$.*

This corollary is easy to justify for $n = 1$ and then in virtue of Theorem 37 it is true for all n.

Definition 40. The space $W_{p,0}^m(\Omega)$ is the completion of $C_0^\infty(\Omega)$ in the metric of $W_p^m(\Omega)$.

Theorem 41 (Friedrichs). *If Ω is a bounded domain in $\mathbf{R}^n$, then there exists a constant $C = C(\Omega)$ such that*

$$\|u\|_{L_p(\Omega)} \le C\|u\|_{W^1_{p,0}(\Omega)} \equiv C\left(\int_\Omega \sum_{j=1}^n \left|\frac{\partial u(x)}{\partial x_j}\right|^p dx\right)^{1/p}$$

for all functions u from $W^1_{p,0}(\Omega)$.

Proof. Let $u \in C_0^\infty(\Omega)$ with Ω lying in the strip $J = \{x \in \mathbf{R}^n,\ a < x_1 < b\}$. Set $u(x) = 0$ outside Ω. Then

$$u(x) = \int_a^{x_1} \frac{\partial u(t, x_2, \dots, x_n)}{\partial t}\, dt.$$

Therefore

$$|u(x)| \le \int_a^b \left|\frac{\partial u(t, x_2, \dots, x_n)}{\partial t}\right| dt$$

and by the Hölder inequality

$$|u(x)|^p \le (b-a)^{p/q} \int_a^b \left|\frac{\partial u(t, x_2, \dots, x_n)}{\partial t}\right|^p dt,$$

where q is such that $1/p + 1/q = 1$. After integrating both parts of the latter inequality over J we obtain the inequality for $u \in C_0^\infty(\Omega)$. Since the space $W^1_{p,0}(\Omega)$ is the completion of $C_0^\infty(\Omega)$, this inequality is valid for all $u \in W^1_{p,0}(\Omega)$. □

Corollary 42. *If Ω is a bounded domain in $\mathbf{R}^n$, then the function $u(x) = 1$ does not belong to the space $W^1_{p,0}(\Omega)$ for $p \ge 1$.*

Remark 43. The constant C in Theorem 41 can be taken equal to $C = b - a$.
This value can sometimes be improved. For instance, if $p = 2$, then

$$\int_\Omega |u(x)|^2\, dx \le \frac{(b-a)^2}{\pi^2} \int_\Omega \sum_{j=1}^n \left|\frac{\partial u(x)}{\partial x_j}\right|^2 dx.$$

The constant in the latter inequality is sharp.

2.16 The averaging and generalized derivatives

Using the averaging we can approximate functions from $W_p^m(\Omega)$ by $C^\infty(\Omega)$ functions.

Theorem 44. *Let Ω be a domain in $\mathbf{R}^n$, $\Omega' \subset \Omega$, and the distance between Ω' and $\partial\Omega$ be equal to $d > 0$. Then for $0 < h < d$ and $x \in \Omega'$*

$$(D^\alpha u)_h(x) = D^\alpha u_h(x).$$

Proof. If $0 < h < d$ and $x \in \Omega'$, we have

$$D^\alpha u_h(x) = D_x^\alpha h^{-n} \int_{R^n} \omega(\frac{x-y}{h}) u(y)\, dy = h^{-n} \int_\Omega (-1)^{|\alpha|} D_y^\alpha \omega(\frac{x-y}{h}) u(y)\, dy$$

$$= h^{-n} \int_\Omega \omega(\frac{x-y}{h}) D_y^\alpha u(y)\, dy = (D^\alpha u)_h(x),$$

since the function $\omega(\frac{x-y}{h})$ for $x \in \Omega'$ belongs to $C_0^\infty(\Omega)$ and one can use the definition of the generalized derivative. □

Corollary 45. *If $\overline{\Omega'} \subset \Omega$ and $u \in W_p^m(\Omega)$, then*

$$\lim_{h \to 0} \|u_h - u\|_{W_p^m(\Omega')} = 0.$$

Proof. The proof follows from Theorems 44 and 26. □

Theorem 46. *Let a sequence u_k of elements of the space $W_p^m(\Omega)$ be bounded: $\|u_k\|_{W_p^m(\Omega)} \le A$ and converging weakly in $L_p(\Omega)$ to $u(x)$ as $k \to \infty$. Then it converges weakly to u in $W_p^m(\Omega)$, $u \in W_p^m(\Omega)$ and, $\|u\|_{W_p^m(\Omega)} \le A$.*

Proof. We have

$$\int_\Omega \varphi(x) D^\alpha u_k(x)\, dx = (-1)^{|\alpha|} \int_\Omega u_k(x) D^\alpha \varphi(x)\, dx,$$

where $\varphi(x) \in C_0^\infty(\Omega)$. This implies that the sequence $\{D^\alpha u_k(x)\}$ is weakly converging in $L_p(\Omega)$ to a function $v_\alpha(x)$. We have

$$\int_\Omega \varphi(x) v_\alpha(x)\, dx = (-1)^{|\alpha|} \int_\Omega u(x) D^\alpha \varphi(x)\, dx$$

for $\varphi(x) \in C_0^\infty(\Omega)$. Therefore, the generalized derivative $D^\alpha u(x)$ does exist and is equal to $v_\alpha(x)$. Moreover,

$$\begin{aligned} \|D^\alpha u(x)\|_{L_p(\Omega)}^p &\le \lim_{k \to \infty} \int |D^\alpha u(x)|^{p-2} D^\alpha u(x) \overline{D^\alpha u_k(x)}\, dx \\ &\le \|D^\alpha u(x)\|_{L_p(\Omega)}^{p-1} \limsup_{k \to \infty} \|D^\alpha u_k(x)\|_{L_p(\Omega)}. \end{aligned}$$

Thus $u \in W_p^m(\Omega)$ and $\|u\|_{W_p^m(\Omega)} \le A$. □

Theorem 47. *If a domain Ω is star-like in $\mathbf{R}^n$, then the space $C^\infty(\overline{\Omega})$ is dense in $W_p^m(\Omega)$.*

Proof. Let the domain Ω be star-like in $\mathbf{R}^n$ with respect to the origin. If $u \in W_p^m(\Omega)$, then the function $u(\lambda x)$ with $\lambda = \text{const} < 1$ is defined in $\Omega_1 = \{x \in \mathbf{R}^n : \lambda x \in \Omega\}$ and $u(\lambda x) \in W_p^m(\Omega_1)$. Moreover,

$$\|u(\lambda x) - u(x)\|_{W_p^m(\Omega)} \le \varepsilon,$$

if $|\lambda - 1| < \delta(\varepsilon)$ by Theorem 25. Corollary 45 implies also that

$$\lim_{h\to 0} \|u_h(\lambda x) - u(\lambda x)\|_{W_p^m(\Omega)} = 0.$$

Therefore

$$\lim_{h\to 0} \|u_h(\lambda x) - u(x)\|_{W_p^m(\Omega)} = 0.$$

Since $u_h(\lambda x) \in C^\infty(\overline{\Omega})$, this implies that the space $C^\infty(\overline{\Omega})$ is dense in $W_p^m(\Omega)$. □

Note that there are domains Ω such that the space $C^\infty(\Omega)$ is not dense in $W_p^m(\Omega)$ (see, for example, [Maz1]).

2.17 Continuation of functions of $W_p^m(\Omega)$

The following theorem describes a construction for continuing functions from $W_p^m(\Omega)$ in a larger domain.

Theorem 48. *Let Q be a parallelepiped in $\mathbf{R}^n$:*

$$Q = \{x \in \mathbf{R}^n : \ -a_j < x_j < a_j,\ j = 1, \dots, n\}$$

and $u \in W_p^m(Q)$, $p \geq 1$. Then there exists a function $u_1 \in W_{p,0}^m(\mathbf{R}^n)$ such that $u_1(x) = u(x)$ for $x \in Q$, the support of u_1 is in

$$Q_1 = \{x \in \mathbf{R}^n : \ -2a_j < x_j < 2a_j,\ j = 1, \dots, n\},$$

and

$$\|u_1(x)\|_{W_p^m(\mathbf{R}^n)} \leq C\|u(x)\|_{W_p^m(Q)},$$

where C does not depend on u.

Proof. Set

$$u_1(x) = \sum_{j=0}^{m-1} \alpha_j u(-a_1 - 2^{-j-1}x_1, x_2, \dots, x_n)$$

for $-2a_1 < x_1 < -a_1, -a_j < x_j < a_j$, $j = 2, \dots, n$. The constants α_j are such that

$$\sum_{j=0}^{m-1} \alpha_j (-1)^k 2^{-k(j+1)} = 1,\ k = 0, 1, \dots, m-1.$$

This is an algebraic system with nonvanishing determinant and therefore it has a solution. Thus we have defined the function u_1 in the parallelepiped

$$Q' = \{x \in \mathbf{R}^n : \ -2a_1 < x_j < a_1,\ -a_j < x_j < a_j,\ j = 2, \dots, n\}.$$

Moreover,

$$\|u_1(x)\|_{W_p^m(Q')} \leq C_1\|u(x)\|_{W_p^m(Q)},$$

where C_1 does not depend on u. Observe that $u_1 \in C^{m-1}(Q')$, if $u \in C^m(Q)$.

The same method can be applied for continuation of u over other sides of Q. This construction has been proposed by Whitney. □

Theorem 49. *Let Ω be a bounded convex domain in $\mathbf{R}^n$ and E be a subset of $\overline{\Omega}$ of positive measure, $u \in W_p^1(\Omega)$, $p > 1$. Then*

$$\int_\Omega |u|^p\,dx \leq C\frac{\mu_n(\Omega)}{\mu_n(E)}\int_E |u|^p\,dx + Cd(\Omega)^{n+p-1}\mu_n(E)^{(1-n)/n}\int_\Omega |\nabla u|^p\,dx,$$

where $d(\Omega)$ is the diameter of the domain Ω, μ_n is the n-dimensional Lebesgue measure, and the constant C does not depend on u.

Proof. It suffices to prove the inequality for functions of $C^\infty(\Omega)$, because the space $C^\infty(\Omega)$ is dense in $W_p^1(\Omega)$. It is obvious that

$$u(x) - u(y) = \int_0^{|x-y|} \frac{\partial u}{\partial \nu}\,dt,$$

where x, y are arbitrary points in Ω, $\partial/\partial\nu$ is the differentiation in the direction $(x-y)/|x-y|$, and dt is a length element. By the Hölder inequality,

$$|u(x)|^p \leq C|u(y)|^p + C|x-y|^{p-1}\int_0^{l(x,y)} |\nabla u(y+t)|^p\,dt,$$

where $l(x,y)$ is the length of the segment connecting the point y and the point of the intersection of $\partial\Omega$ with the ray, having the direction $(x-y)/|x-y|$ and coming from the point y. Integrating this inequality with respect to $x \in \Omega$, we obtain that

$$\int_\Omega |u(x)|^p\,dx \leq C\mu_n(\Omega)|u(y)|^p + Cd(\Omega)^{n+p-1}\int_\Omega \frac{|\nabla u(z)|^p}{|z-y|^{n-1}}\,dz.$$

Integrating now with respect to $y \in E$, we see that

$$\mu_n(E)\int_\Omega |u(x)|^p\,dx \leq C\mu_n(\Omega)\int_E |u(y)|^p\,dy$$

$$+Cd(\Omega)^{n+p-1}\int |\nabla u(z)|^p\,dz\int_E \frac{dy}{|z-y|^{n-1}}.$$

Note that

$$\int_E \frac{dy}{|z-y|^{n-1}} \leq \int_{|z-y|\leq s} \frac{dy}{|z-y|^{n-1}}$$

$$+\int_{|z-y|>s,\ y\in E} \frac{dy}{|z-y|^{n-1}} \leq s\omega_n + s^{1-n}\mu_n\,(E)$$

for any $s > 0$. Setting $s = \mu_n(E)^{1/n}$ we obtain the theorem. □

2.18 The Sobolev integral representation

This representation is used often to study the space $W_p^m(\Omega)$.

Theorem 50. *Let $u(x) \in C_0^\infty(\mathbf{R}^n)$ and $m \in \mathbf{N}$. Then*

$$u(x) = \int_{\mathbf{R}^n} \left(\sum_{|\alpha|=m} a_\alpha(x,y) D^\alpha u(y) \right) |x-y|^{m-n}\, dy,$$

where $a_\alpha(x,y)$ are bounded functions, smooth for $x \neq y$ and independent of u.

Proof. Let e be a unit vector in $\mathbf{R}^n$. Integrating by parts it is easy to verify that

$$u(x) = \frac{(-1)^{m-1}}{(m-1)!} \int_0^\infty t^{m-1} \frac{\partial^m}{\partial t^m} u(x+te)\, dt.$$

Integrating both parts of this equality over the unit sphere with its center at the point x we obtain that

$$u(x) = \int_{\mathbf{R}^n} \frac{1}{|x-y|^{n-m}} \sum_{|\alpha|=m} a_\alpha(x,y) D^\alpha u(y)\, dy. \qquad \square$$

2.19 The space $W_p^1(\Omega, E)$

Let Ω be a bounded domain in $\mathbf{R}^n$ and E be a subset of $\overline{\Omega}$. Let $C_0^\infty(\Omega, E)$ be the space of infinitely differentiable functions with compact supports in Ω, vanishing in a neighborhood of E. The space $W_p^1(\Omega, E)$, $p \geq 1$, is defined as the completion of $C_0^\infty(\Omega, E)$ in the norm of $W_p^1(\Omega)$.

Theorem 51. *Let Ω be a bounded convex domain in $\mathbf{R}^n$, and E be a subset of $\overline{\Omega}$. If a set E is such that its projection E' on the plane $x_n = \text{const}$ has a positive Lebesgue measure, then*

$$\int_\Omega |u(x)|^p\, dx \leq C \int_\Omega |\nabla u(x)|^p\, dx$$

for all $u(x) \in W_p^1(\Omega, E)$, with a constant C independent of u.

Proof. Let x' be a point of E' and $u(x) \in C^1(\overline{\Omega})$. Then

$$u(x', x_n) = \int_{x_n^*}^{x_n} \frac{\partial u}{\partial x_n}\, dx_n,$$

where x_n^* is such that $u(x', x_n^*) = 0$. We have

$$|u^p(x', x_n)| \leq (b(x') - a(x'))^{p-1} \int_{a(x')}^{b(x')} \left| \frac{\partial u}{\partial x_n} \right|^p\, dx_n,$$

where $a(x')$, $b(x')$ are the intersection points of the line $x = (x', t)$, $t \in \mathbf{R}^1$ with $\partial\Omega$. Integrating this inequality with respect to $x' \in E'$ and with respect to x_n from $a(x')$ to $b(x')$, we obtain that

$$\int_{\Omega'} |u(x)|^p \, dx \le C \int_{\Omega} |\nabla u(x)|^p \, dx,$$

where $\Omega' = \{x : \ x' \in E', \ a(x') < x_n < b(x')\}$. By condition, $\operatorname{meas} \Omega' > 0$. The result follows now from Theorem 49. □

2.20 Properties of the space $W^m_{p,0}(\Omega)$

We will discuss here some properties of functions of the Sobolev spaces.

Theorem 52. *If $u(x) \in W^m_p(\Omega)$, $p \ge 1, \operatorname{supp} u \subset\subset \Omega$, then $u(x) \in W^m_{p,0}(\Omega)$.*

Proof. Let $u_h(x)$ be the averaging of a function u. The functions $u_h(x)$ are infinitely differentiable with compact supports and their sequence tends to u in $W^m_p(\Omega)$ as $h \to 0$. □

Theorem 53. *Let $u_k(x) \in W^m_p(\Omega)$, $k = 1, 2, \ldots$, $\|u_k\|_{W^m_{p,0}(\Omega)} \le C$, and let it converge weakly in $L_p(\Omega)$ to $u(x)$ as $k \to \infty$. Then $u(x) \in W^m_{p,0}(\Omega)$, $\|u\|_{W^m_{p,0}(\Omega)} \le C$.*

Proof. It follows from the definition of the space $W^m_{p,0}(\Omega)$ and Theorem 46. □

Theorem 54. *If $u(x) \in W^1_{p,0}(\Omega)$, $f(t)$ satisfies the Lipschitz condition for $-\infty < t < \infty$, and $f(0) = 0$, then $f(u(x)) \in W^1_{p,0}(\Omega)$.*

Proof. Let $u_k \in C_0^\infty(\Omega)$, $\|u_k - u\|_{W^1_p(\Omega)} \to 0$ as $k \to \infty$. Then

$$\|f(u_k) - f(u)\|_{L_p(\Omega)} \to 0$$

as $k \to \infty$ and, by Corollary 38, $f(u_k) \in W^1_{p,0}(\Omega)$. Since $\|\partial f(u_k)/\partial x_j\|_{L_p(\Omega)} \le C$, Theorem 53 implies our statement. □

Example 55. Let $f(x)$ be an absolutely continuous function defined on the segment $\Omega = \{x : \ -1 < x < +1\}$, $f(-1) = f(1) = 0$, and $\int_{-1}^{1} f'^p(x)\, dx < \infty$. Then $f(x) \in W^1_{p,0}(\Omega)$. Indeed, let $\lambda > 1$ and $f_\lambda(x) = f(\lambda x)$ for $-1/\lambda < x < 1/\lambda$, $f_\lambda(x) = 0$ on the segments $[-1, -1/\lambda]$ and $[1/\lambda, 1]$. Then $f_\lambda(x) \in W^1_{p,0}(\Omega)$. In virtue of Theorem 53 the convergence $f(x) = \lim_{\lambda \to 1+} f(\lambda x)$ in the space $L_p(\Omega)$ implies that $f(x) \in W^1_{p,0}(\Omega)$.

Theorem 56. *The spaces $W^m_{p,0}(\mathbf{R}^n)$ and $W^m_p(\mathbf{R}^n)$ coincide.*

Proof. Let $u(x) \in W_p^m(\mathbf{R}^n)$. Let $\theta(t) \in C^\infty(\mathbf{R}^1)$ be a function such that $\theta(t) = 1$ for $t < 1$ and $\theta(t) = 0$ for $t > 2$. Now set $u_k(x) = u(x)\theta(|x| - k)$. It is easy to see that

$$\sum_{|\alpha| \leq m} \int_{\mathbf{R}^n} |D^\alpha(u_k(x) - u(x))|^p \, dx \leq C \int_{|x|>k+1} \sum_{|\alpha|\leq m} |D^\alpha u(x)|^p \, dx \to 0$$

as $k \to \infty$. By Theorem 52 $u_k(x) \in W_{p,0}^m(\mathbf{R}^n)$ for any k. Theorem 53 implies that $u(x) \in W_{p,0}^m(\Omega)$. □

2.21 Sobolev's embedding theorems

The following Sobolev theorem is fundamental for applications.

Theorem 57. *If* $u(x) \in C_0^\infty(\mathbf{R}^n)$, $n > mp$, *then*

$$\left(\int_{\mathbf{R}^n} |u(x)|^{np/(n-mp)} \, dx \right)^{(n-mp)/np} \leq C \left(\int_{\mathbf{R}^n} \sum_{|\alpha|=m} |D^\alpha u|^p \, dx \right)^{1/p}.$$

Proof. Corollary 5 implies that

$$\int_{\mathbf{R}^n} |u(x)|^q \, dx = \int_0^\infty \mu(t^{\frac{1}{q}}) \, dt = \int_0^\infty \mu(t) \, dt^q$$

for any $q > 0$. Here $\mu(t) = \text{meas}_n\{x : \ u(x) > t\}$. The function $\mu(t)$ is nonincreasing, so that

$$\int_0^\infty \mu(t) \, dt^q = q \int_0^\infty [\mu(t)^{1/q} t]^{q-1} \mu(t)^{1/q} \, dt$$

$$\leq q \int_0^\infty [\int_0^t \mu(t')^{1/q} \, dt']^{q-1} \mu(t)^{1/q} \, dt = \left[\int_0^\infty \mu(t')^{1/q} \, dt' \right]^q.$$

Set $q = n/(n-1)$. The isoperimetric inequality (see Section 2.1) implies that

$$\|u\|_{L_q(\mathbf{R}^n)} \leq \int_0^\infty \mu(t)^{1/q} \, dt \leq \int_0^\infty s(t) \, dt (nV_n^{1/n})^{-1} = C \int_0^\infty s(t) \, dt,$$

where $s(t)$ is the $(n-1)$-dimensional Hausdorff measure of the set $E_t = \{x : u(x) = t\}$. Lemma 10 with $\Phi(x) \equiv 1$ implies that

$$\|u(x)\|_{L_{n/(n-1)}(\mathbf{R}^n)} \leq C \|\nabla u\|_{L_1(\mathbf{R}^n)}.$$

So we have proved Theorem 57 for the case $m = p = 1$. Let now $1 \leq p < n$. Set $u(x) = |v(x)|^{p(n-1)/(n-p)}$. Applying the proved inequality and the Hölder inequality, we obtain that

$$\|u(x)\|_{L_{n/(n-1)}(\mathbf{R}^n)} \leq \|v(x)\|^{p(n-1)/(n-p)}_{L_{np/(n-p)}(\mathbf{R}^n)} \leq C \|v^{(pn-n)/(n-p)} \nabla v\|_{L_1(\mathbf{R}^n)}$$

$$\leq C \|v\|^{n(p-1)/(n-p)}_{L_{np/(n-p)}(\mathbf{R}^n)} \|\nabla v\|_{L_p(\mathbf{R}^n)}.$$

Therefore

$$\|v\|_{L_{np/(n-p)}(\mathbf{R}^n)} \le C\|\nabla v\|_{L_p(\mathbf{R}^n)}.$$

Substitute in this inequality the function $D^\alpha u$ with $|\alpha| = m-k$ for $v(x)$ and the number p for $p_1 = np/(n-(k-1)p)$. Then we see that

$$\sum_{|\alpha|=m-k} \|D^\alpha u\|_{L_{pn/(n-pk)}(\mathbf{R}^n)} \le C \sum_{|\alpha|=m-k+1} \|D^\alpha u\|_{L_{pn/(n-(k-1)p)}(\mathbf{R}^n)},$$

if $n > kp$. Writing down all the inequalities for $k = m, m-1, \ldots, 1$, and multiplying them term by term, we obtain the result. □

Remark 58. Theorem 57 has been proved by V.P. Il'in. The estimate

$$\|u\|_{L_q(\Omega)} \le C\|u\|_{W_p^m(\Omega)}$$

with $q < np/(n-mp)$ for a bounded domain Ω, which is star-like with respect to a point, was proved at first by Sobolev. The stated proof is due to V.G.Mazya, see [Maz1]. We proved it for functions defined in $\mathbf{R}^n$ and having compact supports. It is easy to see that it is true for all functions of $W_{p,0}^m(\Omega)$. The class of domains for which Theorem 57 remains true was described by Mazya. It is obvious that it is true, if Ω is a parallelepiped and $q \le np/(n-mp)$, since then $W_p^m(\Omega)$ is the space of restrictions on Ω of functions from $W_{p,0}^m(\mathbf{R}^n)$.

2.22 Poincaré's inequality

The Poincaré inequality can be considered as a very important special case of Sobolev's theorems.

Theorem 59. *Let $Q = \{x : \ a_i < x_i < b_i, \ i = 1, \ldots, n\}$. If $\int_Q u(x)\,dx = 0$, then*

$$\int_Q |u(x)|^p\,dx \le C\int_Q |\nabla u|^p\,dx,$$

where $p \ge 1$ and C is independent of u.

Proof. We will apply induction on n. If $n = 1$ and

$$\int_{a_1}^{b_1} u(x)\,dx = 0,$$

then $u(\xi) = 0$ at a point ξ from (a_1, b_1). Therefore

$$|u(x)| \le |\int_\xi^x u'(t)\,dt|$$

and by the Hölder inequality

$$|u(x)|^p \le (\int_a^b |u'(t)|^p\,dt)(b_1 - a_1)^{p-1}.$$

Integrating this inequality from a_1 to b_1, we obtain the result.

Now let $n > 1$. Since $\int_Q u(x)\,dx = 0$, we have

$$\int_{x\in Q, x_n=\xi} u(x)\,dx_1 \dots dx_{n-1} = 0$$

for some ξ, $a_n < \xi < b_n$. It is easy to see that

$$u(x) = \int_\xi^{x_n} \frac{\partial u}{\partial x_n}\,dx_n + u(x_1, \dots, x_{n-1}, \xi).$$

Therefore

$$\overline{u}(x_n) \equiv \frac{1}{\text{meas}\,Q(x_n)} \int_{Q(x_n)} u(x_1, \dots, x_n)\,dx_1 \dots dx_{n-1} \le C_1 \int_Q |\nabla u|\,dx$$

for any x_n, $a_n < x_n < b_n$, where $Q(x_n)$ is the $(n-1)$-dimensional cube:

$$Q(x_n) = \{(x_1, \dots, x_n) : \; a_i \le x_i \le b_i, \; i = 1, \dots, n-1\}$$

and $\overline{u}(x_n)$ is the mean value of u in $Q(x_n)$. Note that

$$\int_{Q(x_n)} [u(x) - \overline{u}(x_n)]\,dx_1 \dots dx_{n-1} = 0.$$

Using the induction assumption, we have

$$\int_{Q(x_n)} |u(x) - \overline{u}(x_n)|^p\,dx_1 \dots dx_{n-1} \le C_2 \int_{Q(x_n)} |\nabla u|^p\,dx_1 \dots dx_{n-1}.$$

Therefore

$$\int_{Q(x_n)} |u(x)|^p\,dx_1 \dots dx_{n-1} \le C_3 \int_{Q(x_n)} |\nabla u|^p\,dx_1 \dots dx_{n-1}.$$

Integrating this inequality with respect to x_n from a_n to b_n, we obtain our statement. □

Theorem 60. *Let Ω be a bounded domain in $\mathbf{R}^n$ containing a cube Q with edge d_1 and let the diameter of Ω be equal to d_2. Let Ω be star-like with respect to the center of the cube Q. If $u(x) \in W_2^1(\Omega)$, then*

$$\int_\Omega |u(x) - \overline{u}|^2\,dx \le C\frac{d_2^n}{d_1^{n-2}} \int_\Omega |\nabla u|^2\,dx \qquad \text{for } n > 2;$$

$$\int_\Omega |u(x) - \overline{u}|^2\,dx \le C d_2^2(1 + ln\frac{d_2}{d_1}) \int_\Omega |\nabla u|^2\,dx \quad \text{for } n = 2,$$

where C does not depend on u and Ω, and $\overline{u} = \int_\Omega u\,dx/\mu_n(\Omega)$.

Proof. Let the center of Q be at the origin. Set $h = d_1^{-n} \int_Q u\,dx$. By Theorem 59 we have

$$\int_Q |u-h|^2\,dx \le Cd_1^2 \int_Q |\nabla u|^2\,dx.$$

Set $v(x) = u(x) - h$. It suffices to prove our inequality for functions $u(x) \in C^1(\Omega)$, since the space $C^1(\Omega)$ is dense in $W_2^1(\Omega)$, when Ω is convex (see Theorem 47).

Let l be a ray from the origin. Let $P \in l$, $A \in l$, $A \in Q$, and $|A| > d_1/2$. We have

$$v(P) = v(A) + \int_{|A|}^{|P|} \frac{\partial v}{\partial r}\,dr,$$

hence

$$|v^2(P)| \le 2|v^2(A)| + 2\left(\int_{|A|}^{|P|} |\nabla v| r^{(n-1)/2} r^{(1-n)/2}\,dr\right)^2.$$

Integrating this relation, multiplied by $|P|^{n-1}$, with respect to $|P|$ from 0 to $\lambda(l)$, where $\lambda(l)$ is the length of the interval on l from $x=0$ until the point of intersection of l with $\partial\Omega$, we obtain

$$\int_0^{\lambda(l)} |v^2(P)||P|^{n-1}\,d|P| \le C|v(A)|^2 d_2^n + Cd_2^2 \int_0^{\lambda(l)} r^{n-1} |\nabla v(x)|^2\,dr.$$

Now integrating this relation over the unit sphere. We have

$$\int_\Omega |v(x)|^2\,dx \le C_1 \frac{d_2^n}{\rho^{n-1}} \int_{|x|=\rho} |v(x)|^2\,ds + C_1 d_2^2 \int_\Omega |\nabla v(x)|^2\,dx,$$

where $\rho = |A|$. There is ρ_1 such that $d_1/2 \le \rho_1 \le d_1$ and

$$\int_{|x|=\rho_1} |v(x)|^2\,ds \le \frac{4}{d_1} \int_Q |v(x)|^2\,dx.$$

Indeed, otherwise, we have

$$\int_{|x|=\rho} |v(x)|^2\,ds > \frac{4}{d_1} \int_Q |v(x)|^2\,dx$$

for all ρ of this interval. Integrating this inequality with respect to ρ from $d_1/2$ to d_1, we obtain that $\int_Q |v(x)|^2\,dx > 2\int_Q |v(x)|^2\,dx$, which is impossible. Now set

$\rho = \rho_1$. Then

$$\begin{aligned}\int_\Omega |u-h|^2\,dx &\le C_2\left(\frac{d_2}{d_1}\right)^n \int_Q |v(x)|^2\,dx + d_2^2\int_\Omega |\nabla v|^2\,dx\\ &\le C_2\left(\frac{d_2}{d_1}\right)^n d_1^2\int_Q |\nabla u(x)|^2\,dx + d_2^2\int_\Omega |\nabla u|^2\,dx\\ &\le 2C_2\frac{d_2^n}{d_1^{n-2}}\int_\Omega |\nabla u|^2\,dx \le C_2\left[\left(\frac{d_2}{d_1}\right)^n d_1^2 + d_2^2\right]\int_\Omega |\nabla u|^2\,dx\\ &\le 2C_2\frac{d_2^n}{d_1^{n-2}}\int_\Omega |\nabla u|^2\,dx.\end{aligned}$$

This and the Cauchy-Bounyakovsky inequality imply that

$$\begin{aligned}|\overline{u}-h|\operatorname{meas}\Omega = \left|\int_\Omega (u-h)\,dx\right| &\le (\operatorname{meas}\Omega)^{1/2}\left(\int_\Omega |u-h|^2\,dx\right)^{1/2}\\ &\le C_3 d_2^{n/2} d_1^{(2-n)/2}\left(\int_\Omega |\nabla u|^2\,dx\right)^{1/2}(\operatorname{meas}\Omega)^{1/2}.\end{aligned}$$

Therefore

$$\begin{aligned}\left(\int_\Omega |u-\overline{u}|^2\,dx\right)^{1/2} &\le \left(\int_\Omega |u-h|^2\,dx\right)^{1/2} + \left(\int_\Omega |h-\overline{u}|^2\,dx\right)^{1/2}\\ &\le C_4 d_2^{n/2} d_1^{(2-n)/2}\left(\int_\Omega |\nabla u|^2\,dx\right)^{1/2}.\end{aligned}$$

Thus the theorem is proved, if $n > 2$.

Now let $n = 2$. Integrating the relation

$$|v^2(P)| \le 2|v^2(A)| + 2\left(\int_{|A|}^{|P|} |\nabla v| r^{1/2} r^{-1/2}\,dr\right)^2,$$

multiplied by $|P|$, with respect to $|P|$ from 0 to $\lambda(l)$, where $\lambda(l)$ is the length of the interval on l from $x = 0$ until the point of intersection of l with $\partial\Omega$, we obtain

$$\int_0^{\lambda(l)} |v^2(P)||P|\,d|P| \le C|v(A)|^2 d_2^2 + C d_2^2 |\ln(d_2/|P|)| \int_0^{\lambda(l)} r|\nabla v|^2\,dr.$$

Now integrate this relation over the unit circle. We have

$$\int_\Omega |v(x)|^2\,dx \le C_1\frac{d_2^2}{\rho}\int_{|x|=\rho} |v(x)|^2\,ds + C_1 d_2^2 |\ln(d_2/d_1)| \int_\Omega |\nabla v|^2\,dx.$$

The rest of the proof is as above. □

Theorem 61. *Let Ω be a union of domains in $\mathbf{R}^n$, $\Omega = \Omega_1 \cup \Omega_2 \cup \ldots \cup \Omega_k$, the intersection $\Omega_i \cap \Omega_{i+1}$ contain a ball Q_i with its center in a point P_i and the domains Ω_i and Ω_{i+1} be star-like with respect to P_i for $i = 1, \ldots, k-1$. Then there exists a constant $C(\Omega)$ such that for any $u \in W_2^1(\Omega)$*

$$\int_\Omega |u - \overline{u}|^2 \, dx \leq C(\Omega) \int_\Omega |\nabla u|^2 \, dx,$$

where $\overline{u} = (\operatorname{meas} \Omega)^{-1} \int_\Omega u \, dx$.

Proof. Let $h_i = (\operatorname{meas} Q_i)^{-1} \int_{Q_i} u \, dx$. By Theorem 60 we have

$$\int_{\Omega_i} |u - h_i|^2 \, dx + \int_{\Omega_{i+1}} |u - h_i|^2 \, dx \leq C(\Omega) \int_\Omega |\nabla u|^2 \, dx.$$

Moreover

$$\int_{Q_i} |u - h_i|^2 \, dx \leq C(\Omega) \int_\Omega |\nabla u|^2 \, dx.$$

In virtue of the Cauchy-Bounyakovsky inequality we have

$$|h_{i+1} - h_i| \cdot \operatorname{meas} Q_{i+1} = \left| \int_{Q_{i+1}} (u - h_i) \, dx \right|$$

$$\leq \left(\int_{Q_{i+1}} |u - h_i|^2 \, dx \right)^{1/2} (\operatorname{meas} Q_{i+1})^{1/2} \leq C(\Omega) \left(\int_\Omega |\nabla u|^2 \, dx \right)^{1/2}.$$

Therefore

$$|h_{i+1} - h_i| \leq C_1(\Omega) \left(\int_\Omega |\nabla u|^2 \, dx \right)^{1/2}.$$

Hence

$$|h_i - h_1| \leq |h_i - h_{i-1}| + \cdots + |h_2 - h_1| \leq C_2(\Omega) \left(\int_\Omega |\nabla u|^2 \, dx \right)^{1/2}$$

for all $i = 1, \ldots, k$. Hence it follows that

$$\left(\int_{\Omega_i} |u - h_1|^2 \, dx \right)^{1/2} \leq \left(\int_{\Omega_i} |u - h_i|^2 \, dx \right)^{1/2} + \left(\int_\Omega |h_i - h_1|^2 \, dx \right)^{1/2}$$

$$\leq C_3(\Omega) \left(\int_\Omega |\nabla u|^2 \, dx \right)^{1/2}.$$

Summing all these inequalities for $i = 1, \ldots, k$ we obtain that

$$\int_\Omega |u - h_1|^2 \, dx \leq C_4(\Omega) \int_\Omega |\nabla u|^2 \, dx.$$

The Cauchy-Bounyakovsky inequality implies that

$$|\overline{u}-h_1|\cdot \operatorname{meas}\Omega = \Big|\int_\Omega (u-h_1)\,dx\Big| \le C_5(\Omega)\left(\int_\Omega |\nabla u|^2\,dx\right)^{1/2}.$$

Therefore

$$\left(\int_\Omega |u-\overline{u}|^2\,dx\right)^{1/2} \le \left(\int_\Omega |u-h_1|^2\,dx\right)^{1/2} + \left(\int_\Omega |h_1-\overline{u}|^2\,dx\right)^{1/2}$$
$$\le C_6(\Omega)\left(\int_\Omega |\nabla u|^2\,dx\right)^{1/2}.$$ □

Theorem 62. *Let λ be a positive real number and Ω be the domain in $\mathbf{R}^n$ such that*

$$\Omega = \{x:\ \lambda<|x|<2\lambda\}.$$

Then for all $u\in W_2^1(\Omega)$, such that $\int_\Omega u(x)\,dx=0$, the inequality

$$\int_\Omega |u|^2\,dx \le C\lambda^2\int_\Omega |\nabla u|^2\,dx$$

holds and the constant C is independent of λ and u.

Proof. The inequality is true in the case $\lambda=1$ by Theorem 61, since this domain can be covered by a finite number of open cubes. Substituting $\lambda x'=x$, we obtain that

$$\int_\Omega |u|^2\,dx = \lambda^n\int_{1\le|x'|\le 2}|u|^2\,dx'.$$

Since

$$\int_{1<|x'|<2}|u|^2\,dx' \le C\int_{1<|x'|<2}\sum_1^n\left|\frac{\partial u}{\partial x_i'}\right|^2\,dx',$$

we see that in variables x the inequality

$$\int_\Omega |u|^2\,dx \le C\lambda^2\int_\Omega\sum_1^n\left|\frac{\partial u}{\partial x_i}\right|^2\,dx$$

holds. □

Theorem 63. *Let Ω satisfy the conditions of Theorem 60, $p\ge 1$, and*

$$\int_\Omega D^\alpha u\,dx = 0$$

for all α such that $|\alpha|<m$. Then

$$\sum_{|\alpha|=m-k}\int_\Omega |D^\alpha u|^p\,dx \le C\int_\Omega\sum_{|\alpha|=m}|D^\alpha u|^p\,dx$$

for all $k\le m$.

Proof. If $m = 1$, the proof is an easy modification of that of Theorem 60. For any m we will use induction on k. Indeed, it is obvious for $k = 0$. If $k > 0$, then

$$\sum_{|\alpha|=m-k} \int_\Omega |D^\alpha u|^p \, dx \leq C_1 \sum_{|\alpha|=m-k} \int_\Omega |\nabla D^\alpha u|^p \, dx \leq C_2 \sum_{|\alpha|=m-k+1} \int_\Omega |D^\alpha u|^p \, dx.$$

By the inductive hypothesis we have

$$\sum_{|\alpha|=m-k+1} \int_\Omega |D^\alpha u|^p \, dx \leq C_1 \sum_{|\alpha|=m} \int_\Omega |D^\alpha u|^p \, dx,$$

so that

$$\sum_{|\alpha|=m-k} \int_\Omega |D^\alpha u|^p \, dx \leq C_1 C_2 \int_\Omega \sum_{|\alpha|=m} |D^\alpha u|^p \, dx. \qquad \square$$

Theorem 64. *Let Ω satisfy the conditions of Theorem 60, $p \geq 1$, and*

$$\int_\Omega x^\alpha u \, dx = 0$$

for all α with $|\alpha| < m$. Then for $k = 0, 1, \ldots, m$

$$\sum_{|\alpha|=m-k} \int_\Omega |D^\alpha u|^p \, dx \leq C \int_\Omega \sum_{|\alpha|=m} |D^\alpha u|^p \, dx.$$

Proof. Let us find constants $a_\alpha = \text{const}$, $|\alpha| < m$ such that

$$\int_\Omega D^\beta (u - \sum_{|\alpha|<m} a_\alpha x^\alpha) \, dx = 0$$

for all β, $|\beta| < m$. Let us show that the determinant of this algebraic system

$$D = \det\left(\int_\Omega D^\beta x^\alpha \, dx\right)_{|\beta|<m, |\alpha|<m}$$

does not vanish. Indeed, if it vanishes, then there exists a nonzero set of c_α, $|\alpha| < N$, such that

$$\int_\Omega D^\beta \left(\sum_{|\alpha|<m} c_\alpha x^\alpha\right) dx = 0$$

for all β such that $|\beta| < m$. Let k be the maximal value, for which there are $c_\gamma \neq 0$ with $|\gamma| = k$. Setting $\beta = \gamma$, we obtain that

$$\gamma! c_\gamma \operatorname{meas} \Omega = 0,$$

which contradicts our assumption. Thus there are a_α satisfying these relations. By Theorem 63

$$\sum_{|\beta|<m} \int_\Omega |D^\beta(u - \sum_{|\alpha|<m} a_\alpha x^\alpha)|^p \, dx \le C \int_\Omega \sum_{|\alpha|=m} |D^\alpha u|^p \, dx.$$

Note that

$$\int_\Omega \sum_{|\alpha|<m} a_\alpha x^\alpha x^\beta \, dx = \int_\Omega x^\beta \left(\sum_{|\alpha|<m} a_\alpha x^\alpha - u(x) \right) dx, \ \forall \ |\beta| < m.$$

Therefore, for $|\beta| = m$

$$\left| \int_\Omega \sum_{|\alpha|<m} a_\alpha x^\alpha x^\beta \, dx \right| \le C \left(\int_\Omega \sum_{|\alpha|=m} |D^\alpha u|^p \, dx \right)^{1/p}.$$

Let us denote

$$\sum_{|\alpha|<m} \int_\Omega a_\alpha x^\alpha x^\beta \, dx = A_\beta$$

for $|\beta| = m$. Consider the last equality as a linear system of equations for a_α. Its determinant is the Gramian determinant for a set of linearly independent functions and hence does not vanish. Therefore,

$$|a_\alpha| \le C \sum_{|\beta|=m} |A_\beta| = C_1 \left(\int_\Omega \sum_{|\gamma|=m} |D^\gamma u|^p \, dx \right)^{1/p}.$$

The proof is complete. □

2.23 Interpolation inequalities

The following theorem shows that a function from $L_p(\Omega)$ and $W_p^m(\Omega)$ with $m > 1$ belongs to any intermediate space.

Theorem 65. *Let Ω satisfy the conditions of Theorem 60 and $0 \le k \le m$. If $u \in W_p^m(\Omega)$, $p \ge 1$, then*

$$\int_\Omega \sum_{|\alpha|=k} |D^\alpha u|^p \, dx \le C \int_\Omega \left[\sum_{|\alpha|=m} |D^\alpha u|^p + |u^p| \right] dx, \ 0 \le k \le m,$$

where C is independent of u.

Proof. There are constants $c_\alpha = \text{const}$, $|\alpha| < m$ such that

$$\int_\Omega x^\beta [u - \sum_{|\alpha|<m} c_\alpha x^\alpha] \, dx = 0$$

for all β with $|\beta| < m$. These equalities represents an algebraic linear system of equations for c_α with its determinant equal to the determinant of the matrix having as elements $\int_\Omega x^\beta x^\alpha \, dx$ for $|\alpha| < m$, $|\beta| < m$, i.e. to the Gramian determinant of the set of linearly independent functions x^α, $|\alpha| < m$, and hence nonvanishing. Therefore it has a unique solution, i.e. there are constants c_α satisfying this system such that $|c_\alpha| \le c_1 \int_\Omega |u| \, dx$. Hence and by Theorem 64

$$\sum_{|\alpha|=k} \int_\Omega \left| D^\alpha (u - \sum_{|\beta|<m} c_\beta x^\beta) \right|^p dx \le c_2 \int_\Omega \sum_{|\alpha|=m} |D^\alpha u|^p \, dx. \qquad \square$$

Theorem 66. *Let Ω satisfy the conditions of Theorem 60 and $0 < k < m$. For any $\varepsilon > 0$ and $p \ge 1$ there is a constant $C(\varepsilon, p)$ such that for all $u \in W_p^m(\Omega)$ the estimate*

$$\sum_{|\alpha|=k} \int_\Omega |D^\alpha u|^p \, dx \le \varepsilon \int_\Omega \sum_{|\alpha|=m} |D^\alpha u|^p \, dx + C(\varepsilon, p) \int_\Omega |u|^p \, dx$$

holds.

Proof. If Q is a cube with its edge equal to 1, then by Theorem 65 we have

$$\int_Q \sum_{|\alpha|=k} |D^\alpha u|^p \, dx \le C \int_Q \sum_{|\alpha|=m} |D^\alpha u|^p \, dx + C \int_Q |u|^p \, dx.$$

If Q is a cube with its edge equal to δ, then, using the substitution $x = \delta x'$, we obtain that

$$\delta^{kp} \int_Q \sum_{|\alpha|=k} |D^\alpha u|^p \, dx \le C \delta^{mp} \int_Q \sum_{|\alpha|=m} |D^\alpha u|^p \, dx + C \int_Q |u|^p \, dx.$$

Now let Ω be an arbitrary domain in $\mathbf{R}^n$. Let $u(x) \in W_{p,0}^m(\Omega)$. Let us consider a partition of $\mathbf{R}^n$ on congruent closed cubes with its size equal to δ. Summing the inequalities obtained above for each of these cubes, we get the inequality

$$\sum_{|\alpha|=k} \int_\Omega |D^\alpha u|^p \, dx \le C \delta^{(m-k)p} \int_\Omega \sum_{|\alpha|=m} |D^\alpha u|^p \, dx + C \delta^{-kp} \int_\Omega |u|^p \, dx.$$

Now setting $\delta = (\varepsilon C^{-1})^{1/(m-k)p}$, we obtain the result. Observe that

$$C(\varepsilon, p) = c_1 \varepsilon^{-kp/(m-k)p}. \qquad \square$$

2.24 Compactness of the embedding

The following theorems show that the embedding of the Sobolev spaces is often not only continuous but also compact.

Theorem 67. *Let Ω be a bounded domain in $\mathbf{R}^n$ and $p > 1$. Then every set bounded in $W^1_{p,0}(\Omega)$ is compact in $L_p(\Omega)$.*

Proof. In virtue of Theorem 29 it is sufficient to prove the equicontinuity in $L_p(\Omega)$ of a set, bounded in $W^1_{p,0}(\Omega)$. Since

$$f(x_1 + h, x_2, \dots, x_n) - f(x_1, \dots, x_n) = \int_{x_1}^{x_1+h} \frac{\partial f}{\partial x_1}\, dx_1$$

for almost all $(x_2, \dots, x_n)$, we have

$$|f(x_1 + h, x_2, \dots, x_n) - f(x_1, \dots, x_n)|^p \le h^{p/q} \int_{x_1}^{x_1+h} \left|\frac{\partial f}{\partial x_1}\right|^p dx_1,$$

where $1/p + 1/q = 1$. Integrating this inequality over Ω, we see that

$$\int_\Omega |f(x_1 + h, x_2, \dots, x_n) - f(x_1, \dots, x_n)|^p\, dx \le h^p d \int_\Omega \left|\frac{\partial f}{\partial x_1}\right|^p dx,$$

where d is the diameter of the domain Ω. Similar inequalities hold after the substitution of x_1 by x_i with $i = 2, \dots, n$. These inequalities imply the equicontinuity in $L_p(\Omega)$ of a set, bounded in $W^1_{p,0}(\Omega)$. □

Note that the boundedness of Ω is essential. For example, if $\Omega = \mathbf{R}^1$, the family $f_k(x) = f(x + k)$, where $f(x) \in C_0^\infty$, $f(x) \not\equiv 0$, is bounded in $W^1_{p,0}(\Omega)$, but is not compact in $L_p(\Omega)$. Theorem 67 implies compactness of embedding of the space $W^l_{p,0}(\Omega)$ in $W^k_{p,0}(\Omega)$ for $k < l$.

Theorem 68. *Let Ω be a bounded domain in $\mathbf{R}^n$ and*

$$1 \le q < q^* = np/(n - p),\ n > p.$$

Then every bounded set in the space $W^1_{p,0}(\Omega)$ is compact in $L_q(\Omega)$. If $q < np/(n - mp)$, $n > mp$, then every bounded set in $W^m_{p,0}(\Omega)$ set is compact in $L_q(\Omega)$.

Proof. Let a set of functions $u(x)$ be such that $\|u(x)\|_{W^1_{p,0}(\Omega)} \le M = \text{const}$, and let Ω be a bounded domain in $\mathbf{R}^n$. Then

$$\|u(x)\|_{L_q} \le C\|u(x)\|_{L_{q^*}} \le C_1\|u(x)\|_{W^1_{p,0}(\Omega)}$$

by Sobolev's Theorem 57. Let us estimate $\|u(x + y) - u(x)\|_{L_q}$. Observe that $t^q \le \varepsilon t^{q^*} + C(\varepsilon)t^p$ for any $\varepsilon > 0$ and all $t > 0$. Therefore

$$\int_\Omega |u(x + y) - u(x)|^q\, dx \le \varepsilon \int_\Omega |u(x) - u(y)|^{q^*}\, dx + C(\varepsilon) \int_\Omega |u(x + y) - u(x)|^p\, dx.$$

Thus

$$\|u(x+y)-u(x)\|_{L_q} \le \varepsilon C_2 M^{q^*/q} + C(\varepsilon)|y|^{(p-1)/q} M^{p/q} \le C_3\varepsilon,$$

if $|y|$ is small enough. Therefore, the considered set of functions is uniformly bounded and equicontinuous in $L_q(\Omega)$ and by Theorem 29 it is compact. The second statement is proved by induction as in the proof of Theorem 57. □

Theorem 69. *Let Ω be a bounded domain in $\mathbf{R}^n$ and $pm > n$. Then every bounded set in $W^m_{p,0}(\Omega)$ belongs to $C(\Omega)$ and is compact in it.*

Proof. If $u(x) \in W^m_{p,0}(\Omega)$, then by the Sobolev Theorem 50 and the Hölder inequality

$$|u(x)| \le C\left(\sum_{|\alpha|=m}\int_\Omega |D^\alpha u|^p\,dy\right)^{1/p}\left(\int_\Omega \frac{dy}{|y-x|^{p'(n-m)}}\right)^{1/p'},\ p'=\frac{p}{p-1}.$$

If $n-(n-m)p' > 0$, i.e. $n < pm$, then

$$\int_\Omega \frac{dy}{|y-x|^{p'(n-m)}} \le C_1.$$

Therefore

$$|u(x)| \le C\|u\|_{W^m_p}.$$

Every function $u(x) \in W^m_{p,0}(\Omega)$ can be represented as a limit of a sequence of functions from $C_0^\infty(\Omega)$ in the norm of $W^m_p(\Omega)$. The proved inequality also implies that

$$|u_j(x)-u_{j+k}(x)| \le C\|u_j-u_{j+k}\|_{W^m_p(\Omega)}.$$

Therefore, the sequence of continuous functions $u_j(x)$ converges uniformly and its limit $u(x)$ is also a continuous function.

To prove the equicontinuity of functions of a bounded set in $W^m_{p,0}(\Omega)$ we apply Theorem 50. Then

$$|u(x)-u(x')| = \Big|\int_{\mathbf{R}^n}\sum_{|\alpha|=m}(a_\alpha(x,y)|x-y|^{m-n} - a_\alpha(x',y)|x'-y|^{m-n})D^\alpha u(y)\,dy\Big|.$$

Let ε be a small positive number and $B(x,\delta)$ a ball of radius δ with its center at x. If δ is sufficiently small and $x' \in B(x,\delta/2)$, then

$$\Big|\int_{B(x,\delta)}\sum_{|\alpha|=m}(a_\alpha(x,y)|x-y|^{m-n} - a_\alpha(x',y)|x'-y|^{m-n})D^\alpha u(y)\,dy\Big|$$

$$\le \Big(\int_{B(x,\delta)}\sum_{|\alpha|=m}|a_\alpha(x,y)|x-y|^{m-n} - a_\alpha(x',y)|x'-y|^{m-n}|^{p'}\,dy\Big)^{1/p'}$$

$$\times\Big(\int_{B(x,\delta)}\sum_{|\alpha|=m}|D^\alpha u(y)|^p\,dy\Big)^{1/p} \le \varepsilon/2,$$

since a_α are bounded and $p'(n-m) < n$. For fixed δ, we can find for any $\varepsilon_1 > 0$ find a $\rho > 0$ so that $\rho < \delta/2$ and

$$|a_\alpha(x,y)|x-y|^{m-n} - a_\alpha(x',y)|x'-y|^{m-n}|^{p'} \leq \varepsilon_1,$$

if $|x-x'| < \rho$, $|x-y| > \delta$, $|x'-y| > \delta/2$. Therefore

$$|\int_{\mathbf{R}^n \setminus B(x,\delta)} \sum_{|\alpha|=m} (a_\alpha(x,y)|x-y|^{m-n} - a_\alpha(x',y)|x'-y|^{m-n}) D^\alpha u(y)\, dy|$$

$$\leq \varepsilon_1 (\mathrm{meas}(\Omega))^{1/p'} (\int_{\mathbf{R}^n \setminus B(x,\delta)} \sum_{|\alpha|=m} |D^\alpha u(y)|^p \, dy)^{1/p} \leq C\varepsilon_1 \leq \varepsilon/2,$$

for the corresponding choice of ε_1. The proof is complete. □

Using induction it is easy to show that a function $u \in W_p^m(\Omega)$ with $n < p(m-k)$ belongs to the space $C^k(\Omega)$.

2.25 Invariance of $W_p^m(\Omega)$ under change of variables

For applications it is important that the Sobolev spaces are invariant with respect to diffeomorphisms of a domain.

Theorem 70. *Let Ω be a bounded domain in $\mathbf{R}^n$ and $u \in W_p^m(\Omega)$. Consider the transformation $x = g(y)$ of the class $C^m(\overline{\Omega})$, defining a diffeomorphism of Ω on a domain $G \subset \mathbf{R}_y^n$. Suppose that the Jacobian of this transformation does not vanish in $\overline{\Omega}$. Then $u(g(y)) \in W_p^m(G)$.*

Proof. If $v(x) \in C^\infty(\Omega)$, then $v(g(y)) = w(y) \in C^m(G)$ and the relation

$$D^\alpha w(y) = \sum_{0<|\beta|\leq|\alpha|} \psi_{\alpha,\beta}(g(y)) D_x^\beta v$$

is true for $|\alpha| \leq m$ with $\psi_{\alpha,\beta} \in C^{m-|\alpha|+|\beta|-1}(\overline{G})$. Let Ω_1 be a ball, $\Omega_1 \subset\subset \Omega$, and a sequence $u_k(x) \in C^\infty(\Omega_1)$ be converging in $W_p^m(\Omega_1)$ to u as $k \to \infty$. We have

$$D_y^\alpha u_k(g(y)) = \sum_{|\beta|\leq|\alpha|} \psi_{\alpha,\beta} D_x^\beta u_k(x).$$

Passing to the limit as $k \to \infty$, we obtain for generalized derivatives the formula

$$D_y^\alpha u = \sum_{|\beta|\leq|\alpha|} \psi_{\alpha,\beta} D_x^\beta u.$$

Since the functions $\psi_{\alpha,\beta}$ are bounded in $\overline{\Omega}$ and $u \in W_p^m(\Omega)$, we see that $u(g(y)) \in W_p^m(G)$. □

2.26 The spaces $W_p^m(\Omega)$ for a smooth domain Ω

Let Q_j $(j = 1, \dots, s)$ be a set of open balls with their centers in x_j of radii ρ_j and such that $\overline{\Omega} \subset \cup_{j=1}^s Q_j$. Then there is a system of functions $\sigma_j \in C_0^\infty(\mathbf{R}_x^n)$ such that $\operatorname{supp} \sigma_j \subset \overline{Q}_j$ and $\sum_{j=1}^s \sigma_j \equiv 1$ in Ω. Such a system can be constructed in the following way. Let $\varphi_j(x) = e^{1/(|x-x_j|^2-\rho_j^2)}$ for $x \in Q_j$, $\varphi_j = 0$ for $x \notin Q_j$. It is easy to see that the functions $\sigma_j = \varphi_j / \sum_{i=1}^s \varphi_i$ satisfy the required conditions.

Let now Q_j be arbitrary open sets and $\Omega \subset \cup Q_j$. Then in order to construct the partition of unity it suffices to take as $\varphi_j(x)$ any function $C_0^\infty(\mathbf{R}^n)$ positive in Q_j and vanishing outside Q_j. Let us show that such functions exist. Let k be an integer. Consider the set $\Omega_k = \{x : \ x \in Q_j, \ 2^{-k} < \rho(x, \partial Q_j) \le 2^{-k+1}\}$. It is possible that some of Ω_k are empty. However, $Q_j \subset \cup_{-\infty<k<+\infty} \Omega_k$. Let us consider the covering of $\overline{Q}_j$ by balls of the radius 2^{-k-1} with their centers at the points of Ω_k. By the Heine-Borel lemma one can choose a finite subcovering of Q_j. Let this subcovering consist of the balls K_j, $j = 1, \dots, l$, having the radii r_j and the centers y_j. Set now $\psi_{k1}(x) = \sum_{j=1}^l e^{1/(|x-y_j|^2-r_j^2)}$. These functions are of the class $C_0^\infty(\mathbf{R}^n)$ and $\operatorname{supp} \psi_{k1} \in \Omega_{k+1} \cup \Omega_k \cup \Omega_{k-1}$. Set

$$M_k = \sum_{|\alpha| \le k} \max |D^\alpha \psi_{k1}|$$

and $\psi_k = 2^{-k} \psi_{k1} / M_k$. The function $\varphi_j(x) = \sum_{-\infty}^{+\infty} \psi_k(x)$ satisfies the required conditions.

If we set

$$\sigma_j(x) = \varphi_j(x) / \sum_{i=1}^s \varphi_i(x),$$

we obtain a partition of unity $\sum_{j=1}^s \sigma_j(x) \equiv 1$.

A partition of unity $\sum_{j=1}^s \sigma_j \equiv 1$ such that $\operatorname{supp} \sigma_j \subset \overline{Q}_j$ is called *subordinated to the covering* of Ω by sets Q_j.

We considered before the Sobolev spaces $W_p^m(\Omega)$, $p > 1$, for a bounded star-like domain. Now let Ω be a bounded domain with an infinitely smooth boundary.

Theorem 71. *The space $W_p^m(\Omega)$ is the restriction of $W_{p,0}^m(\mathbf{R}_x^n)$ on Ω.*

Proof. For any point $P \in \partial\Omega$ let S_P be a ball such that $S_P \cap \Omega$ can be mapped with the help of a smooth invertible transformation $y = F_P x$ in the half-space $y_n > 0$ in such a way that $S_P \cap \partial\Omega$ lies in the plane $y_n = 0$. Choose from the set of S_P a finite subsystem, covering $\partial\Omega$. Let this subsystem consist of the balls $S_{P_1}, \dots, S_{P_N}$. The set $\overline{\Omega}$ is covered by the domains $S_{P_1}, S_{P_2}, \dots, S_{P_N}$, $\Omega = S_{P_{N+1}}$. Let $1 = \sum_1^{N+1} \varphi_j(x)$ be a partition of unity subordinated to this covering. Let $u(x) \in W_p^l(\Omega)$, $u_j(x) = u(x)\varphi_j(x)$, $1 \le j \le N+1$. Set now $v_j(y) = u_j(F_{P_j}^{-1} y)$ if $y \in F_{P_j}(S_{P_j}) \cap \mathbf{R}_+^n$, and $v_j(y) = 0$ if $y \in \mathbf{R}_+^n \setminus F_{P_j}(S_{P_j})$. We have

$$\|v_j(y)\|_{W_p^m(\mathbf{R}_+^n)} \le C \|u\|_{W_p^m(\Omega)}.$$

Let $w_j(y)$ be the Whitney continuation of $v_j(y)$ in $\mathbf{R}^n$. Using cut-off functions, we can assume that $\operatorname{supp} v_j \subset F_{P_j} S_{P_j}$. Set now $u_j^0(x) = 0$, if $x \in \mathbf{R}^n \setminus S_{P_j}$; $u_j^0(x) = w_j(F_{P_j} x)$, if $x \in S_{P_j}$ for $j \le N$. It is obvious that $\operatorname{supp} u_j^0 \subset S_{P_j}$, $u_j^0 \in W_p^m(\Omega)$. Then the function $U(x) = \sum_{j=1}^{N+1} u_j^0(x)$ coincides with $u(x)$ in Ω, has a compact support, and

$$\|U(x)\|_{W_p^m(\mathbf{R}^n)} \le C\|u(x)\|_{W_p^m(\Omega)}. \qquad \square$$

Theorem 71 allows us to relate the above-mentioned statements on the properties of the spaces $W_p^m(\Omega)$ to general domains. For example, the interpolation inequality

$$\sum_{|\alpha|<m} \int_\Omega |D^\alpha u|^p \, dx \le \varepsilon \sum_{|\alpha|=m} \int_\Omega |D^\alpha u|^p \, dx + C(\varepsilon) \int_\Omega |u|^p \, dx$$

holds, since it is true for functions of $W_{p,0}^l(\mathbf{R}^n)$. The space $W_{p,0}^m(\Omega)$ is embedded in $L_q(\Omega)$ compactly if $\partial\Omega$ is a smooth surface. The same is true for the embedding of $W_p^m(\Omega)$ in $W_p^k(\Omega)$ if $k < m$.

Theorem 72. *Let Ω, Ω' be domains in $\mathbf{R}^n$ and $\overline{\Omega'} \subset \Omega$. Let $u(x) \in L_p(\Omega)$, $p > 1$, be such a function that*

$$\frac{u(x_1, \dots, x_{k-1}, x_k + h, x_{k+1}, \dots, x_n) - u(x)}{h} \in L_p(\Omega')$$

for $|h| \le h_0$ and

$$\left\| \frac{u(x_1, \dots, x_{k-1}, x_k + h, x_{k+1}, \dots, x_n) - u(x)}{h} \right\|_{L_p(\Omega')} \le C,$$

where C is independent of h. Then the function $u(x)$ has the generalized derivative $\partial u/\partial x_k \in L_p(\Omega')$ and $\|\partial u/\partial x_k\|_{L_p(\Omega')} \le C$.

Proof. Since a bounded set in the space $L_p(\Omega')$ is weakly compact, there is a sequence $\{h_j\}$ tending to 0 as $j \to \infty$ and such that the sequence of functions

$$\left\{ \frac{u(x_1, \dots, x_{k-1}, x_k + h_j, x_{k+1}, \dots, x_n) - u(x)}{h_j} \right\}$$

is weakly converging in $L_p(\Omega')$ to a function $u_k(x) \in L_p(\Omega')$ and $\|u_k(x)\|_{L_p(\Omega')} \le C$. The definition of weak convergence implies that for any function $\psi(x) \in C_0^\infty(\Omega')$ we have

$$\begin{aligned}\int_{\Omega'} u_k(x)\psi(x)dx &= \lim_{j\to\infty} \int_{\Omega'} \frac{u(x_1,\dots,x_{k-1},x_k+h_j,x_{k+1},\dots,x_n)-u(x)}{h_j} \psi(x)dx \\ &= \lim_{j\to\infty} \int_{\Omega'} u(x) \frac{\psi(x_1,\dots,x_{k-1},x_k-h_j,x_{k+1},\dots,x_n)-\psi(x)}{h_j} dx \\ &= -\int_{\Omega'} u(x) \frac{\partial \psi}{\partial x_k} dx,\end{aligned}$$

i.e. $u_k(x) = \partial u/\partial x_k$. $\square$

The inverse statement is also true, as shown by the following theorem.

Theorem 73. *Let Ω, Ω' be domains in $\mathbf{R}^n$ and $\overline{\Omega'} \subset \Omega$. Let $u(x) \in L_p(\Omega)$, $p > 1$, have a generalized derivative $\partial u/\partial x_k \in L_p(\Omega)$. Then*

$$\frac{u(x_1, \ldots, x_{k-1}, x_k + h, x_{k+1}, \ldots, x_n) - u(x)}{h} \in L_p(\Omega')$$

for $|h| \leq h_0$ and

$$\|\frac{u(x_1, \ldots, x_{k-1}, x_k + h, x_{k+1}, \ldots, x_n) - u(x)}{h}\|_{L_p(\Omega')} \leq \|\frac{\partial u}{\partial x_k}\|_{L_p(\Omega)}.$$

Proof. It is sufficient to prove our statement for functions $u(x) \in C^\infty(\Omega'')$, where Ω'' is such a domain that $\overline{\Omega}' \subset \Omega''$, $\overline{\Omega}'' \subset \Omega$, and to use averaging. If $u(x) \in C^\infty(\Omega'')$, then the Newton-Leibniz formula implies that

$$\left\| \frac{u(x_1, \ldots, x_{k-1}, x_k + h, x_{k+1}, \ldots, x_n) - u(x)}{h} \right\|^p_{L_p(\Omega')}$$

$$= \left\| \frac{1}{h} \int_{x_k}^{x_k+h} \frac{\partial u(x_1, \ldots, x_{k-1}, t, x_{k+1}, \ldots, x_n)}{\partial t} dt \right\|^p_{L_p(\Omega')}$$

$$\leq \int_{\Omega'} \frac{1}{h} (\int_{x_k}^{x_k+h} \left| \frac{\partial u(x_1, \ldots, x_{k-1}, t, x_{k+1}, \ldots, x_n)}{\partial t} \right|^p dt)\, dx.$$

Changing the integration order in the latter expression, we obtain the statement. □

Remark 74. If $u \in L_p(\Omega)$ and $\operatorname{supp} u \subset \Omega$, then in Theorems 72 and 73 one can take $\Omega' = \Omega$. This is also true in the case when Ω has the form: $x_k \geq x_k^0$, $\sum_{i=1}^n (x_i - x_i^0)^2 \leq \rho^2$, and $u(x) = 0$ for $\sum_{i=1}^n (x_i - x_i^0)^2 \geq \rho'^2$, $\rho' < \rho$.

2.27 The traces of functions of $W_p^1(\Omega)$

The space of traces of functions from $W_p^1(\Omega)$ with $p > 1$ on $\partial\Omega$ is defined as the quotient space $W_p^1(\Omega)/W_{p,0}^1(\Omega)$ with the corresponding norm. This space is denoted by $W^{1-1/p}(\partial\Omega)$. Suppose that $\partial\Omega$ is a surface of the class C^1. Let $\vec{e}_P$ be the unit vector of the inner normal to $\partial\Omega$ at a point P. Set for $\tau > 0:\ S_\tau = \{x : x = P + \tau\vec{e}_P\}$. If $\tau \leq \tau_0 = \text{const}$, then the set S_τ is a surface of the class C^1, and the surfaces S_τ, corresponding to different values of τ, do not intersect.

Theorem 75. *If $u(x) \in W_p^1(\Omega)$, then $u(x) \in L_p(S_\tau)$ for almost all τ, $0 < \tau \leq \tau_0$.*

Proof. Let us introduce in the set $S \times [0, \tau_0]$ a coordinate system $(P(x), \tau(x))$, where $P(x) \in \partial\Omega$, $\tau(x) \in [0, \tau_0]$ are such that $x = P(x) + \tau \vec{e}_P$. Then

$$\int_{S\times[0,\tau_0]} |u(x)|^p \, dx = \int_0^{\tau_0} \int_{S_\tau} |u(x)|^p \, dS_\tau d\tau,$$

where dS_τ is the element of the surface S_τ. By the Fubini theorem,

$$\int_0^{\tau_0} d\tau \int_{S_\tau} |u(x)|^p dS_\tau \le C \int_{S\times[0,\tau_0]} |u(x)|^p \, dx < \infty$$

and the integral $\int_{S_\tau} |u(x)|^p \, dS_\tau$ exists for almost all τ. Besides, there is a τ_1, $0 \le \tau_1 \le \tau_0$, such that

$$\int_{S_{\tau_1}} |u(x)|^p dS_\tau \le \frac{1}{\tau_0} \int_{S\times[0,\tau_0]} |u(x)|^p \, dx \le \frac{1}{\tau_0} \|u\|^p_{W^1_p}. \qquad \square$$

Theorem 76. *If $u(x) \in W^1_p(\Omega)$, then for all τ, $0 < \tau \le \tau_0$, the estimate*

$$\|u\|_{L_p(S_\tau)} \le C\|u\|_{W^1_p(\Omega)}$$

holds.

Proof. Since for smooth domains the space $W^1_p(\Omega)$ is the completion of $C^\infty(\Omega)$ in the norm of $W^1_p(\Omega)$, it suffices to prove the estimate for smooth functions u. However, if $x = \tau\vec{e}_P + P$, $y = \tau_1\vec{e}_P + P$, then

$$u(x) - u(y) = \int_{\tau_1}^{\tau} \frac{\partial u}{\partial \vec{e}_P} \, d\tau \quad \text{and} \quad |u(x)| \le |u(y)| + \int_{\tau_1}^{\tau} |\nabla u| \, d\tau.$$

Therefore

$$\int_{S_\tau} |u(x)|^p \, dS_\tau \le C_p \int_{S_{\tau_1}} |u(y)|^p \, dS_{\tau_1} + C_p|\tau_1 - \tau|^{p/p'} \int_{S\times[0,\tau_0]} |\nabla u|^p \, dx,$$

where $1/p + 1/p' = 1$. Using Theorem 75, let us choose the value of τ_1 in such a way that

$$\int_{S_{\tau_1}} |u|^p \, dS_\tau \le \frac{1}{\tau_0} \int_{S\times[0,\tau_0]} |u|^p \, d\tau \le \frac{1}{\tau_0}\|u\|^p_{W^1_p}.$$

Then we obtain the desired estimate. $\square$

The proved inequality allows to define the trace of a function $u(x) \in W^1_p(\Omega)$ on S_τ in another way. Let $u_k \in C^1(\overline{\Omega})$ and the sequence $\{u_k\}$ converge to u in $W^1_p(\Omega)$. Then the sequence $u_k|_{S_\tau}$ is fundamental in $L_p(S_\tau)$. Its limit is called the trace of u on S_τ.

Theorem 76 implies the continuous dependence of the trace on S_τ with respect to τ. Indeed, if $1/p + 1/p' = 1$, then

$$\int |u(P + \tau e_P) - u(P + \tau_1 e_P)|^p \, dS \le C|\tau - \tau_1|^{p/p'} \int_{S\times[0,\tau_0]} |\nabla u|^p \, dx \to 0,$$

as $\tau, \tau_1 \to 0$. If $u(x) \in W_{p,0}^1(\Omega)$, then

$$\int_{S_\tau} |u|^p \, dS \to 0$$

as $\tau \to 0$. The inverse statement is also true, as seen from the following theorem.

Theorem 77. *If* $u(x) \in W_p^1(\Omega)$ *and*

$$\int_{S_\tau} |u(x)|^p \, dS \to 0$$

as $\tau \to 0$, *then* $u(x) \in W_{p,0}^1(\Omega)$.

Proof. We have for $0 < \tau \le \tau_0$:

$$\int_{S_\tau} |u(x)|^p \, dS_\tau \le C_p \tau^{p/p'} \int_\Omega |\nabla u(x)|^p \, dx,$$

where $1/p + 1/p' = 1$. Integrating this inequality with respect to τ from $\tau = 0$ to $\tau = \rho = \text{const} \le \tau_0$, we obtain that

$$\int_{S\times[0,\rho]} |u|^p \, dx \le C_1 \rho^p \int_\Omega |\nabla u|^p \, dx.$$

Let $\sigma(x)$ be the characteristic function of the set $\Omega \setminus (S \times [0, \rho/2])$ and $\sigma_h(x)$ be its averaging, i.e.

$$\sigma_h(x) = h^{-n} \int_{|x-y|\le h} \omega\left(\frac{x-y}{h}\right) \sigma(y) \, dy.$$

It is obvious that $|\sigma_h| \le 1$, $|\nabla \, \sigma_h| \le Ch^{-1}$. Set $u_\rho(x) = u(x)\sigma_{\rho/4}(x)$. Since $\sigma_{\rho/4}(x) \equiv 0$ in $S \times [0, \rho/8]$, we have $u_\rho(x) \in W_{p,0}^1(\Omega)$.

To prove our statement it suffices to show that

$$\|u_\rho(x) - u(x)\|_{W_p^1(\Omega)} \to 0$$

as $\rho \to 0$. This implies that $u(x) \in W_{p,0}^1(\Omega)$. We have

$$\int_\Omega |u_\rho(x) - u(x)|^p \, dx + \int_\Omega |\nabla(u_\rho - u)(x)|^p \, dx$$

$$\le C(\int_{S\times[0,\rho]} |u(x)|^p \, dx + \rho^{-p} \int_{S\times[0,\rho]} |u(x)|^p \, dx + \int_{S\times[0,\rho]} |\nabla u(x)|^p \, dx)$$

$$\le C \int_{S\times[0,\rho]} (|u(x)|^p + |\nabla u(x)|^p) \, dx \to 0$$

as $\rho \to 0$. Therefore, $\|u - u_\rho\|_{W_p^1(\Omega)} \to 0$ as $\rho \to 0$ and u belongs to $W_{p,0}^1(\Omega)$. □

2.28 The space H^s

Let us define H^s as the completion of $C_0^\infty(\mathbf{R}^n)$ in the norm

$$\|u\|_s = \left(\int_{\mathbf{R}^n} (1+|\xi|^2)^s |\tilde{u}(\xi)|^2 \, d\xi \right)^{1/2}, \quad s \in \mathbf{R}.$$

In particular, $H^0 = L_2(\mathbf{R}^n)$.

Theorem 78.
1. *$H_{s_1} \subset H_{s_2}$, if $s_1 > s_2$.*
2. *Let $u \in H^s$ and $D^\alpha u$ be a generalized derivative of u of order α, $|\alpha| \le s$. Then $D^\alpha u \in H^{s-|\alpha|}$ and $\|D^\alpha u\|_{s-|\alpha|} \le \|u\|_s$.*
3. *If $s > 0$ is an integer, then $H^s = W_2^s(\mathbf{R}^n)$.*
4. *If $0 < s < 1$, then the norm of the space H^s is equivalent to*

$$\left[\int_{\mathbf{R}^n} u^2 \, dx + \int\!\!\int_{\mathbf{R}^n \times \mathbf{R}^n} \frac{|u(x+h) - u(x)|^2}{|h|^{2s+n}} \, dh \, dx \right]^{1/2}.$$

If $s = m + \delta$, $0 < \delta < 1$, and m is a positive integer, then

$$\|u\|_s^2 \sim \int \sum_{|\alpha| \le m} |D^\alpha u|^2 \, dx + \int\!\!\int_{\mathbf{R}^n \times \mathbf{R}^n} \sum_{|\alpha| = m} \frac{|D^\alpha u(x+h) - D^\alpha u(x)|^2}{|h|^{2\delta+n}} \, dx \, dh.$$

5. *If $u \in H^s$, $a \in C^{s_1}(\mathbf{R}^n)$ with $0 < s < s_1 \le 1$, then $au \in H^s$.*
If $a \in C^{s_1}(\mathbf{R}^n)$, $u \in H^s$, $s = m + \delta$, $0 < \delta < 1$, and m is an integer, $s_1 = m + \delta_1$, and $\delta_1 > \delta$, then $au \in H^s$.
6. *If $s_1 < s_2 < s_3$ and $\varepsilon > 0$, then there is a constant $C(s_1, s_2, s_3, \varepsilon)$ such that*

$$\|u\|_{s_2} \le \varepsilon \|u\|_{s_3} + C\|u\|_{s_1}.$$

7. *A bounded set of functions from H_{s_1}, $s_1 > 0$ vanishing outside a ball $|x| \le R$ is compact in H^{s_2} if $s_2 < s_1$.*
8. *If $2s > n$, $u \in H^s$, then u is continuous and $|u(x)| \le C\|u\|_s$ for all $x \in \mathbf{R}^n$.*
9. *If $u \in H^s(\mathbf{R}^n)$, $2s > n + 2\alpha$, $0 < \alpha < 1$, then*

$$|u(x) - u(y)| \le C|x-y|^\alpha \|u\|_s$$

for any x, y from $\mathbf{R}^n$.
10. *The space H^{-s} can be identified with the space of linear bounded functionals over H^s.*
11. *The space H^s is invariant with respect to the change of variables $x = f(y)$ such that $f \in C^\infty(\mathbf{R}^n)$, $\det|\partial f_i/\partial y_j| > C = \text{const} > 0$, $\sum_{|\alpha| \le [s]+1} |D^\alpha f| \le M < \infty$.*

Proof.

1. It follows from the definition.

2. The statement is true for $u \in \mathcal{S}(\mathbf{R}^n)$ and hence for all $u \in H^s$.

3. Let $u \in H^s$, $u_k \in \mathcal{S}$ be a Cauchy sequence in H^s, defining u. Since $L_2 = H_0^0 \subset H^s$, we have $u(x) \in L_2(\mathbf{R}^n)$. Let $|\alpha| \le s$. The sequence $\{D^\alpha u_k\}$ is fundamental in L_2 and therefore it has a limit in $L_2(\mathbf{R}^n)$, coinciding with the generalized derivative $D^\alpha u \in L_2(\mathbf{R}^n)$. Therefore, $H^s \subset W_2^s(\mathbf{R}^n)$.

Now let $u \in W_2^s(\mathbf{R}^n)$. Since $W_2^s(\mathbf{R}^n) = W_{2,0}^s(\mathbf{R}^n)$, there is a sequence $u_k \in C_0^\infty(\mathbf{R}^n)$ such that $\|u - u_k\|_{W_2^s(\mathbf{R}^n)} \to 0$ as $k \to \infty$. The sequence u_k is fundamental in H^s, since

$$\begin{aligned}\|u_k - u_{k+p}\|_s^2 &= C_{n,s}\int_{\mathbf{R}^n} |\tilde{u}_k(\xi) - \tilde{u}_{k+p}(\xi)|^2\,d\xi + C_{n,s}\int_{\mathbf{R}^n} |\xi|^{2s}|\tilde{u}_k(\xi) - \tilde{u}_{k+p}(\xi)|^2\,d\xi\\ &= C_{n,s}\int_{\mathbf{R}^n} |u_k - u_{k+p}|^2\,dx + C_{n,s}\sum_{|\alpha|=s}\int_{\mathbf{R}^n} |D^\alpha(u_k - u_{k+p})|^2\,dx\\ &\le C_{n,s}\|u_k - u_{k+p}\|_{W_2^s}^2 \to 0\end{aligned}$$

as $k \to \infty$. Therefore, it defines an element $u \in H^s(\mathbf{R}^n)$.

4. Let $u \in H^s$, where $0 < s < 1$. Then

$$\int_{\mathbf{R}^n} |u(x+h) - u(x)|^2\,dx = (2\pi)^{-n}\int_{\mathbf{R}^n} |\tilde{u}(\xi)|^2|e^{-i\xi h} - 1|^2\,d\xi\,,$$

$$\int_{\mathbf{R}^n} \frac{dh}{h^{n+2s}}\int_{\mathbf{R}^n} |u(x+h) - u(x)|^2\,dx = (2\pi)^{-n}\int_{\mathbf{R}^n} |\tilde{u}(\xi)|^2\,d\xi\int_{\mathbf{R}^n} \frac{|e^{-i\xi h} - 1|^2}{|h|^{n+2s}}\,dh.$$

It is easy to see that

$$C_1|\xi|^{2s} \le \int_{\mathbf{R}^n} \frac{|e^{-i\xi h} - 1|^2}{|h|^{n+2s}}\,dh \le C_2|\xi|^{2s}.$$

Therefore

$$\|u\|_s^2 \le C_3\int_{\mathbf{R}^n} |u|^2\,dx + C_3\int\int_{\mathbf{R}^n\times\mathbf{R}^n} \frac{|u(x+h) - u(x)|^2}{|h|^{2s+n}}\,dh\,dx;$$

$$\|u\|_s^2 \ge C_4\int_{\mathbf{R}^n} |u|^2\,dx + C_4\int\int_{\mathbf{R}^n\times\mathbf{R}^n} \frac{|u(x+h) - u(x)|^2}{|h|^{2s+n}}\,dh\,dx,$$

i.e.

$$\|u\|_s^2 \sim \int_{\mathbf{R}^n} u^2\,dx + \int\int_{\mathbf{R}^n\times\mathbf{R}^n} \frac{|u(x+h) - u(x)|^2}{|h|^{2s+n}}\,dh\,dx.$$

The second statement follows readily from the proved one and the definition of H^s.

5. We have

$$|a(x)| \le A_1,\ |a(x+h) - a(x)| \le A_2|h|^{s_1},\ x \in \mathbf{R}^n,\ h \in \mathbf{R}^n.$$

By the property 4

$$\begin{aligned}
&\|a(x)u(x)\|^2_{H^s(\mathbf{R}^n)} \\
&\leq C\int_{\mathbf{R}^n} |a^2u^2(x)|\,dx + C\int_{\mathbf{R}^n}\int_{\mathbf{R}^n} \frac{|a(x+h)u(x+h)-a(x)u(x)|^2}{|h|^{2s+n}}\,dh\,dx \\
&\leq C\int_{\mathbf{R}^n} |a^2u^2(x)|\,dx + C_1\int\int_{\mathbf{R}^n\times\mathbf{R}^n} \frac{|a(x+h)|^2|u(x+h)-u(x)|^2}{|h|^{2s+n}}\,dh\,dx \\
&\quad + C_1\int\int_{\mathbf{R}^n\times\mathbf{R}^n} \frac{|u(x)|^2|a(x+h)-a(x)|^2}{|h|^{2s+n}}\,dh\,dx \\
&\leq C_2\int_{\mathbf{R}^n} |u|^2\,dx + C_3\int\int_{\mathbf{R}^n\times\mathbf{R}^n} \frac{|u(x+h)-u(x)|^2}{|h|^{2s+n}}\,dh\,dx.
\end{aligned}$$

The second statement follows readily from the proved one and the definition of H^s.

6. The definition of the norm in H^s implies that

$$\begin{aligned}
\|u\|^2_{s_2} &= (2\pi)^{-n}\int_{\mathbf{R}^n} (1+|\xi|^2)^{s_2}|\tilde{u}(\xi)|^2\,d\xi \\
&= \int_{x\in\mathbf{R}^n,|\xi|>K} (1+|\xi|^2)^{s_2-s_3}\cdot(1+|\xi|^2)^{s_3}|\tilde{u}(\xi)|^2\,d\xi \\
&\quad + \int_{x\in\mathbf{R}^n,|\xi|<K} (1+|\xi|^2)^{s_2-s_1}\cdot(1+|\xi|^2)^{s_1}|\tilde{u}(\xi)|^2\,d\xi \\
&\leq K^{2(s_2-s_3)}\int_{\mathbf{R}^n} (1+|\xi|^2)^{s_3}|\tilde{u}(\xi)|^2 d\xi + (1+K^2)^{s_2-s_1}\int_{\mathbf{R}^n} (1+|\xi|^2)^{s_1}|\tilde{u}|^2\,d\xi.
\end{aligned}$$

If K is large enough, for example, $K = \varepsilon^{\frac{1}{2(s_2-s_3)}}$, the statement follows.

7. Consider the set $\mathcal{A}$ of functions of H^{s_1}, $s_1 > 0$, vanishing for $|x| \geq R$ and such that $\|u(x)\|_{s_1} \leq M$. Since $s_1 > 0$, we have $u(x) \in L_2(\Omega)$ and the Fourier transform $\tilde{u}(\xi)$ is defined. We have

$$\int_{\mathbf{R}^n} |\tilde{u}(\xi)|^2(1+|\xi|^2)^{s_1}\,dx \leq M_1.$$

Besides,

$$\frac{\partial}{\partial\xi_k}\tilde{u}(\xi) = (2\pi)^{-n/2}\int_{|x|<R} e^{-i\xi x}(-ix_k)u(x)\,dx$$

and therefore, $|\partial\tilde{u}(\xi)/\partial\xi_k| + |\tilde{u}(\xi)| \leq M_2$ for all $k = 1,2,\dots,n$ with a constant M_2 depending on M_1 and R only. Let $\varepsilon > 0$. There exists h, depending on ε and u and such that $\int_{|\xi|>h}(1+|\xi|^2)^{s_2}|\tilde{u}(\xi)|^2\,d\xi \leq \varepsilon$. In the ball $K_h = \{\xi : |\xi| < h\}$ the set of functions $\tilde{u}(\xi)$ is uniformly bounded and equicontinuous. By the Arzela theorem this set is compact in the space of continuous functions and therefore in the space $L_2(K_h)$. Set $\varepsilon_1 = (1+h^2)^{-s_2/2}\varepsilon^{1/2}$. There is an ε_1-net in $L_2(K_h)$ for the set of

$\{\tilde{u}(\xi)\}$. It consists of functions $v_1, v_2, \ldots, v_N$. Set $v_j = 0$, $j = 1, 2, \ldots, N$, outside of K_h^0. Then these functions form a $\sqrt{2}\varepsilon$-net in H_{s_2}. Indeed,

$$\|u - v_k\|_{H_{s_2}}^2 = (2\pi)^{-n} \int_{\mathbf{R}^n} (1+|\xi|^2)^{s_2} |\tilde{u}(\xi) - v_k(\xi)|^2 \, d\xi$$

$$\leq \int_{|\xi|<h} (1+h^2)^{s_2} |\tilde{u}(\xi) - v_k(\xi)|^2 \, d\xi + \int_{|\xi|>h} (1+|\xi|^2)^{s_2} |\tilde{u}(\xi)|^2 \, d\xi \leq 2\varepsilon^2$$

for each $k = 1, \ldots, N$. This proves the compactness of the set $\mathcal{A}$.

8. Let $u(x) \in C_0^\infty$. Then

$$|u(x)| \leq \frac{1}{(2\pi)^n} \int_{\mathbf{R}^n} |\tilde{u}(\xi)| \, d\xi$$

$$\leq \frac{1}{(2\pi)^n} \left(\int_{\mathbf{R}^n} \frac{d\xi}{(1+|\xi|^2)^s} \right)^{1/2} \left(\int_{\mathbf{R}^n} (1+|\xi|^2)^s |\tilde{u}(\xi)|^2 \, d\xi \right)^{1/2} = C\|u\|_s.$$

If $u(x) \in H^s$, $u_m(x) \in C_0^\infty$, $u_m \to u$ in H^s, then by the same inequality

$$|u_m(x) - u_{m+k}(x)| \leq C\|u_m - u_{m+k}\|_s \to 0$$

as $m, k \to \infty$. The Cauchy criterion implies that the sequence of $u_m(x)$ is uniformly converging to $u(x)$ and $u(x)$ is continuous. Since $|u_m(x)| \leq C\|u_m\|_s$, we have $|u(x)| \leq C\|u\|_s$.

Remark that if $2s = n$, then the space H^s is not embedded in C. Let, for instance $n = 2$, $s = 1$. The function $u(x) = \ln^\alpha |x| \cdot g(x)$, $0 < \alpha < \frac{1}{2}$, where $g(x) \in C_0^\infty(\mathbf{R}^2)$, $g(x) \equiv 1$ for $|x| < \frac{1}{2}$; $g(x) \equiv 0$ for $|x| > 1$, is unbounded, but $u \in H^1$. Indeed,

$$\|u(x)\|_{H^1(\mathbf{R}^2)}^2 = (2\pi)^n \Big(\int_{\mathbf{R}^2} u^2 \, dx + \int_{\mathbf{R}^2} |\nabla u|^2 \, dx \Big)$$

by the Parseval equality. Since

$$\int_{\mathbf{R}^2} |\nabla u|^2 \, dx \leq \int_{|x|<1} r \Big(\frac{\partial}{\partial r} u\Big)^2 \, dr < \infty,$$

one can see that $u(x) \in W_2^1(\mathbf{R}^2)$.

9. Indeed,

$$u(x) - u(y) = \frac{1}{(2\pi)^n} \int_{\mathbf{R}^n} \tilde{u}(\xi) [e^{i\xi x} - e^{i\xi y}] \, d\xi.$$

Therefore

$$|u(x) - u(y)| \leq (2\pi)^{-n} \int_{\mathbf{R}^n} |\tilde{u}(\xi)| \cdot |x - y|^\alpha \cdot |\xi|^\alpha \, d\xi.$$

Hence

$$|u(x) - u(y)| \leq (2\pi)^{-n}|x-y|^{\alpha}\Big(\int_{\mathbf{R}^n} |\tilde{u}(\xi)|^2(1+|\xi|)^{2s}\,d\xi\Big)^{1/2}$$

$$\cdot\left(\int_{\mathbf{R}^n} \frac{|\xi|^{2\alpha}}{1+|\xi|^{2s}}\,d\xi\right)^{1/2} \leq C_2|x-y|^{\alpha}\|u\|_{H_s}.$$

10. If $u \in C_0^\infty(\mathbf{R}^n)$, then the functional

$$T_u(v) = \int_{R^n} \overline{u(x)}v(x)\,dx$$

is defined for all $v \in H^s$ and

$$|T_u(v)| \leq (2\pi)^{-n}\int_{\mathbf{R}^n} |\tilde{v}(\xi)\tilde{u}(\xi)|\,d\xi \leq \Big(\int_{\mathbf{R}^n} |\tilde{u}(\xi)|^2(1+|\xi|^2)^{-s}\,d\xi\Big)^{1/2}$$

$$\cdot\Big(\int_{\mathbf{R}^n} |\tilde{v}(\xi)|^2(1+|\xi|^2)^{s}\,d\xi\Big)^{1/2} \leq (2\pi)^{-n}\|u\|_{-s}\|v\|_s.$$

This inequality allows us to define the functional $T_u(v)$ for all $u \in H^{-s}$. Indeed, let $u \in H^{-s}$ and a sequence of $u_k \in C_0^\infty(\mathbf{R}^n)$ be a Cauchy sequence in H^{-s}, defining the element u. The sequence of the functionals $T_k(v) = \int_{\mathbf{R}^n} u_k v\,dx$ is such that $\lim_{k\to\infty} T_k(v) = T(v)$ exists for each $v \in H^s$. Therefore, each element $u \in H^{-s}$ defines a functional T_u over H^s. It also gives sense to the integral $\int_{R^n} \overline{u(x)}v(x)\,dx$, when $u \in H^{-s}$ and $v \in H^s$.

Now let T be a functional in $H^s(s > 0)$. One can introduce in H^s an inner product

$$(f,g)_s = (2\pi)^{-n}\int_{\mathbf{R}^n} \tilde{f}(\xi)\bar{\tilde{g}}(\xi)(1+|\xi|^2)^s d\xi.$$

So H_s is a Hilbert space. By the Riesz theorem there is an element $w \in H^s$ such that

$$T(v) = (v,w)_s.$$

Since $w \in H^s$, there exists a sequence w_k in $\mathcal{S}$ converging in the norm of H^s to w. The functions $u_k(x)$ such that $\tilde{u}_k = \tilde{w}_k(1+|\xi|^2)^s$ belong to H^{-s} and form a Cauchy sequence there. This sequence defines an element $u \in H^{-s}$. Therefore, every functional T over H^s defines an element $u \in H^{-s}$. Moreover, $\|u\|_{-s} = \|T\|$. If $v \in H^s$, then

$$T(v) = \int_{R^n} \overline{u(x)}v(x)\,dx.$$

11. If s is natural, it follows from the formulas for derivatives in new coordinates. If $s = m + s'$, $m \geq 0$, $0 < s' < 1$, we use the definition of the space, given in step 4. Here it is important that

$$C_1|x_1 - x_2| \leq |y_1 - y_2| \leq C_2|x_1 - x_2|,$$

if $x_i = f(y_i)$, $x \in \mathbf{R}^n$, $y \in \mathbf{R}^n$. To prove the statement in the case $s < 0$ one can use the duality $H^{-s} = (H^s)^*$ for $s > 0$, proved above (see 10) and the invariance of H^{-s}. □

The last property allows us to define the space $H^s(M)$, where M is a smooth (C^∞) surface of dimension $n-1$ in $\mathbf{R}^n$. For each point $P \in M$ there are a neighborhood $U_P \subset \mathbf{R}^n$ and a map $T_P : \mathbf{R}^n_x \to \mathbf{R}^n_y$ such that $T:\ M \cap U_P \to \mathbf{R}^{n-1}_y$. We can choose a finite covering of M by such neighborhoods and a finite partition of unity $\sum \varphi_j(x) \equiv 1$, subordinated to this covering, i.e. such that $\operatorname{supp}\varphi_j(x) \subset \overline{U}_P$ for some $P \in M$. Let $u(x) \in C^\infty(M)$. Set $\|u(x)\|_s = \sum_{1\le j\le N} \|\varphi_j u\|_{H^s(\mathbf{R}^{n-1}_y)}$ for $s>0$. That defines a norm in $C^\infty(M)$ and the completion of C^∞ in this norm is denoted by $H^s(M)$. It is easy to verify that the definition does not depend on the choice of the covering and the partition of unity, although the norm does depend on them. The properties of the space $H^s(M)$ are induced by the properties of the space $H^s(\mathbf{R}^n)$. In particular, one can induce in $H^s(M)$ $(0<s<1)$ another norm

$$\|u\|^2_{H^s(M)} = \int_M u^2\, dS + \int_{M\times M} \frac{|u(x)-u(y)|^2}{|x-y|^{2s+n-1}}\, dS_x\, dS_y.$$

2.29 The traces of functions of $W_2^k(\mathbf{R}^n)$

The following theorem can be considered as a partial case of Theorems 75 and 76.

Theorem 79. *Let $u(x) \in W_2^k(\mathbf{R}^n_+)$, $k \ge 1$, where $\mathbf{R}^n_+ = \{x:\ x \in \mathbf{R}^n,\ x_n > 0\}$. Then for each fixed $x_n \ge 0$ the function $u(x', x_n)$, where $x' = (x_1, \dots, x_{n-1})$, belongs to $L_2(\mathbf{R}^{n-1})$ and $\|u(x',x_n)\|_{L_2(\mathbf{R}^{n-1})} \le C\|u(x)\|_{W_2^1(\mathbf{R}^n)}$.*

Proof. Since for $t \ge 0$

$$\int\int_{x'\in\mathbf{R}^{n-1},\, t<x_n<t+1} |u(x',x_n)|^2\, dx'\, dx_n \le \|u\|^2_{W_2^k(\mathbf{R}^n)},$$

by the Fubini theorem the function $\int |u(x',x_n)|^2\, dx'$ of x_n is summable over the segment $t \le x_n \le t+1$ and

$$\int_t^{t+1} [\int_{\mathbf{R}^{n-1}} |u(x',x_n)|^2\, dx']\, dx_n \le \|u\|^2_{W_2^k(\mathbf{R}^n_+)}.$$

Therefore, there is a $t' \in [t, t+1]$ such that $\int_{\mathbf{R}^{n-1}} |u(x',t')|^2\, dx' \le \|u\|^2_{W_2^k(\mathbf{R}^n_+)}$. The formula $u(x',t) = u(x',t') + \int_{t'}^t \partial u/\partial x_n\, dx_n$ implies that

$$|u(x',t)|^2 \le 2|u(x',t')|^2 + 2\int_t^{t+1} \left|\frac{\partial u}{\partial x_n}\right|^2\, dx_n.$$

Integrating this inequality with respect to $x' \in \mathbf{R}^{n-1}$ and applying the above stated inequality, we obtain that $\|u(x',t)\|^2_{L_2(\mathbf{R}^{n-1})} \le 2\|u\|^2_{W_2^k(\mathbf{R}^n_+)}$. □

Theorem 80. *Let* $u(x) \in W_2^k(\mathbf{R}_+^n)$, $k \geq 1$, *where*

$$\mathbf{R}_+^n = \{x : \ x \in \mathbf{R}^n, \ x_n > 0\}.$$

Then the function $u(x',0)$ *belongs to* $H_{k-1/2}(\mathbf{R}^{n-1})$ *and*

$$\|u(x',0)\|^2_{H^{k-1/2}(\mathbf{R}^{n-1})} \leq C_1 \|u\|_{W_2^k(\mathbf{R}_+^n)}.$$

Proof. It suffices to prove the theorem for $u \in C_0^\infty(\overline{\mathbf{R}_+^n})$. Let $\tilde{u}(\xi, x_n)$ be the Fourier transform of $u(x', x_n)$ with respect to x'.

The Parseval identity implies that

$$\|u\|^2_{W_2^k(\mathbf{R}_+^n)} = (2\pi)^{n-1} \int_{\xi \in \mathbf{R}^{n-1}, 0<x_n<\infty} \sum_{|\alpha|+j\leq k} |\xi^\alpha|^2 \left| \frac{d^j \tilde{u}(\xi, x_n)}{dx_n^j} \right|^2 d\xi \, dx_n.$$

Let us consider the function

$$I(\xi) = \sum_{j=0}^k \int_0^\infty |\xi|^{2(k-j)} \left| \frac{d^j \tilde{u}(\xi, x_n)}{dx_n^j} \right|^2 dx_n.$$

After substituting $x_n = t|\xi|$ we obtain

$$|\xi|^{-2k+1} I(\xi) = \sum_{j=0}^k \int_0^\infty \left| \frac{d^j \tilde{u}(\xi, x_n)}{dt_n^j} \right|^2 dt.$$

Each segment $[\alpha, \alpha+1]$ contains a point t_α such that $|\tilde{u}(\xi, t_\alpha)|^2 = \int_\alpha^{\alpha+1} |\tilde{u}(\xi, t)|^2 \, dt$. Thus for every $t \in (\alpha, \ \alpha+1)$ we have

$$|\tilde{u}(\xi, t)|^2 \leq 2\left(\int_\alpha^{\alpha+1} |\frac{d}{ds}\tilde{u}(\xi, s)|^2 \, ds + \int_\alpha^{\alpha+1} |\tilde{u}(\xi, s)|^2 \, ds\right)$$

and therefore $|\tilde{u}(\xi, x_n)|^2 \leq 2|\xi|^{-2k+1} I(\xi)$ or $|\xi|^{2k-1} |\tilde{u}(\xi, x_n)|^2 \leq 2I(\xi)$. Integrating this relation with respect to $\xi \in \mathbf{R}^{n-1}$, we obtain the result. □

In the same way one can define the trace of a function of $W_2^k(\mathbf{R}^n)$ on any plane $x_n = t$. Moreover, $u|_{x_n=t} \in W_2^{k-1/2}(\mathbf{R}^{n-1})$ and

$$\|u(x',t)\|_{H^{k-1/2}(\mathbf{R}_{x'}^{n-1})} \leq C\|u(x)\|_{W_2^k(\mathbf{R}_+^n)}.$$

Theorem 81. *If* $u(x) \in W_2^1(\mathbf{R}_+^n)$, $t \geq 0$, *then the function* $u(x', t)$ *of* y *is "continuous in mean", i.e.* $\|u(x', t+\varepsilon) - u(x', t)\|_{L_2(\mathbf{R}^{n-1})} \to 0$ *as* $\varepsilon \to 0$.

Proof. If $u(x) \in C^1(\mathbf{R}^n_+)$, then $u(x', t+\varepsilon) - u(x', t) = \int_t^{t+\varepsilon} \partial u/\partial x_n \, dx_n$ and hence $|u(x', t+\varepsilon) - u(\tilde{x}, t)|^2 \le \varepsilon \int_t^{t+\varepsilon} |\partial u/\partial x_n|^2 \, dx_n$. Hence

$$\int_{\mathbf{R}^{n-1}} |u(x', t+\varepsilon) - u(x', t)|^2 \, dx'$$

$$\le \varepsilon \int\int_{x'_n \in \mathbf{R}^{n-1}, t<x_n<t+\varepsilon} \left| \frac{\partial u}{\partial x_n} \right|^2 dx \le \varepsilon \int\int_{x' \in \mathbf{R}^{n-1}, t<x_n<t+\varepsilon} |\nabla u|^2 \, dx.$$

If $u(x) \in W_2^1(\mathbf{R}^n_+)$, then there exists a sequence $u_m(x) \in C^\infty(\mathbf{R}^n_+)$, such that $\|u_m - u\|_{W_2^1(\mathbf{R}^n_+)} \to 0$ as $m \to \infty$. Each function $u_m(x)$ satisfies the proved inequality, therefore it is valid for $u(x)$, too. □

Theorem 82. *If $u \in W_2^1(\mathbf{R}^n_+)$, $u(x', 0) = 0$, then $u \in W_{2,0}^1(\mathbf{R}^n_+)$.*

Proof. Let $\sigma_1(x) \in C_0^\infty(\mathbf{R}^n)$, $\sigma_1(x) \equiv 1$ for $|x| < 1$, $\sigma_1(x) \equiv 0$ for $|x| > 2$, and $\sigma_2(t) \in C^\infty(\mathbf{R}^1)$, $\sigma_2(t) \equiv 0$ for $0 < t < 1/2$, $\sigma_2(t) \equiv 1$ for $t > 1$. Consider the sequence

$$u_m(x) = \sigma_1(\frac{x}{m})\sigma_2(mx_n)u(x),$$

where $m = 1, 2, \ldots$. Each function $u_m(x)$ has a support, compactly embedded in $\mathbf{R}^n_+$, and belongs to $W_2^1(\mathbf{R}^n_+)$. Therefore, it belongs to $W_{2,0}^1(\mathbf{R}^n_+)$. Let us show that $\|u(x) - u_m(x)\|_{W_2^1(\mathbf{R}^n_+)} \to 0$ as $m \to \infty$.

Observe that

$$|\nabla \sigma_1(\frac{x}{m})| \le \frac{C_1}{m}, \ |\nabla \sigma_2(mx_n)| \le C_2 m,$$

where C_1, C_2 are independent of m. Hence

$$\|u - u_m\|^2_{W_2^1(\mathbf{R}^n_+)} \le C_3 \int_{|x|>m, x\in\mathbf{R}^n_+} u^2 \, dx + C_3 \int_{|x|>m, x\in\mathbf{R}^n_+} |\nabla u|^2 \, dx$$

$$+ C_3 \int_{m<|x|<2m, x\in\mathbf{R}^n_+} \frac{2}{m^2} u^2 \, dx + C_3 m^2 \int_{0<x_n<\frac{1}{m}, x\in\mathbf{R}^n_+} u^2 \, dx.$$

The first three terms on the right-hand side tend to 0 as $m \to \infty$, since $\int_{\mathbf{R}^n_+} u^2 \, dx + \int_{\mathbf{R}^n_+} |\nabla u|^2 \, dx < \infty$. Since $u(x', 0) = 0$, we have for $\tau \in [0, 1/m]$

$$\int_{\mathbf{R}^{n-1}} |u^2(x', \tau)|^2 \, dx' \le \tau \int_{0<x_n<\frac{1}{m}, x'\in\mathbf{R}^{n-1}} |\nabla u(x)|^2 \, dx.$$

Integrate this relation with respect to τ from $\tau = 0$ to $\tau = \frac{1}{m}$. We obtain

$$\int\int_{0<x_n<\frac{1}{m}, x\in\mathbf{R}^n_+} u^2 \, dx \le \frac{1}{2m^2} \int\int_{0<x_n<\frac{1}{m}, x\in\mathbf{R}^n_+} |\nabla u|^2 \, dx.$$

Therefore, the last term also tends to zero as $m \to \infty$. □

Theorem 83. *If* $\varphi \in H^{k-1/2}(\mathbf{R}^{n-1})$, $k \geq 1$, *then there exists a function* $u(x) \in W_2^k(\mathbf{R}_+^n)$ *such that* $u|_{x_n=0} = \varphi$ *and*

$$\|u\|_{W_2^k(\mathbf{R}_+^n)} \leq C\|\varphi\|_{H^{k-1/2}(\mathbf{R}^{n-1})}.$$

Proof. Let us define the function $u(x)$ by the formula

$$u(x) = (2\pi)^{-n+1} \int_{\mathbf{R}^{n-1}} e^{i\xi x'} \tilde{\varphi}(\xi) e^{-|\xi| x_n} e^{-|x_n|} \, d\xi,$$

where $\xi = (\xi_1, \dots, \xi_{n-1})$, $x' = (x_1, \dots, x_{n-1})$, $x = (x', x_n)$. Let at first $\varphi(x') \in S$. Then $\tilde{\varphi}(\xi) \in S$ and the integral converges absolutely and uniformly with respect to $x = (x', x_n)$ in $\mathbf{R}_+^n$. The same is true for all integrals obtained by differentiation of the integral with respect to x. Therefore, $u(x) \in C^\infty(\mathbf{R}_+^n)$. Moreover, by the Parseval equality

$$\|u(x)\|^2_{W_2^k(\mathbf{R}_+^n)} = \sum_{|\alpha| \leq k} \int_{\mathbf{R}_+^n} |D^\alpha u(x)|^2 \, dx$$

$$\leq C_1 \sum_{|\alpha| \leq k} \int_{\mathbf{R}^{n-1}} (1 + |\xi|)^{2|\alpha|-1} |\tilde{\varphi}(\xi)|^2 \, d\xi \leq C_2 \|\varphi(x')\|^2_{H^{k-1/2}(\mathbf{R}^{n-1})}.$$

If $\varphi(x') \in H^{k-1/2}$ one can consider a sequence $\{\varphi_m(x')\}$ of functions belonging to S and defining $\varphi(x')$ as an element of $H^{k-1/2}(\mathbf{R}^{n-1})$. For each $\varphi_m(x)$ there is a function $u_m(x) \in C^\infty(\mathbf{R}_+^n)$, satisfying the inequality

$$\|u_m(x)\|^2_{W_2^k(\mathbf{R}_+^n)} \leq \|\varphi_m(x')\|^2_{H^{k-1/2}(\mathbf{R}^{n-1})}$$

and such that

$$u_m|_{x_n=0} = \varphi_m(x').$$

We also have

$$\|u_m - u_{m+p}\|^2_{W_2^k(\mathbf{R}_+^n)} \leq C\|\varphi_m(x') - \varphi_{m+p}(x')\|^2_{H^{k-1/2}(\mathbf{R}^{n-1})}.$$

Therefore, the sequence $\{u_m(x)\}$ is a Cauchy sequence in $W_2^k(\mathbf{R}_+^n)$ and hence there is an element $u(x) \in W_2^k(\mathbf{R}_+^n)$ such that

$$\|u_m(x) - u(x)\|_{W_2^k(\mathbf{R}_+^n)} \to 0$$

as $m \to \infty$. Passing to the limit we obtain that

$$\|u(x)\|_{W_2^k(\mathbf{R}_+^n)} \leq C\|\varphi(x)\|_{H^{k-1/2}(\mathbf{R}^{n-1})}.$$

By the definition of the trace of a function from W_2^k, the trace of u on the plane $x_n = 0$ is equal to $\lim_{m\to\infty} u_m(x', 0)$ in the space $H^{k-1/2}(\mathbf{R}^{n-1})$, i.e. to the function $\varphi(x)$. □

Theorem 83 allows us to give another definition of the norm in $H^{k-1/2}(\mathbf{R}^{n-1})$, namely

$$\|\varphi\|'_{H^{k-1/2}(\mathbf{R}^{n-1})} = \inf \|u(x)\|_{W_2^k(\mathbf{R}^n_+)},$$

where the inf is taken over the set of functions $u(x) \in W_2^k(\mathbf{R}^n_+)$, such that $u(x', 0) = \varphi(x)$. This means that the space $H^{k-1/2}(\mathbf{R}^{n-1})$ is isomorphic to the quotient space $W_2^k(\mathbf{R}^n_+)/W_{2,0}^k(\mathbf{R}^n_+)$.

Now let Ω be a bounded domain in $\mathbf{R}^n$ and its boundary $\partial\Omega$ be a smooth surface. By definition, two functions $u_1(x) \in W_2^k(\Omega)$ and $u_2(x) \in W_2^k(\Omega)$ coincide on $\partial\Omega$, if $u_1 - u_2 \in W_{2,0}^1(\Omega)$. Therefore, the space of traces on $\partial\Omega$ of functions from $W_2^k(\Omega)$ can be identified with the quotient space $W_2^k(\Omega)/W_{2,0}^k(\Omega)$.

Let us show that the space of traces is the space $H^{k-1/2}(\partial\Omega)$. For each point $P \in \partial\Omega$ there exists a neighborhood $U_P \in \mathbf{R}^n$ and a smooth map $\varphi_P : U_P \to \mathbf{R}^n$ with nonvanishing Jacobian and such that $\varphi_P : U_P \cap \Omega \to \mathbf{R}^n_+$. Moreover, $\varphi_P : \partial\Omega \cap U_P \to \mathbf{R}^{n-1}$. Let a partition of unity

$$1 = \sum_{i=1}^{N} g_i(x)$$

be such that $\operatorname{supp} g_i$ lies in U_{P_i}, $i = 1, \ldots, N$, $g_i \in C_0^\infty(\mathbf{R}^n)$. The definition of trace implies that

$$u|_\Gamma \equiv \sum_{i=1}^{N} g_i u|_\Gamma = \sum_{i=1}^{N} u_i|_\Gamma,$$

where $u_i = g_i u$. Since the spaces W_2^k, $W_{2,0}^k$ are invariant with respect to smooth transformations, we have

$$u_i(x)|_\Gamma = u_i(x(y))|_{y_n=0},$$

if $(y_1, \ldots, y_n)$ is the image of a point x under the map φ_{P_i}. The trace $u_i(x)|_\Gamma$ can be defined as an element of the space $W^{k-1/2}(\partial\Omega)$ and we can set

$$\|u\|_{W^{k-1/2}(\partial\Omega)} = \sum_{i=1}^{N} \|u_i(x(y))\|_{H^{k-1/2}(\mathbf{R}^{n-1})}.$$

Theorem 83 implies that this definition of the norm in the space $W^{k-1/2}(\partial\Omega)$ is equivalent to the following one:

$$\|u\|_{W^{k-1/2}(\partial\Omega)} = \inf \|U\|_{W^k(\Omega)},$$

where U is an arbitrary function in $W^k(\Omega)$ such that $U|_{\partial\Omega} = u|_{\partial\Omega}$.

The problem of characterization of functions $\varphi(x)$, given on the boundary of a domain and allowing a continuation $\Phi(x) \in W_2^1(\Omega)$ on the domain, is very important for variational calculus and for other analyses. Hadamard has found a continuous function $\varphi(x)$ such that there is no a function $\Phi(x) \in W_2^1(\Omega)$ whose trace on $\partial\Omega$ coincides with $\varphi(x)$. The above theorems show that the condition, necessary and sufficient for $\varphi(x)$ to be a trace of a function $\Phi(x) \in W_2^1(\Omega)$, is that $\varphi(x) \in H^{1/2}(\partial\Omega)$, if $\partial\Omega$ is smooth. The space $H^s(\partial\Omega)$ was first introduced by Slobodetzky [Sl].

We now give an example of a function, continuous on $\partial\Omega$ and not belonging to the space $H^{1/2}(\partial\Omega)$.

Let $\Omega = \mathbf{R}^2_+ = \{(x,y) : \ y > 0, \ \infty < x < +\infty\}$. Let $\varphi(x) \equiv 0$ for $x < 0, \varphi(x) = 1/|\ln\ x|^{1/2}$ for $0 < x < 1/2$. For $x > 1/2$ this is an arbitrary smooth function, vanishing for $|x| > 1$. Let us verify that $\varphi(x) \notin H^{1/2}(\partial\Omega)$. If this is not true, then there exists a function $u(x,y) \in W_2^1(\Omega)$ such that $u|_{y=0} = \varphi(x)$. It is clear that for almost all r the function u is an absolutely continuous function of θ in the polar coordinates r, θ. Set $t = \ln \frac{1}{r}$. The domain Ω passes in the strip

$$\Pi = \{(t,\theta) : \ -\infty < t < +\infty, \ 0 < \theta < \pi\}.$$

Moreover, the function $v(t,\theta)$, which is the image of u, belongs to the space $W_2^1(\Pi')$, where Π' is an arbitrary bounded subdomain of Π. Since v is absolutely continuous on almost all lines $t =$ const, the function u is absolutely continuous on almost all lines $r =$ const. Therefore,

$$-u(r,0) = \int_0^\pi \frac{\partial u}{\partial\theta}\, d\theta,$$

and since $|\partial u/\partial\theta| \le r|\nabla u|$, we have

$$|u(r,0)|^2 \le \pi r^2 \int_0^\pi |\nabla u|^2\, d\theta, \quad \text{so that} \quad \frac{|u(r,0)|^2}{r} \le \pi \int_0^\pi |\nabla u|^2 r\, d\theta.$$

Integrating this inequality with respect to r, we obtain that

$$\int_0^{1/2} \frac{|u(x,0)|^2}{x}\, dx \le \pi \int_{\mathbf{R}^2} |\nabla u|^2\, dx.$$

Since $u(x,0) = (\ln\ x)^{-1/2}$ for $0 < x \le 1/2$, the last inequality is impossible.

Observe that for any $\alpha < 1/2$ there is a function $\varphi(x)$, belonging to C^α, and not belonging to the space $H^{1/2}(\partial\Omega)$.

2.30 The Hardy inequalities

Now we consider the classical Hardy inequality and some of its generalizations.

Theorem 84 (Hardy's inequality). *If $y(x)$ is absolutely continuous on $[0,\infty)$ and $y(0)=0$, then*

$$\int_0^\infty \frac{|y(x)|^2}{x^2}\,dx \le 4\int_0^\infty |y'(x)|^2\,dx.$$

Proof. We have

$$\int_0^\infty \left(\frac{y(x)}{x}+2y'(x)\right)^2 dx \ge 0.$$

Therefore

$$\int_0^\infty \left(\frac{y(x)^2}{x^2}+4\frac{y(x)y'(x)}{x}+4y'(x)^2\right) dx \ge 0.$$

Observe that $|y(x)|^2 \le x\int_0^x y'(t)^2dt = o(x)$ as $x\to 0$. Integrating by parts we obtain that

$$\int_0^\infty \left(-\frac{y(x)^2}{x^2}+4y'(x)^2\right) dx \ge 0. \qquad \square$$

The following inequality generalizes Theorem 84.

Theorem 85. *If $y(0)=0$ and $y(x)$ is absolutely continuous on $[0,\infty)$, then the inequality*

$$\int_0^\infty \frac{|y|^p}{x^p}\,dx \le \left(\frac{p}{p-1}\right)^p \int_0^\infty |y'|^p\,dx \tag{2}$$

is true for all $p>1$.

Proof. Note that

$$|y(x)| = |\int_0^x y'(t)\,dt| \le (\int_0^x |y'(t)|^p\,dt)^{1/p}x^{1/p'},\ \frac{1}{p}+\frac{1}{p'}=1$$

and hence $|y(x)| = o(x^{1/p'})$ as $x\to 0$. Moreover, for $x>N$ we have

$$|y(x)| \le |y(N)| + \int_N^x |y'(t)|\,dt \le |y(N)| + (\int_N^x |y'(t)|^p\,dt)^{1/p}(x-N)^{1/p'}.$$

Let ε be a positive number. If $N>0$ is so large that

$$\int_N^\infty |y'(x)|^p\,dx < \varepsilon^p,$$

then $|y(x)| \le |y(N)| + \varepsilon(x-N)^{1/p'}$ and for $x > N + (|y(N)|/\varepsilon)^{p'}$ we have $|y(x)| \le 2\varepsilon x^{1/p'}$. Thus $|y(x)| = o(x^{1/p'})$ as $x\to+\infty$. Using the obtained results and integrating by parts we can see that

$$\int_0^\infty \frac{|y(x)|^p}{x^p}\,dx = \int_0^\infty \frac{px^{1-p}}{p-1} y(x)|y(x)|^{p-2}y'(x)\,dx.$$

Using the Hölder inequality, we obtain

$$\int_0^\infty \frac{|y(x)|^p}{x^p}\,dx \le \frac{p}{p-1}(\int_0^\infty |y'(x)|^p\,dx)^{1/p}\left(\int_0^\infty \frac{|y(x)|^{(p-1)p'}}{x^{(p-1)p'}}\,dx\right)^{1/p'},$$

and since $(p-1)p' = p$, we have

$$\int_0^\infty \frac{|y(x)|^p}{x^p}\,dx \le \left(\frac{p}{p-1}\right)^p \int_0^\infty |y'(x)|^p\,dx. \qquad \square$$

Observe that the constant $(p/(p-1))^p$ in this inequality cannot be replaced by $(p/(p-1))^p - \varepsilon$ with any $\varepsilon > 0$. To see this, one can consider the function such that $y(x) = x^\alpha$, $\alpha = (p-1+\varepsilon)/p$ on $[0,N]$, $y(x) = N^\alpha$ on $[N,\infty)$.

Note also another generalization of the Hardy inequality:

Theorem 86. *The inequality*

$$\int_0^\infty x^{\alpha-p}|y(x)|^p\,dx \le \frac{p^p}{|\alpha+1-p|^p}\int_0^\infty x^\alpha|y'(x)|^p\,dx \tag{3}$$

is true if $p>1, \alpha \ne p-1$, and $y(x)$ is absolutely continuous on $[0,\infty)$ and satisfies the following boundary condition: $y(0) = 0$ when $\alpha < p-1$, and $\lim_{x\to+\infty} y(x) = 0$ when $\alpha > p-1$.

Proof. It suffices to change the variable $x = t^{1+\alpha/(1-p)}$ and apply inequality (2). $\square$

Inequality (3) can be also stated in the following form:

Theorem 87. *If y is an absolutely continuous function on $[0,\infty)$ and*

$$\int_0^\infty x^\alpha|y'(x)|^p\,dx < \infty,$$

for some α, p, such that $\alpha \ne p-1$, $p>1$, then $y(x) = c + y_0(x)$, where $c = y_0(0)$ and

$$\int_0^\infty x^{\alpha-p}|y_0(x)|^p\,dx \le \frac{p^p}{|\alpha+1-p|^p}\int_0^\infty x^\alpha|y'(x)|^p\,dx, \tag{4}$$

$$|c|^p \le C\int_1^2 |y(x)|^p\,dx + C\int_0^\infty x^\alpha|y'(x)|^p\,dx, \tag{5}$$

where C does not depend on y.

Proof. Let for example, $\alpha < p-1$. Let us show that there exists $\lim_{x\to0} y(x)$. We have by the Hölder inequality

$$|y(x_1) - y(x_2)|^p = |\int_{x_1}^{x_2} y'\,dx|^p \le \int_{x_1}^{x_2} x^\alpha|y'|^p\,dx(\int_{x_1}^{x_2} x^{-\alpha p'/p}\,dx)^{p/p'},$$

where $1/p + 1/p' = 1$. Therefore, if $x_1 \to 0,\ x_2 \to 0$, then $|y(x_1) - y(x_2)| \to 0$. The existence of the limit $\lim_{x\to+0} y(x)$ follows now from the Cauchy criterion. Set $c = \lim_{x\to+0} y(x)$. Then the function $y_0(x) = y(x) - c$ satisfies inequality (3) and since $|c|^p \le C_p[|y_0(x)|^p + |y(x)|^p]$, we have

$$\int_1^2 c^p\,dx \le C_{1p}\int_1^2 |y(x)|^p\,dx + C_{2p\alpha}\int_1^2 |y_0(x)|^p x^{\alpha-p}\,dx. \qquad \square$$

Theorem 88. *If* $p > 1,\ \alpha = p-1$, *then the following inequality is true*

$$\int_1^\infty x^{-1}|\ln\, x|^{-p}|y(x)|^p\,dx \le (\frac{p}{p-1})^p\int_1^\infty x^{p-1}|y'(x)|^p\,dx, \tag{6}$$

if the function y *is absolutely continuous on* $[0,\infty)$ *and* $y(1) = 0$.

Proof. In fact, after the change $x = e^t$ we obtain using (2) that

$$\int_1^\infty x^{-1}|\ln\, x|^{-p}|y(x)|^p\,dx = \int_0^{+\infty} t^{-p}|y(t)|^p\,dt$$
$$\le (\frac{p}{p-1})^p\int_0^{+\infty}\left|\frac{dy}{dt}\right|^p dt = (\frac{p}{p-1})^p\int_1^\infty x^{p-1}|y'(x)|^p\,dx. \qquad \square$$

As above, we can state (6) in the following form: if a function y is absolutely continuous on $(0,\infty)$ and $y(1) = c$, then $y(x) = y_0(x) + c$ and

$$\int_1^\infty \frac{|y_0(x)|^p}{x|\ln\, x|^p}\,dx \le \left(\frac{p}{p-1}\right)^p\int_1^\infty x^{p-1}|y'(x)|^p\,dx\ ,$$
$$|c| \le C(\int_2^3 |y(x)|^p\,dx + \int_1^\infty x^{p-1}|y'(x)|^p\,dx).$$

At last, note one more analogue of the Hardy inequality:
Theorem 89. *The inequality*

$$(\int_0^\infty x^\alpha|y(x)|^q\,dx)^{1/q} \le C_{\alpha,\beta,q,p}(\int_0^\infty x^\beta|y'(x)|^p\,dx)^{1/p} \tag{7}$$

is valid for absolutely continuous functions y *such that* $\lim_{x\to+\infty} y(x) = 0,\ p > 1,\ q \ge p,\ (1+\alpha)/q = (1+\beta-p)/p,\ \beta > p-1$.

Proof. Indeed, we have

$$|y(x)| = |\int_x^\infty y'(t)\,dt| \le (\int_x^\infty |y'(t)|^p t^\beta\,dt)^{1/p}\cdot(\int_x^\infty t^{-\frac{\beta p'}{p}}\,dt)^{1/p'}$$
$$= \left(\frac{x^{1-\beta p'/p}}{1-\beta p'/p}\right)^{1/p'}\left(\int_0^\infty |y'(x)|^p x^\beta\,dx\right)^{1/p}. \tag{8}$$

Using (8) and (3), we obtain that

$$\int_0^\infty x^\alpha |y(x)|^q\,dx \equiv \int_0^\infty x^\alpha |y(x)|^p \cdot |y(x)|^{q-p}\,dx$$
$$\le C_1 \int_0^\infty x^{\alpha+(q-p)(1/p'-\beta/p)}|y(x)|^p\,dx \cdot \Big(\int_0^\infty x^\beta |y'(x)|^p\,dx\Big)^{(q-p)/p}$$
$$= C_1 \int_0^\infty x^{\beta-p}|y(x)|^p\,dx \Big(\int_0^\infty x^\beta |y'(x)|^p\,dx\Big)^{(q-p)/p}$$
$$\le C\Big(\int_0^\infty x^\beta |y'(x)|^p\,dx\Big)^{q/p}.$$

□

Observe that the sharp value of the constant C in (7) is

$$\left(\frac{p}{p-\beta p'}\right)^{(q-p)/p'} \cdot \frac{p^p}{|\beta+1-p|^p}.$$

This inequality holds also for $\beta < p-1$, if $y(0) = 0$. The proof is similar.

Now we consider multi-dimensional analogues of the Hardy inequality.

Theorem 90. *If* $u(x) \in C^1(\mathbf{R}^n)$, $\int_{\mathbf{R}^n} |x|^\alpha |\nabla u|^2\,dx < \infty$, *and* $\alpha + n \ne 2$, *then* $u(x) = c_0 + u_0(x)$, $c_0 = \text{const}$, *where*

$$\int_{\mathbf{R}^n} |x|^{\alpha-2}|u_0(x)|^2\,dx \le C \int_{\mathbf{R}^n} |x|^\alpha |\nabla u(x)|^2\,dx$$

and

$$|c_0|^2 \le C \int_{1<|x|<2} |u(x)|^2\,dx + C \int_{\mathbf{R}^n} |x|^\alpha |\nabla u(x)|^2\,dx.$$

Proof. Let us consider the function

$$\overline{u}(r) = \frac{1}{\omega_n r^{n-1}} \int_{|x|=r} u(x)\,ds,$$

where ω_n is the area of the unit sphere. It is evident that $\overline{u}(r) = \int_{|x|=1} u(xr)\,dS/\omega_n$, where dS is the surface element of the unit sphere, so that $ds = r^{n-1}dS$. It is easy to see that

$$\overline{u}'(r) = \frac{1}{\omega_n} \int_{|x|=1} \frac{\partial}{\partial r} u(rx)\,dS = \frac{1}{\omega_n r^{n-1}} \int_{|x|=r} \frac{\partial u(x)}{\partial r}\,ds.$$

The Cauchy inequality implies that

$$|\overline{u}'(r)| \le \frac{1}{\omega_n r^{n-1}} \Big(\int_{|x|=r} |\nabla u(x)|^2\,ds\Big)^{1/2} (\omega_n r^{n-1})^{1/2}.$$

Therefore
$$r^{\alpha+n-1}|\overline{u}'(r)|^2 \le \frac{r^\alpha}{\omega_n}\int_{|x|=r}|\nabla u(x)|^2\,ds.$$

Integrating the latter inequality with respect to r from 0 to ∞, we obtain that
$$\int_0^\infty r^{\alpha+n-1}|\overline{u}'(r)|^2\,dr \le \omega_n^{-1}\int_{\mathbf{R}^n} r^\alpha|\nabla u|^2\,dx.$$

Inequalities (4), (5) proved above imply that $\overline{u}(r) = c_0 + v(r)$, where $c_0 = const$ and
$$\int_0^\infty r^{\alpha+n-3}|v(r)|^2 dr \le \frac{4}{(\alpha+n-2)^2}\int_0^\infty r^{\alpha+n-1}|\overline{u}'(r)|^2\,dr,$$
$$|c_0|^2 \le C[\int_1^2 \overline{|}u(x)|^2\,dr + \int_0^\infty r^{\alpha+n-1}|\overline{u}'(r)|^2\,dr].$$

Therefore
$$\int_{\mathbf{R}^n} r^{\alpha-2}|v(r)|^2\,dx \le \frac{4}{(\alpha+n-2)^2\omega_n}\int_{\mathbf{R}^n} r^\alpha|\nabla u(x)|^2\,dx,$$
$$|c_0|^2 \le C_1[\int_{\mathbf{R}^n} r^\alpha|\nabla u(x)|^2\,dx + \int_{1<|x|<2}|u(x)|^2\,dx].$$

Let us consider the function $u_1(x) = u(x) - \overline{u}(|x|)$. By the definition of $\overline{u}(r)$ we have
$$\int_{|x|=r} u_1(x)\,ds = 0$$
for any $r > 0$. The Poincaré inequality (Theorem 62) implies that
$$\int_{\lambda<|x|<2\lambda}|u_1(x)|^2\,dx \le C\lambda^2\int_{\lambda<|x|<2\lambda}|\nabla u_1(x)|^2\,dx.$$

Therefore
$$\int_{\lambda<|x|<2\lambda}\lambda^{\alpha-2}|u_1(x)|^2\,dx \le C\int_{\lambda<|x|<2\lambda}\lambda^\alpha|\nabla u_1|^2\,dx.$$

Hence
$$2^{-|\alpha-2|}\int_{\lambda<|x|<2\lambda}|x|^{\alpha-2}|u_1(x)|^2\,dx \le C\int_{\lambda<|x|<2\lambda}|x|^\alpha|\nabla u_1|^2\,dx.$$

Setting $\lambda = 2^{-k}$, $k = 0, \pm 1, \ldots$, and summing the inequalities over all values of k, we obtain that
$$\int_{\mathbf{R}^n}|x|^{\alpha-2}|u_1(x)|^2\,dx \le C_1\int_{\mathbf{R}^n}|x|^\alpha|\nabla u_1(x)|^2\,dx \le C_2\int_{\mathbf{R}^n}|x|^\alpha|\nabla u(x)|^2\,dx.$$

Therefore, $u = \overline{u} + u_1 = c_0 + v(r) + u_1 = c_0 + u_0$, where u_0 and c_0 satisfy the conditions

$$\int_{\mathbf{R}^n} |x|^{\alpha-2}|u_0(x)|^2\,dx \le C\int_{\mathbf{R}^n} |x|^{\alpha}|\nabla u(x)|^2\,dx$$

and

$$|c_0|^2 \le C\int_{1<|x|<2} |u(x)|^2\,dx + C\int_{\mathbf{R}^n} |x|^{\alpha}|\nabla u(x)|^2\,dx. \qquad \square$$

Lemma 91. *If a function $f(t)$ is absolutely continuous on $[0,\infty)$, $f(t)\equiv 0$ for $t<2$, and*

$$\int_0^\infty \frac{t^\alpha}{|\ln\ t|^\beta}|f'(t)|^2\,dt < \infty,\ \alpha<1,\ \beta>0,$$

then

$$\int_0^\infty \frac{t^{\alpha-2}}{|\ln\ t|^\beta}|f(t)|^2\,dt \le C\int_0^\infty \frac{t^\alpha}{|\ln\ t|^\beta}|f'(t)|^2\,dt.$$

Proof. It suffices to prove the inequality for real-valued functions f. The following estimate is obvious if $t > t_1 \ge 2$:

$$|f(t)-f(t_1)| = |\int_{t_1}^t f'(s)\,ds| \le (\int_{t_1}^t \frac{s^\alpha f'(s)^2}{|\ln\ s|^\beta}\,ds)^{1/2}(\int_{t_1}^t \frac{|\ln\ s|^\beta}{s^\alpha}\,ds)^{1/2}$$

$$\le (\int_{t_1}^\infty \frac{s^\alpha f'(s)^2}{|\ln\ s|^\beta}\,ds)^{1/2} t^{(1-\alpha)/2}|\ln\ t|^{\beta/2}.$$

Therefore

$$|f(t)| = o(t^{(1-\alpha)/2}(\ln\ t)^{\beta/2})$$

as $t\to+\infty$. Integrating by parts, we obtain that

$$\int_0^T \frac{t^{\alpha-2}}{|\ln\ t|^\beta} f(t)^2\,dt = \int_0^T \frac{f(t)^2}{|\ln\ t|^\beta}\frac{dt^{\alpha-1}}{\alpha-1} = \frac{T^{\alpha-1}}{(\alpha-1)(\ln\ T)^\beta}f^2(T)$$

$$-2\int_0^T \frac{t^{\alpha-1}f(t)f'(t)}{(\alpha-1)|\ln\ t|^\beta}\,dt + \beta\int_0^T \frac{t^{\alpha-2}f^2(t)}{(\alpha-1)|\ln\ t|^{\beta+1}}\,dt.$$

The first term on the right-hand side tends to 0 as $T\to\infty$, as was proved above. The latter term is negative, since $\alpha<1$.

Therefore, for any $\varepsilon>0$ we have as $T\to\infty$

$$\int_0^T \frac{t^{\alpha-2}f(t)^2}{|\ln\ t|^\beta}\,dt \le \frac{2}{1-\alpha}\left[\int_0^T \frac{t^{\alpha-2}f(t)^2}{|\ln\ t|^\beta}\,dt\right]^{1/2}\cdot[\int_0^T \frac{t^\alpha f'(t)^2}{|\ln\ t|^\beta}\,dt]^{1/2} + o(1)$$

$$\le \varepsilon\int_0^T \frac{t^{\alpha-2}f(t)^2}{|\ln\ t|^\beta}\,dt + C(\varepsilon)\int_0^T \frac{t^\alpha f'(t)^2}{|\ln\ t|^\beta}\,dt + o(1).$$

Setting $\varepsilon=1/2$ and letting $T\to\infty$, we obtain the result. $\square$

Theorem 92. *If a function u is such that $u \equiv 0$ for $|x| < 2$, $u(x) \in C^1(\mathbf{R}^n)$, and $(\ln|x|)^{-\beta/2}|x|^{\alpha/2}|\nabla u(x)| \in L_2(\mathbf{R}^n)$ with $\alpha + n < 2$, $\beta > 0$, then*

$$\int_{\mathbf{R}^n} (\ln|x|)^{-\beta}|x|^{\alpha-2}|u(x)|^2\, dx \le C \int_{\mathbf{R}^n} (\ln|x|)^{-\beta}|x|^{\alpha}|\nabla u(x)|^2\, dx,$$

where the constant C is independent of u.

Proof. Consider the function $\overline{u}(r) = \omega_n^{-1} r^{1-n} \int_{|x|=r} u\, ds$, where ds is the element of area of a sphere of radius r. It is easy to see that

$$|\overline{u}'(r)|^2 \le r^{1-n}\omega_n^{-1} \int_{|x|=r} |\nabla u|^2\, ds$$

and

$$\int_0^\infty \int_{|\phi|=1} r^{-1+\alpha+n}|\ln\, r|^{-\beta}|\overline{u}'(r)|^2\, dr\, d\phi \le \int_{\mathbf{R}^n} |\ln\, r|^{-\beta} r^{\alpha}|\nabla u|^2\, dx.$$

Using Lemma 91 we can state that

$$\int_0^\infty |\ln\, r|^{-\beta} r^{\alpha+n-3}\overline{u}(r)^2\, dr \le C \int_{\mathbf{R}^n} |\ln\, r|^{-\beta} r^{\alpha}|\nabla u(x)|^2\, dx.$$

Consider the function $u(x) - \overline{u}(r)$ in the domain $S_\lambda = \{x :\ \lambda < |x| < 2\lambda\}$, $\lambda > 2$. Since

$$\int_{S_\lambda} [u(x) - \overline{u}(r)]\, dx = 0,$$

we obtain, using the Poincaré inequality, that

$$\int_{S_\lambda} [u(x) - \overline{u}(r)]^2\, dx \le C\lambda^2 \int_{S_\lambda} |\nabla u|^2\, dx.$$

This gives the relation

$$\int_{S_\lambda} \frac{|u(x) - \overline{u}(r)|^2|x|^{\alpha-2}}{(\ln|x|)^\beta}\, dx \le C_1 \int_{S_\lambda} \frac{|x|^\alpha|\nabla u|^2}{(\ln|x|)^\beta}\, dx,$$

since for points in the domain S_λ we have the inequalities:

$$\ln|\lambda| \le \ln|x| \le \ln 2\lambda = \ln 2 + \ln\lambda \le 2\ln\lambda.$$

Setting $\lambda = 2^k$, $k = 1, 2, \ldots$, and summing over all k, we obtain

$$\int_{\mathbf{R}^n} \frac{|u(x) - \overline{u}(r)|^2|x|^{\alpha-2}}{(\ln|x|)^\beta}\, dx \le C_1 \int_{\mathbf{R}^n} \frac{|x|^\alpha|\nabla u|^2}{(\ln|x|)^\beta}\, dx.$$

Therefore

$$\int_{\mathbf{R}^n} |u(x)|^2 (\ln|x|)^{-\beta}|x|^{\alpha-2}\,dx \le 2\int_{\mathbf{R}^n} |u(x)-\overline{u}|^2(\ln|x|)^{-\beta}|x|^{\alpha-2}\,dx$$
$$+2\int_{\mathbf{R}^n} |\overline{u}(r)|^2(\ln|x|)^{-\beta}|x|^{\alpha-2}\,dx \le C_2\int_{\mathbf{R}^n} |x|^{\alpha}(\ln|x|)^{-\beta}|\nabla u|^2\,dx. \qquad \square$$

Theorem 93. *If* $|x|^{1-n/2}u_{x_i}(x) \in L_2(\mathbf{R}^n)$, $i=1,\dots,n$, *and*

$$\int_{1<|x|<2} |u(x)|^2\,dx < \infty, \ \alpha+n=2,$$

then

$$\int_{\mathbf{R}^n} \frac{|u(x)|^2}{|x|^n[\ln^2|x|+1]}\,dx \le C\int_{\mathbf{R}^n} |x|^{2-n}|\nabla u(x)|^2\,dx + C\int_{1<|x|<2} |u(x)|^2\,dx,$$

where $C=\text{const}$ *does not depend on* u.

Proof. The proof repeats the proof of Theorem 90, but instead of (4),(5) we use inequality (6):

$$\int_0^\infty \frac{|y_0(x)|^p}{x|\ln x|^p}\,dx \le \left(\frac{p}{p-1}\right)^p \int_0^\infty x^{p-1}|y'(x)|^p\,dx. \qquad \square$$

Theorem 94. *If*

$$\int_{\mathbf{R}^n} \frac{|x|^\alpha|\nabla u(x)|^2}{(1+|\ln|x||)^\beta}\,dx < \infty$$

$$\int_{2<|x|<3} |u(x)|^2\,dx < \infty, \ \alpha+n<2, \ \beta>0,$$

then

$$\int_{|x|>3} \frac{|x|^{\alpha-2}|u(x)|^2}{(1+|\ln|x||)^\beta}\,dx \le C\int_{\mathbf{R}^n} \frac{|x|^\alpha|\nabla u(x)|^2}{(1+|\ln|x||)^\beta}\,dx + C\int_{2<|x|<3} |u(x)|^2\,dx,$$

where the constant C *does not depend on* u.

Proof. Let $\sigma(x) \in C_0^\infty(\mathbf{R}^n)$, $\sigma(x)\equiv 1$ for $|x|>3$, $\sigma(x)\equiv 0$ for $|x|<2$. Applying Theorem 92 to the function $\sigma(x)u$, we obtain that

$$\int_{\mathbf{R}^n} \frac{|x|^{\alpha-2}|\sigma^2u^2(x)|}{(1+|\ln|x||)^\beta}\,dx \le C_1\int_{\mathbf{R}^n} \frac{|x|^\alpha|\nabla(\sigma u)(x)|^2}{(1+|\ln|x||)^\beta}\,dx$$
$$\le C_2\int_{\mathbf{R}^n} \frac{|x|^\alpha|\nabla u(x)|^2}{(1+|\ln|x||)^\beta}\,dx + C_2\int_{\mathbf{R}^n} \frac{|x|^\alpha u(x)^2|\nabla\sigma|^2}{(1+|\ln|x||)^\beta}\,dx$$
$$\le C_2\int_{\mathbf{R}^n} \frac{|x|^\alpha|\nabla u(x)|^2}{(1+|\ln|x||)^\beta}\,dx + C_3\int_{|x|\le 3} \frac{|x|^\alpha u(x)^2}{(1+|\ln|x||)^\beta}\,dx,$$

since $\nabla\sigma\equiv 0$ for $|x|>3$ and for $|x|<2$. We can set $C=\max(C_2,C_3)$. $\square$

There are also analogues of the Hardy inequality, using the higher order derivatives. For example, if $y(x) \in C^m[0,\infty)$ and the support of y is lying on $(0,\infty)$, then

$$\int_0^\infty x^{\alpha-2m}|y(x)|^2\,dx \le C_{m,\alpha}\int_0^\infty x^\alpha |y^{(m)}(x)|^2\,dx,$$

if α is not equal to an odd integer from $[1,m]$. This inequality follows from the inequality

$$\int_0^\infty x^{\alpha-2}|y(x)|^2\,dx \le \frac{4}{|\alpha-1|^2}\int_0^\infty x^\alpha |y'(x)|^2\,dx,\ \alpha \ne 1,$$

proved above, after its application to the functions $y^{(m-1)},\dots,y',y$.

Theorem 95. *If*

$$\sum_{|\alpha|=k}\int_{\mathbf{R}^n}|x|^\gamma |D^\alpha u(x)|^2\,dx < \infty,\ u(x)\in C^\infty(\mathbf{R}^n\setminus 0),$$

where $\gamma \ne 2j+2-n$ for all integers j such that $0\le j<k$, then there is a polynomial $P(x)=\sum_{|\beta|<k} c_\beta x^\beta$ of degree $k-1$ such that $u(x)=P(x)+u_0(x)$ and

$$\sum_{|\alpha|<k}\int_{\mathbf{R}^n}|x|^{\gamma-2k+2|\alpha|}|D^\alpha u_0(x)|^2\,dx \le C\sum_{|\alpha|=k}\int_{\mathbf{R}^n}|x|^\gamma |D^\alpha u(x)|^2\,dx,$$

$$|c_\beta|^2 \le C\sum_{|\alpha|=k}\int_{\mathbf{R}^n}|x|^\gamma|D^\alpha u(x)|^2\,dx + \int_{1<|x|<2}|u(x)|^2\,dx.$$

Proof. We will use induction on k. If $k=1$, then the statement follows from Theorem 90. Let $l>1$ and the statement be true for $k<l$. Each function $D^\alpha u$ with $|\alpha|=l-1$ satisfies the inequality:

$$\int_{\mathbf{R}^n}|x|^\gamma|\nabla(D^\alpha u)|^2\,dx \le \sum_{|\beta|=l}\int_{\mathbf{R}^n}|x|^\gamma|D^\beta u|^2\,dx.$$

Theorem 90 implies that for each α with $|\alpha|=l-1$ the representation $D^\alpha u(x)=a_\alpha+v_\alpha(x)$, where $a_\alpha=\text{const}$ such that

$$\int_{\mathbf{R}^n}|x|^{\gamma-2}|v_\alpha(x)|^2\,dx \le C\sum_{|\beta|=l}\int_{\mathbf{R}^n}|x|^\gamma|D^\beta u(x)|^2\,dx,$$

$$|a_\alpha|^2 \le \sum_{|\beta|=l}\int_{\mathbf{R}^n}|x|^\gamma|D^\beta u(x)|^2\,dx + C\int_{1<|x|<2}|D^\alpha u(x)|^2\,dx.$$

Consider the function $u_1(x) = u(x) - P_1(x)$, where

$$P_1(x) = \sum_{|\alpha|=l-1} a_\alpha x^\alpha/\alpha!.$$

Since $D^\alpha u_1 = D^\alpha u - a_\alpha$ if $|\alpha| = l-1$, we have

$$\sum_{|\alpha|=l-1} \int_{\mathbf{R}^n} |x|^{\gamma-2}|D^\alpha u_1(x)|^2\,dx \le C \sum_{|\beta|=l} \int_{\mathbf{R}^n} |x|^\gamma |D^\beta u(x)|^2\,dx.$$

By the induction conjecture we have $u_1 = P_2(x)+u_0$, where $P_2(x)=\sum_{|\alpha|\le l-2} b_\alpha x^\alpha$ is a polynomial of degree $l-2$ and the function $u_0(x)$ is such that

$$\sum_{|\alpha|\le l-1} \int_{\mathbf{R}^n} |x|^{\gamma-2l+2}|D^\alpha u_0(x)|^2\,dx \le C \sum_{|\alpha|=l-1} \int_{\mathbf{R}^n} |x|^{\gamma-2}|D^\alpha u_1(x)|^2\,dx,$$

$$|b_\alpha|^2 \le C \sum_{|\alpha|=l-1} \int_{\mathbf{R}^n} |x|^{\gamma-2}|D^\alpha u_1(x)|^2\,dx + C\int_{1<|x|<2} |u_1(x)|^2\,dx.$$

Therefore $u(x) = u_1(x) + P_1(x) = u_0(x) + P(x)$, where $P(x) = P_1(x) + P_2(x)$ is a polynomial of degree $l-1$. The proof is complete. □

Theorem 96. *If $u(x) \in C^\infty(\mathbf{R}^n)$,*

$$I(u) = \sum_{|\alpha|=k} \int_{\mathbf{R}^n} |x|^\gamma |D^\alpha u(x)|^2\,dx + \int_{1<|x|<2} |u(x)|^2\,dx < \infty,$$

and for some integer s, $0 \le s < k$, we have the equality $\gamma + n = 2s+2$, then there exists a polynomial $P(x) = \sum_{|\alpha|<k} c_\alpha x^\alpha$, such that

$$u(x) = P(x) + v(x)$$

and

$$\sum_{|\alpha|<k} |c_\alpha|^2 + \sum_{|\beta|<k-s} \int_{\mathbf{R}^n} \frac{(1+|x|)^{\gamma-2k+2|\beta|}|D^\beta v(x)|^2}{1+\ln^2|x|}\,dx$$
$$+ \sum_{k-s\le|\beta|<k} \int_{\mathbf{R}^n} |x|^{\gamma-2k+2|\beta|}|D^\beta v(x)|^2\,dx \le CI(u),$$

where C does not depend on u.

Proof. Let β be a multi-index such that $\beta = (\beta_1,\dots,\beta_n)$ and $|\beta| + s = k$. Set $u_\beta(x) = D^\beta u(x)$. Theorem 95 implies that

$$u_\beta(x) = P_\beta(x) + u_{\beta 0}(x),$$

where $P_\beta(x) = \sum_{|\alpha|<s} a_{\alpha\beta} x^\alpha$ and

$$\int_{\mathbf{R}^n} |x|^{\gamma-2s}|u_{\beta 0}(x)|^2\,dx + \sum_{|\alpha|<s} |a_{\alpha\beta}|^2 \le C \sum_{|\alpha|=s} \int_{\mathbf{R}^n} |x|^\gamma |D^\alpha u_\beta(x)|^2\,dx \le CI(u).$$

Consider the function

$$v(x) = u(x) - \sum_{\substack{|\alpha|<s\\|\beta|=k-s}} a_{\alpha\beta} x^{\alpha+\beta} \alpha!/(\alpha+\beta)!$$

for $s > 0$; if $s = 0$, we set $v(x) = u(x)$. We have

$$D^\beta v(x) = D^\beta u(x) - \sum_{|\alpha|<s} a_{\alpha\beta} x^\alpha = u_\beta(x) - P_\beta(x) = u_{\beta 0}(x).$$

Therefore

$$\sum_{j=0}^{k-s} \sum_{|\beta|=k-s+j} \int_{\mathbf{R}^n} |x|^{\gamma-2s+2j}|D^\beta v(x)|^2\,dx \le C \sum_{|\alpha|=k} \int_{\mathbf{R}^n} |x|^\gamma |D^\alpha u(x)|^2\,dx.$$

By Theorem 93 we have

$$\sum_{|\alpha|=k-s-1} \int_{\mathbf{R}^n} \frac{|x|^{\gamma-2s-2}|D^\alpha v(x)|^2}{(1+\ln|x|)^2}\,dx$$

$$\le C \sum_{|\alpha|=k-s-1} \int_{\mathbf{R}^n} |x|^{\gamma-2s}|\nabla D^\alpha v(x)|^2\,dx + C\int_{1<|x|<2} \sum_{|\alpha|=k-s-1} |D^\alpha u(x)|^2\,dx$$

and therefore

$$\sum_{|\alpha|=k-s-1} \int_{\mathbf{R}^n} \frac{|x|^{\gamma-2s-2}|D^\alpha v(x)|^2}{(1+\ln|x|)^2}\,dx$$

$$\le C_1\Big(\sum_{|\alpha|=k-s} \int_{\mathbf{R}^n} |x|^{\gamma-2s}|D^\alpha v(x)|^2\,dx + \int_{1<|x|<2} \sum_{|\alpha|=k-s-1} |D^\alpha u(x)|^2\,dx\Big)$$

$$\le C_2 \sum_{|\alpha|=k} \int_{\mathbf{R}^n} |x|^\gamma |D^\alpha u(x)|^2\,dx + C_1 \int_{1<|x|<2} \sum_{|\alpha|\le k} |D^\alpha u(x)|^2\,dx.$$

Set $w(x) = v(x)\theta(x)$, where $\theta(x) \in C^\infty(\mathbf{R}^n)$ is a function such that $\theta(x) \equiv 0$ for $|x| < e$, $\theta(x) \equiv 1$ for $|x| > 3$, $0 \le \theta(x) \le 1$. If $|\alpha| = k - s - 1$, then Theorem 93 implies that

$$\int_{\mathbf{R}^n} \frac{|x|^{\gamma-2s-2}|D^\alpha w(x)|^2}{\ln^2|x|}\,dx \le 4\int_{|x|>e} \frac{|x|^{\gamma-2s-2}|D^\alpha w(x)|^2}{(1+|\ln|x||)^2}\,dx$$

$$\le C_3 \int_{\mathbf{R}^n} \sum_{|\alpha|=k} |x|^\gamma |D^\alpha w(x)|^2\,dx.$$

Let us prove that for $j < k - s$

$$\sum_{|\alpha|=j} \int_{\mathbf{R}^n} \frac{|x|^{\gamma-2k+2j}|D^\alpha w(x)|^2}{\ln^2|x|}\,dx$$
$$\leq C_4 \int_{\mathbf{R}^n} \sum_{|\alpha|=k} |x|^\gamma |D^\alpha u(x)|^2\,dx + C_4 \int_{1<|x|<2} \sum_{|\alpha|<k} |D^\alpha u(x)|^2\,dx.$$

If $j = k - s - 1$, the statement has been proved. Let now $j < k - s - 1$. Then by Theorem 92

$$\sum_{|\alpha|=j} \int_{\mathbf{R}^n} \frac{|x|^{\gamma-2k+2j}|D^\alpha w(x)|^2}{\ln^2|x|}\,dx$$
$$\leq C_5\Big(\sum_{|\alpha|=j+1} \int_{\mathbf{R}^n} \frac{|x|^{\gamma-2k+2+2j}|D^\alpha w(x)|^2}{\ln^2|x|}\,dx + \int_{1<|x|<2} \sum_{|\alpha|<k} |D^\alpha w(x)|^2\,dx\Big).$$

Since $|w(x)| \leq |v(x)| \leq C_6|u(x)| + C_6 \sum |a_{\alpha\beta}|$ for $|x| < 2$, and the coefficients $a_{\alpha\beta}$ were estimated above, we have

$$\sum_{|\alpha|=j} \int_{\mathbf{R}^n} \frac{|x|^{\gamma-2k+2j}|D^\alpha w(x)|^2}{\ln^2|x|}\,dx \leq C_8 \sum_{|\alpha|=j+1} \int_{\mathbf{R}^n} \frac{|x|^{\gamma-2k+2+2j}|D^\alpha w(x)|^2}{\ln^2|x|}\,dx$$
$$+ C_8 \sum_{|\alpha|=k} \int_{\mathbf{R}^n} |x|^\gamma |D^\alpha u(x)|^2\,dx + C \int_{1<|x|<2} |u(x)|^2\,dx.$$

Applying induction on j for $j = 1, \ldots, k - s$, we obtain the result. Indeed, if it is true for some $j = j_0 = k - s - i_0$, then

$$\sum_{|\alpha|=j_0-1} \int_{\mathbf{R}^n} \frac{|x|^{\gamma-2k+2j_0-2}|D^\alpha w(x)|^2}{\ln^2|x|}\,dx \leq C \sum_{|\alpha|=j_0} \int_{\mathbf{R}^n} \frac{|x|^{\gamma-2k+2j_0}|D^\alpha w(x)|^2}{\ln^2|x|}\,dx$$
$$+ C \int_{\mathbf{R}^n} |x|^\gamma \sum_{|\alpha|=k} |D^\alpha u(x)|^2\,dx + C \int_{1<|x|<2} |u(x)|^2\,dx,$$

where we used the inductive conjecture that

$$\int_{\mathbf{R}^n} |x|^{\gamma-2k+2j_0} \sum_{|\alpha|=j_0} |D^\alpha w|^2\,dx \leq C \sum_{|\alpha|=k} |D^\alpha u(x)|^2\,dx + C \int_{1<|x|<2} |u(x)|^2\,dx.$$

Therefore, the inequality

$$\sum_{|\alpha|=j} \int_{\mathbf{R}^n} \frac{|x|^{\gamma-2k+2j}|D^\alpha w(x)|^2}{\ln^2|x|}\,dx$$
$$\leq C \int_{\mathbf{R}^n} \sum_{|\alpha|=k} |D^\alpha u(x)|^2\,dx + C \int_{1<|x|<2} \sum_{|\alpha|<k} |D^\alpha u(x)|^2\,dx$$

holds for $j \leq k-s-1$ and

$$\int_{|x|>2} \frac{|x|^{\gamma-2k+2|\alpha|}|D^\alpha v(x)|^2}{\ln^2|x|}\,dx \leq CI(u).$$

Moreover,

$$\sum_{k-s\leq|\alpha|<k} \int_{|x|>2} |x|^{\gamma-2k+2|\alpha|}|D^\alpha w(x)|^2\,dx \leq CI(u)$$

and

$$\sum_{|\gamma|<k} \int_{|x|<2} |D^\gamma v(x)|^2\,dx \leq 2\sum_{|\gamma|<k}\int_{|x|<2} |D^\gamma u(x)|^2\,dx$$
$$+ 2\sum_{|\gamma|<k}\int_{|x|<2} |\sum D^\gamma a_{\alpha\beta}x^{\alpha+\beta}\alpha!/(\alpha+\beta)!|^2\,dx.$$

Therefore

$$\sum_{|\gamma|<k}\int_{|x|<2} |D^\gamma v(x)|^2\,dx \leq I(u)$$

and

$$\sum_{|\alpha|\leq k-s-1}\int_{\mathbf{R}^n} \frac{(1+|x|)^{\gamma-2k+2|\alpha|}|D^\alpha v(x)|^2}{1+\ln^2|x|}\,dx$$
$$= \int_{|x|<2}\frac{(1+|x|)^{\gamma-2k}}{1+\ln^2|x|}|v(x)|^2\,dx + \int_{|x|>2}\frac{(1+|x|)^{\gamma-2k}}{1+\ln^2|x|}|v(x)|^2\,dx$$
$$\leq \int_{|x|<2}|v(x)|^2\,dx + \int_{|x|>2}\frac{|x|^{\gamma-2k}}{\ln^2|x|}|v(x)|^2\,dx \leq CI(u). \qquad \square$$

Theorems 95 and 96 are true if $\gamma = 0$.

The condition $u \in C^\infty$ can be eliminated. Indeed, let $u \in C^k(\mathbf{R}^n\backslash 0)$. Consider its averaging u_h. In virtue of Theorem 95 we have $u_h(x) = P(h,x)+u_0(h,x)$, where $P(h,x) = \sum_{|\alpha|<k} c_{\alpha h}x^\alpha$, and

$$\int_{\mathbf{R}^n} |u_0(h,x)|^2|x|^{-2k}\,dx \leq C\sum_{|\alpha|=k}\int_{\mathbf{R}^n} |D^\alpha u_h(x)|^2\,dx,$$

$$|c_{\alpha h}|^2 \leq C\sum_{|\alpha|=k}\int_{\mathbf{R}^n} |D^\alpha u_0(h,x)|^2\,dx + C\int_{1<|x|<2} |u_h(x)|^2\,dx.$$

Therefore

$$\int_{\varepsilon<|x|<1/\varepsilon} |u_0(x)|_h^2|x|^{-2k}\,dx \leq C\sum_{|\alpha|=k}\int_{\mathbf{R}^n} |D^\alpha u_h(x)|^2\,dx.$$

Since $D^\alpha u_h \to D^\alpha u$ as $h \to 0$, we see that $\|u_0\|_{L_2(\Omega_\varepsilon)} \le C$, where $\Omega_\varepsilon = \{x : \varepsilon < |x| < 1/\varepsilon\}$ and C is independent of h. Since $D^\alpha u_0 = D^\alpha u_h$ for $|\alpha| = k$, we see that the set of u_0 is bounded in $W_2^k(\Omega_\varepsilon)$. Therefore, there exists a sequence of u_{0h} converging in $L_2(\Omega_\varepsilon)$ to a function u_0. The latter function is such that

$$\int_{\varepsilon<|x|<1/\varepsilon} |x|^{-2k}|u_0(x)|^2\,dx \le C\int_{\mathbf{R}^n}\sum_{|\alpha|=k}|D^\alpha u(x)|^2\,dx$$

for $\varepsilon = 2^{-l}$, $l = 1, 2, \ldots$. Using diagonalization, we can choose a subsequence u_h converging to u_0 in $L_2(\Omega_\varepsilon)$ for all ε. Thus we obtain that

$$\int_{\mathbf{R}^n} |x|^{-2k}|u_0(x)|^2\,dx \le C\int_{\mathbf{R}^n}\sum_{|\alpha|=k}|D^\alpha u(x)|^2\,dx.$$

Furthermore, there is a subsequence of $c_{\alpha h}$ converging to c_α and

$$u(x) = P(x) + u_0(x) = \sum_{|\alpha|<k} c_\alpha x^\alpha + u_0(x).$$

Moreover c_α, u_0 satisfy the inequalities stated in Theorem 95. This shows that the requirement $u(x) \in C^\infty$ is superfluous.

Theorem 97. *If $u(x) \in C_0^\infty(\mathbf{R}^n)$, $n > 2$, and $\int_{\mathbf{R}^n}|\nabla u(x)|^2\,dx < \infty$, then*

$$\Big(\int_{\mathbf{R}^n}|x|^{-n+p(n-2)/2}|u(x)|^p\,dx\Big)^{2/p} \le C\int_{\mathbf{R}^n}|\nabla u(x)|^2\,dx$$

for $2 \le p \le 2n/(n-2)$.

Proof. The Hölder inequality implies that

$$\Big(\int_{\mathbf{R}^n}|x|^{-n+p(n-2)/2}|u(x)|^p\,dx\Big)^{2/p}$$

$$\le \Big(\int_{\mathbf{R}^n}|x|^{-2}|u(x)|^2\,dx\Big)^{n/p-n/2+1}\Big(\int_{\mathbf{R}^n}|u(x)|^{2n/(n-2)}\,dx\Big)^{(n-2)(1/2-1/p)}.$$

The first factor on the right-hand side is estimated with the help of the Hardy inequality. Since $u(x) \in C_0^\infty(\mathbf{R}^n)$, we have $c_0 = 0$ and $u(x) \equiv u_0(x)$. The second factor is estimated with the help of the Sobolev embedding theorem 57. □

Theorem 98. *If $u(x) \in C_0^\infty(\mathbf{R}^n)$, $\varphi(x) = (\sum_{i=1}^n |x_i|^{\gamma_i})^{-1}$, $\gamma_i > 0$, $n > 2$, and $\sum_1^n 1/\gamma_i = n/2$, then*

$$\int_{\mathbf{R}^n}\varphi(x)|u(x)|^2\,dx \le C\int_{\mathbf{R}^n}\Big(\sum_1^n |u_{x_i}(x)|^2\Big)\,dx.$$

Proof. The inequality

$$\int_{\lambda<|x|<2\lambda} \varphi(x)|u(x)|^2\,dx$$

$$\leq \left[\int_{\lambda<|x|<2\lambda} |u(x)|^{2n/(n-2)}\,dx\right]^{(n-2)/n} \cdot \left[\int_{\lambda<|x|<2\lambda} \varphi(x)^{n/2}\,dx\right]^{2/n}$$

and the Sobolev theorem 57 imply that for $\lambda > 0$

$$\int_{\lambda<|x|<2\lambda} \varphi(x)|u(x)|^2\,dx \leq C[\int_{\lambda<|x|<2\lambda} |\nabla u(x)|^2\,dx$$

$$+\int_{\lambda<|x|<2\lambda} \frac{|u(x)|^2}{|x|^2}\,dx] \cdot \left[\int_{\lambda<|x|<2\lambda} \varphi(x)^{n/2}\,dx\right]^{2/n},$$

where C does not depend on λ. Let us estimate the integral

$$S(\lambda) = \int_{\lambda<|x|<2\lambda} \varphi(x)^{n/2}\,dx.$$

The domain $\lambda < |x| < 2\lambda$ can be represented as a finite sum of domains $\Omega_{j\lambda}$ such that $|x_j| > \frac{\lambda}{2}$ if $x \in \Omega_{j\lambda}$ for $j = 1, \ldots, n$. Let, for instance, $|x_1| > \frac{\lambda}{2}$. Then

$$\int_{\Omega_{j\lambda}} \varphi(x)^{n/2}\,dx \leq \frac{3\lambda}{2} \int_{\lambda<|x|<2\lambda} \frac{dx_2 \ldots dx_n}{((\lambda/2)^{\gamma_1} + |x_2|^{\gamma_2} + \cdots + |x_n|^{\gamma_n})^{n/2}}.$$

The substitution $x_j = t_j(\lambda/2)^{\gamma_1/\gamma_j}$ gives the relations

$$S(\lambda) \leq C_1 \int_{\mathbf{R}^{n-1}} \frac{dt_2 \ldots dt_n}{(1 + |t_2|^{\gamma_2} + \ldots + |t_n|^{\gamma_n})^{n/2}} \leq C_2,$$

since the integral is converging. To see this, set $t_s = \tau_s^{1/\gamma_s}$. Then

$$\int_{\mathbf{R}^{n-1}} \frac{dt_2 \ldots dt_n}{(1 + |t_2|^{\gamma_2} + \cdots + |t_n|^{\gamma_n})^{n/2}}$$

$$\leq C_3 \int_{\mathbf{R}^{n-1}} \frac{|\tau|^{1/\gamma_2+\ldots+1/\gamma_n-(n-1)}}{(1+|\tau|)^{n/2}} d\tau_2 \ldots d\tau_n \leq C_4 \int_0^\infty \frac{d|\tau|}{(1+|\tau|)^{1+1/\gamma_1}} < \infty.$$

Therefore

$$\int_{\lambda<|x|<2\lambda} |u(x)|^2\varphi(x)\,dx \leq C_5[\int_{\lambda<|x|<2\lambda} |\nabla u(x)|^2\,dx + \int_{\lambda<|x|<2\lambda} r^{-2}|u(x)|^2\,dx].$$

Setting $\lambda = 2^k$, $k = 0, \pm 1, \pm 2, \ldots$ and summing these inequalities over all k, we obtain that

$$\int_{\mathbf{R}^n} |u(x)|^2\varphi(x)\,dx \leq C_6 \left(\int_{\mathbf{R}^n} |\nabla u(x)|^2\,dx + \int_{\mathbf{R}^n} r^{-2}|u(x)|^2\,dx\right)$$

and by the Hardy inequality

$$\int_{\mathbf{R}^n} |u(x)|^2 \varphi(x)\, dx \le C_7 \int_{\mathbf{R}^n} |\nabla u(x)|^2\, dx. \qquad \square$$

We can find the sharp value of the constant C in Theorem 90.

Theorem 99. *If* $u \in C_0^\infty(\mathbf{R}^n)$, $\gamma + n \neq 2$, *and* $u(0) = 0$, *then*

$$\int_{R^n} |x|^{\gamma-2}|u(x)|^2\, dx \le \frac{4}{(\gamma+n-2)^2}\int_{R^n} |x|^{\gamma}|\nabla u(x)|^2\, dx.$$

Proof. Theorem 90 implies that for almost all rays the integrals

$$\int_0^\infty r^{\alpha+n-3}|u(x)|^2\, dr \quad \text{and} \quad \int_0^\infty r^{\alpha+n-1}|\frac{\partial u(x)}{\partial r}|^2 dr$$

are finite. Applying to u inequalities (4), (5), we see that

$$\int_0^\infty r^{\gamma+n-3}|u(x)|^2\, dr \le \frac{4}{(\gamma+n-2)^2}\int_0^\infty r^{\alpha}|\frac{\partial u(x)}{\partial r}|^2\, dr.$$

Integrating this inequality over the unit sphere, we obtain the result. The sharpness can be verified by considering the functions u depending on r only, as was done for inequality (2). $\square$

Generalizing the Hardy inequality we come to the following problem: Under which conditions on functions $V(x) \ge 0$, $W(x) \ge 0$, does the inequality

$$\int_{\mathbf{R}^n} V(x)|u(x)|^2\, dx \le C \int_{\mathbf{R}^n} W(x)|\nabla u(x)|^2\, dx \tag{9}$$

hold? Here $u(x) \in C_0^\infty(\mathbf{R}^n)$ and the constant C is independent of u. Such inequalities are essential in the study of the spectrum of the Schrödinger operator. In the case $W(x) \equiv 1$ a necessary and sufficient condition has been obtained by Mazya. His result uses as a base the notion of capacity. Recall its definition.

Let Ω be a domain in $\mathbf{R}^n$ and F be its subset. Set

$$\operatorname{cap}(F, \Omega) = \inf\{\int_\Omega |\nabla u(x)|^2\, dx\},$$

where inf is taken in the class of function $u(x) \in C_0^\infty(\Omega)$ such that $u \equiv 1$ in a neighborhood of F and $0 \le u(x) \le 1$.

Mazya proved that inequality (9) with $n \ge 3$ and $W(x) = 1$ is true if and only if

$$\int_F V\, dx \le \beta \operatorname{cap}(F, \Omega)$$

for any compact F with a constant β independent of F.

Inequality (9) was considered by many mathematicians, e.g. Makenhaupt, Kufner, Dyn'kin, Fefferman, Schechter, Sawyer, and others. (See the review of results in [Dy].) The following theorem is an exact result for $n = 1$.

Theorem 100. *The inequality*

$$\int_0^\infty q(x)u(x)^2\,dx \le C\int_0^\infty u'(x)^2\,dx,\ u(0) = 0 \tag{10}$$

is valid if and only if there is a constant C_1 such that

$$\sup_{x\in(0,\infty)} x\int_x^\infty q(x)\,dx \le C_1. \tag{11}$$

Proof. Let us show that (10) implies (11). Let $u(x) = x$ for $x \le r$ and $u(x) = r$ for $x > r$. Then

$$\int_0^r q(x)x^2\,dx + r^2\int_r^\infty q(x)\,dx \le Cr,$$

i.e. (11) is true.

Inversely, let (11) be valid. Integrating by parts we obtain

$$\int_0^\infty q(x)u(x)^2\,dx = 2\int_0^\infty u(x)u'(x)\int_x^\infty q(t)\,dt\,dx - \lim_{T\to\infty} u^2(T)\int_T^\infty q(x)\,dx.$$

If $\int_0^\infty u'(x)^2\,dx < \infty$, then $u^2(T) \le T\int_0^\infty u'(x)^2dx$, so that

$$\int_0^\infty q(x)u(x)^2\,dx \le C_2\Big(\int_0^\infty u'(x)^2\,dx\Big)^{1/2}\Big(\int_0^\infty \frac{u(x)^2\,dx}{x^2}\Big)^{1/2} + C_1\int_0^\infty u'(x)^2dx \le C_3\int_0^\infty u'(x)^2\,dx.$$

□

2.31 The Morrey embedding theorem

Morrey introduced the spaces $M_{m,\alpha}(\Omega)$ of functions $u(x) \in W_{2,0}^m(\Omega)$ such that

$$\sup_{\substack{x_0\in\mathbf{R}^n\\ 0<\rho<\infty}} \frac{1}{\omega_n\rho^{n\alpha}}\int_{|x-x_0|\le\rho}\sum_{|\beta|=m}|D^\beta u|^2\,dx < \infty.$$

Here ω_n is the area of the unit sphere, $\alpha > 0$ and $u = 0$ outside Ω. The norm in $M_{m,\alpha}$ is defined in the following way:

$$\|u\|^2_{M_{m,\alpha}(\Omega)} = \|u(x)\|^2_{W^m_{2,0}(\Omega)} + \sup_{\substack{x_0\in\mathbf{R}^n\\ 0<\rho<\infty}} \frac{1}{\omega_n\rho^{n\alpha}}\int_{|x-x_0|\le\rho}\sum_{|\beta|=m}|D^\beta u|^2\,dx.$$

It is easy to see that $M_{m,\alpha}(\Omega)$ is a Banach space. Indeed, let $\{u_k\}$ be a Cauchy sequence. The definition implies that it is a Cauchy sequence in $W_{2,0}^m(\Omega)$ and therefore has a limit u in $W_{2,0}^m(\Omega)$. Besides, for any $x_0 \in \mathbf{R}^n$, $\rho > 0$

$$\frac{1}{\omega_n \rho^{n\alpha}} \int_{|x-x_0|\leq\rho} \sum_{|\beta|=m} |D^\beta u_k - D^\beta u_{k+p}|^2 \, dx \leq \|u_k - u_{k+p}\|^2_{M_{m,\alpha}(\Omega)} \leq \varepsilon,$$

if $k \geq K, p \geq 0$. Since for all $\rho > 0$

$$\frac{1}{\omega_n \rho^{n\alpha}} \int_{|x-x_0|\leq\rho} |D^\beta u_k(x)|^2 \, dx \leq C,$$

we see passing to the limit as $x \to \infty$, that $u \in M_{m,\alpha}(\Omega)$.

Theorem 101. *If $u(x) \in W_{2,0}^m(\mathbf{R}^n)$, $n \geq 2m$, $u(x) \equiv 0$ outside a domain Ω having a compact closure, and*

$$\int_{|x-x_0|\leq\rho} \sum_{|\alpha|=m} |D^\alpha u(x)|^2 \, dx \leq A\rho^a$$

for any $x_0 \in \mathbf{R}^n$ and some a such that $n \geq a > n - 2m$, where A and a do not depend on x_0, ρ, and u, then for any $x \in \Omega$

$$|u(x)| \leq C[A^{1/2} + \|u\|_{W_2^m(\mathbf{R}^n)}],$$

where C is independent of u.

Proof. Let at first $u(x) \in C_0^m(\mathbf{R}^n)$. For any $\gamma \in (n-2m, a)$ we have

$$\begin{aligned}
&\int_{\mathbf{R}^n} |x-x_0|^{-\gamma} \sum_{|\alpha|=m} |D^\alpha u(x)|^2 \, dx \\
&= \sum_{k=-\infty}^{+\infty} \int_{2^k\leq|x-x_0|\leq 2^{k+1}} |x-x_0|^{-\gamma} \sum_{|\alpha|=m} |D^\alpha u(x)|^2 \, dx \\
&\leq \sum_{k=-\infty}^{N} \int_{2^k\leq|x-x_0|\leq 2^{k+1}} 2^{-k\gamma} \sum_{|\alpha|=m} |D^\alpha u(x)|^2 \, dx \\
&\leq \sum_{k=-\infty}^{N} 2^{-k\gamma} \cdot A \cdot 2^{(k+1)a} \leq C_2 A.
\end{aligned}$$

Theorem 95 implies that u can be represented as

$$u(x) = \sum_{|\alpha|\leq m} c_\alpha (x-x_0)^\alpha + v(x),$$

where

$$\int_{\mathbf{R}^n} |x - x_0|^{-\gamma-2m}|v(x)|^2\,dx \le C_3A, \quad \text{and} \quad |c_\alpha|^2 \le C_4A + \int_\Omega |u(x)|^2\,dx.$$

Since $n - 2m \le \gamma < a$, we obtain that $v(x_0) = 0$ and therefore,

$$|u(x_0)|^2 = |c_0|^2 \le C_4A + \int_\Omega |u(x)|^2\,dx.$$

Now let $u(x) \in W_0^m(\mathbf{R}^n)$. Consider its averaging u_h for $h < 1$. Observe that the inequality

$$\int_{|x-x_0|\le\rho} |f(x)|^2\,dx \le A\rho^a$$

implies that

$$\int_{|x-x_0|\le\rho} |f_h(x)|^2\,dx \le CA\rho^a. \tag{12}$$

Indeed, we have

$$|f_h(x)|^2 \le C_1h^{-n}\int_{|x-y|\le h} |f(y)|^2\,dy < C_1Ah^{-n+a}$$

and therefore

$$\int_{|x-x_0|\le\rho} |f_h(x)|^2\,dx \le C_2Ah^{-n+a}\rho^n. \tag{13}$$

Besides,

$$\int_{|x-x_0|\le\rho} |f_h(x)|^2\,dx \le \int_{|x-x_0|\le\rho+h} |f(x)|^2\,dx \le C_3A(\rho+h)^a. \tag{14}$$

Inequality (12) follows from (13) if $h \ge \rho$ and from (14) if $h \le \rho$.

Since $D^\alpha u_h = (D^\alpha u)_h$, we have

$$\int_{|x-x_0|\le\rho} \sum_{|\alpha|=m} |D^\alpha u_h(x)|^2\,dx \le CA\rho^a, \ h \le 1.$$

Therefore

$$|u_h(x)| \le C_7(A^{1/2} + \|u\|_{W_2^m(\mathbf{R}^n)}).$$

Since the sequence $\{u_h\}$ contains a subsequence converging to u almost everywhere in Ω, the proof is complete. □

Chapter 3

Elliptic Operators

3.1 Strongly elliptic equations

We begin with some definitions.
Definition 1. A differential operator

$$L(x, D) = \sum_{|\alpha| \le m} a_\alpha(x) D^\alpha$$

is *strongly elliptic* in a domain Ω if

$$\mathrm{Re} \sum_{|\alpha|=m} a_\alpha(x)\xi^\alpha \neq 0 \quad \text{for} \quad \xi \in \mathbf{R}^n \setminus 0,\ x \in \Omega.$$

Here, as usual, $\xi^\alpha = \xi_1^{\alpha_1} \dots \xi_n^{\alpha_n}$.

Consider a differential equation of the form

$$\sum_{\substack{|\alpha| \le m \\ |\beta| \le m}} D^\alpha a_{\alpha\beta}(x) D^\beta u + \sum_{|\alpha| \le m} a_\alpha(x) D^\alpha u = \sum_{|\alpha| \le m} D^\alpha F_\alpha(x),$$

where

$$\alpha = (\alpha_1, \dots, \alpha_n), \qquad \beta = (\beta_1, \dots, \beta_n),$$

$$D^\alpha = \partial^{|\alpha|}/\partial x_1^{\alpha_1} \dots \partial x_n^{\alpha_n}, \quad |\alpha| = \sum_{i=1}^n \alpha_i, \qquad x = (x_1 \dots, x_n).$$

We assume that $a_{\alpha\beta}(x)$, $a_\alpha(x)$ are measurable bounded in a domain Ω functions with complex values, $F_\alpha(x) \in L_2(\Omega)$; $a_{\alpha\beta} \equiv \overline{a_{\beta\alpha}}$ for $|\alpha| = |\beta| = m$ and $a_{\alpha\beta}(x)$ are continuous functions in $\overline{\Omega}$, if $|\alpha| = |\beta| = m$. We will consider the first boundary value problem (*the Dirichlet problem*).

We will assume that Ω is a bounded domain, but we do not require any regularity of its boundary.

Definition 2. A *generalized solution* of the first boundary problem is a function $u(x) \in W_{2,0}^m(\Omega)$ such that for any $\psi(x) \in W_{2,0}^m(\Omega)$ the equality

$$\sum_{\substack{|\alpha|\le m\\|\beta|\le m}} (-1)^{|\alpha|} \int_\Omega a_{\alpha\beta}(x)\overline{D^\alpha\psi(x)}D^\beta u(x)\,dx + \sum_{|\alpha|\le m}\int_\Omega a_\alpha(x)\overline{\psi(x)}D^\alpha u(x)\,dx$$
$$= \sum_{|\alpha|\le m}\int_\Omega (-1)^{|\alpha|}F_\alpha(x)\overline{D^\alpha\psi(x)}\,dx$$

holds.

We assume that the considered equation is *strongly elliptic*, i.e.

$$\operatorname{Re}\sum_{|\alpha|=|\beta|=m} a_{\alpha\beta}(x)\xi^\alpha\xi^\beta \ne 0$$

for all real $\xi = (\xi_1, \dots, \xi_n)$, $\xi \ne 0$. It is easy to see that this condition is equivalent to the following one:

$$|\operatorname{Re}\sum_{|\alpha|=|\beta|=m} a_{\alpha\beta}(x)\xi^\alpha\xi^\beta| \ge a_0|\xi|^{2m} \quad \text{for all} \quad \xi \in \mathbf{R}^n, \tag{1}$$

where a_0 is a positive constant. One of the most important results in the theory of strongly elliptic operators is the *Fredholm alternative*: either a generalized solution exists for any F_α, or there exists a nontrivial solution for $F_\alpha = 0$. This statement can be proved on the basis of the following *Gårding inequality*.

Theorem 3. *Let* (1) *be satisfied. For any function* $u(x) \in W_{2,0}^m(\Omega)$ *the inequality*

$$\left|\sum_{|\alpha|=|\beta|=m}\int_\Omega a_{\alpha\beta}(x)D^\alpha u\overline{D^\beta u}\,dx\right| \ge k_1\int_\Omega \sum_{|\alpha|=m}|D^\alpha u|^2\,dx - k_2\int_\Omega |u|^2\,dx$$

holds, where $k_1 = \text{const} > 0$, $k_2 = \text{const}$.

Proof. 1. Let at first the coefficients $a_{\alpha\beta}(x)$ be constant: $a_{\alpha\beta}(x) = a_{\alpha\beta}$. Let $\tilde u(\xi)$ be the Fourier transform of u. Then the Parseval equality implies that

$$(2\pi)^n\left|\sum_{\substack{|\alpha|=m\\|\beta|=m}}\int_\Omega a_{\alpha\beta}D^\alpha u\overline{D^\beta u}\,dx\right| = \left|\sum_{\substack{|\alpha|=m\\|\beta|=m}}\int_{\mathbf{R}^n} a_{\alpha\beta}(i\xi)^\alpha\cdot\overline{(i\xi)^\beta}|\tilde u(\xi)|^2\,d\xi\right|$$
$$= \left|\int_{\mathbf{R}^n}\Big(\sum_{\substack{|\alpha|=m\\|\beta|=m}} a_{\alpha\beta}\xi^\alpha\xi^\beta\Big)|\tilde u(\xi)|^2\,d\xi\right| \ge \left|\int_{\mathbf{R}^n}\operatorname{Re}\sum_{\substack{|\alpha|=m\\|\beta|=m}} a_{\alpha\beta}\xi^\alpha\xi^\beta|\tilde u(\xi)|^2\,d\xi\right|$$
$$\ge a_0\int_{\mathbf{R}^n}|\xi|^{2m}|\tilde u(\xi)|^2\,d\xi = a_0(2\pi)^n\sum_{|\alpha|=m}\int_\Omega|D^\alpha u|^2\,dx.$$

2. Now let $|a_{\alpha\beta}(x) - a_{\alpha\beta}| \le \varepsilon$, where $a_{\alpha\beta}$ are constants, and $\varepsilon > 0$ is a small number, $|\alpha| = |\beta| = m$. Then

$$\begin{aligned} &|\sum_{|\alpha|=|\beta|=m} \int_\Omega a_{\alpha\beta}(x) D^\alpha u \overline{D^\beta u}\, dx| \\ &\ge |\sum_{|\alpha|=|\beta|=m} \int_\Omega a_{\alpha\beta} D^\alpha u \overline{D^\beta u}\, dx| - |\int_\Omega \sum_{|\alpha|=|\beta|=m} (a_{\alpha\beta}(x) - a_{\alpha\beta}) D^\alpha u \overline{D^\beta u}\, dx| \\ &\ge a_1 \sum_{|\alpha|=m} \int_\Omega |D^\alpha u|^2\, dx - C\varepsilon \sum_{|\alpha|=m} \int_\Omega |D^\alpha u|^2\, dx, \end{aligned}$$

where the constant C is independent of u and ε. If ε is sufficiently small, for example, if $C\varepsilon < a_1/2$, then

$$|\sum_{|\alpha|=|\beta|=m} \int_\Omega a_{\alpha\beta} D^\alpha u \overline{D^\beta u}\, dx| \ge a_2 \sum_{|\alpha|=m} \int_\Omega |D^\alpha u|^2\, dx, \ a_2 > 0.$$

3. Let the coefficients $a_{\alpha\beta}(x)$ be continuous for $|\alpha| = |\beta| = m$. For each point $x_0 \in \Omega$ there is a ball Q_{x_0} with its center at x_0 such that $|a_{\alpha\beta}(x) - a_{\alpha\beta}(x_0)| \le \varepsilon$ for $|\alpha| = |\beta| = m$, where ε is such that the inequality proved in step 2, is true. Choose from the covering of Ω by such balls Q_{x_0} a finite subcovering $Q_1, \dots, Q_N$ and consider a partition of unity $1 \equiv \sum_{j=1}^N \sigma_j^2(x)$ such that $\operatorname{supp} \sigma_j \subset \overline{Q}_j$, $\sigma_j \in C^\infty(\overline{Q}_j)$. Then we obtain

$$\begin{aligned} &|\sum_{\substack{|\alpha|=m\\|\beta|=m}} \int_\Omega a_{\alpha\beta}(x) D^\alpha u \overline{D^\beta u}\, dx| = |\sum_{\substack{|\alpha|=m\\|\beta|=m}} \int_\Omega \sum_{j=1}^N a_{\alpha\beta}(x) \sigma_j^2(x) D^\alpha u \overline{D^\beta u}\, dx| \\ &\ge |\sum_{\substack{|\alpha|=m\\|\beta|=m}} \sum_{j=1}^N \int_\Omega a_{\alpha\beta}(x) D^\alpha \sigma_j u \overline{D^\beta \sigma_j u}\, dx| \\ &- C_1 \int_\Omega \sum_{|\alpha|<m} |D^\alpha u|^2\, dx - C_1 \int_\Omega \sum_{\substack{|\alpha|=m\\|\beta|<m}} |D^\alpha u||D^\beta u|\, dx. \end{aligned}$$

The last integral can be estimated by using the Cauchy inequality:

$$|\int fg\, dx| \le \varepsilon/2 \int |f|^2\, dx + 1/2\varepsilon \int |g|^2\, dx,$$

so that

$$\int_\Omega \sum_{\substack{|\alpha|=m\\|\beta|<m}} |D^\alpha u||D^\beta u|\, dx \le \varepsilon/2 \sum_{|\alpha|=m} |D^\alpha u|^2\, dx + 1/2\varepsilon \int_\Omega \sum_{|\alpha|<m} |D^\alpha u|^2\, dx.$$

We have

$$|\sum_{\substack{|\alpha|=m\\|\beta|=m}}\int_\Omega a_{\alpha\beta}(x)D^\alpha u\overline{D^\beta u}\,dx| \geq |\sum_{\substack{|\alpha|=m\\|\beta|=m}}\sum_{j=1}^N\int_{Q_j} a_{\alpha\beta}(x)D^\alpha\sigma_j u\overline{D^\beta\sigma_j u}\,dx|$$

$$-C_2\int_\Omega\sum_{|\alpha|<m}|D^\alpha u|^2\,dx - \varepsilon\int_\Omega\sum_{|\alpha|=m}|D^\alpha u|^2\,dx.$$

This implies that

$$|\sum_{|\alpha|=|\beta|=m}\int_\Omega a_{\alpha\beta}(x)D^\alpha u\overline{D^\beta u}\,dx| \geq C_3\sum_{|\alpha|=m}\int_\Omega|D^\alpha u|^2\,dx - C_4\int_\Omega\sum_{|\alpha|<m}|D^\alpha u|^2\,dx,$$

where $C_3 > 0$. Using the interpolation inequalities (see Chapter 2, Theorem 66), we obtain the result. □

Theorem 4. *Let L be a strongly elliptic operator and, for $|\alpha| = |\beta| = m$, $a_{\alpha\beta}(x)$ be real continuous functions such that $a_{\alpha\beta} = a_{\beta\alpha}$. Let all other coefficients of L be bounded in Ω. Then the space of solutions to the Dirichlet problem is finite dimensional and this dimension is independent of F. If the Dirichlet problem with $F_\alpha = 0$ has no nontrivial solutions, then this problem has a generalized solution for any $F_\alpha \in L_2(\Omega)$.*

Proof. The Gårding inequality implies that there is k_0 such that

$$(-1)^m\sum_{\substack{|\alpha|=m\\|\beta|=m}}\int_\Omega a_{\alpha\beta}(x)D^\alpha u\overline{D^\beta u}\,dx + k_0\int_\Omega|u(x)|^2\,dx > 0,$$

(this may occur after change of the signs of all $a_{\alpha\beta}$), if $u(x) \in W^m_{2,0}(\Omega)$ and $u(x) \not\equiv 0$. Consider the bilinear form

$$[v,w] = (-1)^m\sum_{\substack{|\alpha|=m\\|\beta|=m}}\int_\Omega a_{\alpha\beta}(x)D^\alpha v\overline{D^\beta w}\,dx + k_0\int_\Omega v\overline{w}\,dx$$

for $v \in W^m_{2,0}(\Omega)$, $w \in W^m_{2,0}(\Omega)$. This bilinear form can be considered as a scalar product in the Hilbert space $W^m_{2,0}(\Omega)$. Using the interpolation inequality (see Chapter 2, Theorem 66), it is easy to see that $[v,v]^{1/2}$ is equivalent to the standard norm in $W^m_{2,0}(\Omega)$. For any $F_\alpha \in L_2(\Omega)$ the expression

$$\sum_{|\alpha|\leq m}\int_\Omega(-1)^{|\alpha|}F_\alpha(x)\overline{D^\alpha\psi(x)}\,dx$$

defines a linear continuous functional of $\overline{\psi}$ in the Hilbert space $W^m_{2,0}(\Omega)$. The Riesz theorem states that this functional can be represented in the form $[AF,\psi]$, where A

is a linear bounded operator transferring F_α in an element of $W^m_{2,0}(\Omega)$. Therefore, the definition of a generalized solution can be written as

$$[u,\psi] = \sum_{\substack{|\alpha|=m\\|\beta|=m}} (-1)^m \int_\Omega a_{\alpha\beta}(x)\overline{D^\alpha\psi}D^\beta u\,dx + k_0\int_\Omega u(x)\overline{\psi(x)}\,dx$$

$$= -\sum_{\substack{|\alpha+\beta|<2m\\|\alpha|\le m,|\beta|\le m}} (-1)^{|\alpha|}\int_\Omega a_{\alpha\beta}(x)\overline{D^\alpha\psi}D^\beta u\,dx + k_0\int_\Omega u(x)\overline{\psi(x)}\,dx$$

$$-\sum_{|\alpha|\le m}\int_\Omega a_\alpha(x)\overline{\psi(x)}D^\alpha u\,dx + \sum_{|\alpha|\le m}\int_\Omega (-1)^{|\alpha|}F_\alpha(x)\overline{D^\alpha\psi}\,dx$$

$$= [AF,\psi] + [Tu,\psi]$$

and therefore

$$u = AF + Tu.$$

It is easy to see that T is a bounded operator in $W^m_{2,0}(\Omega)$. Let us show that it is compact. Let a sequence $u_k \in W^m_{2,0}(\Omega)$, $k = 1,2,\dots$, weakly converge to u_0. We will show that then Tu_k converges strongly to Tu_0. We have

$$[Tu_k,w] = -\sum_{\substack{|\alpha+\beta|<2m\\|\alpha|\le m,|\beta|\le m}} (-1)^{|\alpha|}\int_\Omega a_{\alpha\beta}(x)D^\beta u_k(x)\overline{D^\alpha w(x)}\,dx$$

$$-\int_\Omega \sum_{|\alpha|\le m} a_\alpha(x)D^\alpha u_k\overline{w}\,dx + k_0\int_\Omega u_k\overline{w}\,dx$$

for any $w \in W^m_{2,0}(\Omega)$. Set $w = Tu_k$. Then

$$[Tu_k,Tu_k] = \sum_{\substack{|\alpha+\beta|<2m\\|\alpha|\le m,|\beta|\le m}} \int_\Omega b_{\alpha\beta}(x)D^\alpha u_k\overline{D^\beta Tu_k}\,dx,$$

where $b_{\alpha\beta}(x)$ are bounded functions. We can suppose that the sequences u_k and Tu_k converge to u_0 and Tu_0 respectively, weakly in $W^m_{2,0}(\Omega)$ and strongly in $W^{m-s}_{2,0}(\Omega)$ for any $s > 0$, since the embedding $W^{m-s}_{2,0}(\Omega)$ in $W^m_{2,0}(\Omega)$ is compact. Therefore, we can pass to the limit as $k\to\infty$, if $|\alpha| < m$, $|\beta| < m$. Assume that $|\alpha|$ or $|\beta|$ is equal to m. Let, for instance, $|\beta| = m$. Then $|\alpha| < m$ and

$$\int_\Omega b_{\alpha\beta}(x)D^\alpha u_k\overline{D^\beta Tu_k}\,dx$$

$$= \int_\Omega b_{\alpha\beta}(x)\overline{D^\beta Tu_k}(D^\alpha u_k - D^\alpha u_0)\,dx + \int_\Omega b_{\alpha\beta}(x)\overline{D^\beta Tu_k}D^\alpha u_0\,dx.$$

We can pass to the limit in the first term on the right-hand side as $k \to \infty$ since $\|D^\alpha u_k - D^\alpha u_0\|_{L_2} \to 0$ and in the second one since the sequence $D^\beta T u_k$ is weakly converging in $L_2(\Omega)$ to $D^\beta T u_0$. Therefore, $\lim_{k\to\infty}[Tu_k, Tu_k] = [Tu_0, Tu_0]$, which implies the strong convergence of Tu_k to Tu_0 in $W_{2,0}^m(\Omega)$ (see Theorem 21 of Chapter 1). Therefore, T is a compact operator and the Fredholm theorem is applicable. □

Theorem 5. *Let the coefficients of L be measurable and bounded in Ω, $F_\alpha \in L_2(\Omega)$, $a_{\alpha\beta}(x) = a_{\beta\alpha}(x)$ for $|\alpha| = |\beta| = m$, and $a_{\alpha\beta}(x)$ be real. Suppose also that for all real $v(x) \in C_0^\infty(\Omega)$ the inequality*

$$|\sum_{|\alpha|=|\beta|=m} \int_\Omega a_{\alpha\beta}(x) D^\alpha v D^\beta v \, dx| \geq \gamma \sum_{|\alpha|=m} \int_\Omega |D^\alpha v|^2 \, dx, \ \gamma = \text{const} > 0$$

holds. Then the statement of Theorem 4 holds.

Proof. The proof is similar to that of Theorem 4. It is even simpler, because it does not require the Gårding inequality, using the continuity of the leading coefficients $a_{\alpha\beta}(x)$ with $|\alpha| = |\beta| = m$, but is based on the condition replacing this inequality. □

Note that the condition of Theorem 5 is fulfilled, for example, if $a_{\alpha\beta}$ are real constants and

$$\sum_{|\alpha|=|\beta|=m} a_{\alpha\beta} \xi^\alpha \xi^\beta \geq \gamma \sum_{|\alpha|=m} |\xi^\alpha|^2$$

for all real ξ. In particular, if $m = 1$, then this condition takes the form

$$\sum_{i,j=1}^n a_{ij} \xi_i \xi_j \geq \gamma \sum_{i=1}^n \xi_i^2.$$

Until now we have been considering the Dirichlet problem with zero data. In the general case the problem consists in looking for a function $u(x) \in W_2^m(\Omega)$ satisfying the equality

$$\sum_{\substack{|\alpha|\leq m \\ |\beta|\leq m}} (-1)^{|\alpha|} \int_\Omega a_{\alpha\beta}(x) D^\alpha \psi D^\beta u \, dx$$

$$+ \sum_{|\alpha|\leq m} \int_\Omega a_\alpha(x) \psi(x) D^\alpha u \, dx = \sum_{|\alpha|\leq m} \int_\Omega (-1)^{|\alpha|} F_\alpha(x) D^\alpha \psi \, dx$$

for any $\psi(x) \in W_{2,0}^m(\Omega)$ and such that $u(x) - v(x) \in W_{2,0}^m(\Omega)$, where $v(x) \in W_2^m(\Omega)$ is a given function.

Denoting $z = u - v$, we see that $z(x) \in W^m_{2,0}(\Omega)$ and

$$\sum_{\substack{|\alpha|\le m\\|\beta|\le m}} D^\alpha a_{\alpha\beta}(x)D^\beta z + \sum_{|\alpha|\le m} a_\alpha(x)D^\alpha z$$

$$= \sum_{|\alpha|\le m} D^\alpha F_\alpha(x) - \sum_{\substack{|\alpha|\le m\\|\beta|\le m}} D^\alpha a_{\alpha\beta}(x)D^\beta v - \sum_{|\alpha|\le m} a_\alpha(x)D^\alpha v$$

$$= \sum_{\substack{|\alpha|\le m\\ \alpha\neq 0}} D^\alpha(F_\alpha - \sum_{|\beta|\le m} a_{\alpha\beta}D^\beta v) + (F_0 - \sum_{|\alpha|\le m} a_\alpha(x)D^\alpha v).$$

This equation can be studied in the same way as above.

The creation of the theory of boundary value problems for strongly elliptic equations and the systems of such equations has been initiated by Vishik in [Vi]. The notion of a generalized solution for the Laplace equation $\Delta u = 0$ and for the polyharmonic equation $\Delta^m u = 0$ has been introduced by Sobolev [So]. The existence of generalized solutions to the Laplace and Poisson equations $\Delta u = f$ was proved by Weyl. The Weyl method served as a basis for further investigations.

We emphasize that in the theory of strongly elliptic equations there are no regularity conditions for the boundary of the domain. If the coefficients of the operator and the second term are smooth, then the solution is smooth inside the domain. In what follows we will consider more general equations, which are called elliptic. The theory of boundary problems for such equations was developed only for domains with regular boundaries.

3.2 Elliptic equations

Let us consider the equation

$$L(x, D)u \equiv \sum_{|\alpha|\le m} a_\alpha(x)D^\alpha u = f(x).$$

Here as always, $\alpha = (\alpha_1, \dots, \alpha_n)$, $|\alpha| = \sum_{i=1}^n \alpha_i$, $x = (x_1, \dots, x_n) \in \mathbf{R}^n$, and $D^\alpha = \partial^{|\alpha|}/\partial x_1^{\alpha_1} \dots \partial x_n^{\alpha_n}$. The coefficients $a_\alpha(x)$ are measurable complex-valued functions. The *characteristic polynomial* of the operator L is

$$L_0(x, \xi) = \sum_{|\alpha|=m} a_\alpha(x)\xi^\alpha, \text{ where } \xi^\alpha = \xi_1^{\alpha_1} \dots \xi_n^{\alpha_n}.$$

This is a polynomial of ξ with its coefficients depending on x. If $a_\alpha \not\equiv 0$ for some α with $|\alpha| = m$, then the *order* of L is equal to m. The operator L is *elliptic* at a point x, if

$$L_0(x, \xi) \neq 0 \quad \text{for all} \quad \xi \in \mathbf{R}^n \setminus 0.$$

The operator L is *elliptic in a domain* if it is elliptic at each point of the domain.

Remark that the condition of ellipticity can be rewritten in the form

$$|L_0(x,\xi)| \geq \gamma_1 |\xi|^m,$$

where $\gamma_1 = \text{const} > 0$, since on the unit sphere we have $|L_0(x,\xi)| \geq \gamma_1$ and L_0 is a homogeneous function in ξ of order m. The constant γ_1 is called the *ellipticity constant.* An equation $Lu = f$ is called *elliptic* in a domain Ω, if L is elliptic in Ω.

The ellipticity is invariant with respect to the change of variables with nonvanishing Jacobian. Indeed, if an equation $Lu = f$ is elliptic at a point x_0, then after the change $y_i = g_i(x)$, $i = 1, \ldots, n$, with $\det \|\partial g_i(x)/\partial x_j\| \neq 0$, $g_i(x) \in C^m(\Omega)$ it is transformed in the equation

$$L'u \equiv \sum_{|\alpha|\leq m} b_\alpha(y) D_y^\alpha u = f, \; D_y^\alpha = \frac{\partial^{|\alpha|}}{\partial y_1^{\alpha_1} \ldots \partial y_n^{\alpha_n}},$$

which is elliptic at the point $y_0 = g(x_0)$. To show this we can note that the characteristic polynomial of L' has the form

$$L_0'(y,\eta) = L_0\left(x, \sum_1^n \eta_i \frac{\partial g_i}{\partial x_1}, \ldots, \sum_1^n \eta_i \frac{\partial g_i}{\partial x_n}\right) = \sum_{|\alpha|=m} b_\alpha(x)\eta^\alpha.$$

Set $\xi_k = \sum_1^n \eta_i \partial g_i/\partial x_k$. Since $\det \|\partial g_i/\partial x_k\| \neq 0$, the vanishing of $L_0'(y,\eta)$ at a point (y_0, η_0) with $\eta_0 \neq 0$ is equivalent to the vanishing of $L_0(x,\xi)$ at the corresponding point (x_0, ξ_0), such that $y_0 = g(x_0)$ and

$$\xi_{0k} = \sum_{j=1}^n \eta_{0j} \partial g_j(x_0)/\partial x_k.$$

The Laplace equation is elliptic. Note that all strongly elliptic equations are elliptic, but not inversely.

If $n > 2$ and an equation is elliptic, then its order is even. Indeed, let $L_0(x_0,\xi) \neq 0$ for all $\xi \in \mathbf{R}^n \setminus 0$. Let us consider the polynomial of a variable τ: $p(\tau) = L_0(x_0, \tau, \xi_2, \xi_3, 0, \ldots, 0)$. The ellipticity of the operator L implies that the equation $p(\tau) = 0$ has no real roots, if $\xi_2^2 + \xi_3^2 = 1$, $(\xi_2, \xi_3) \in \mathbf{R}^2$. Let the number of the roots with positive imaginary parts be equal to m_1 and the number of the roots with negative imaginary parts be equal to $m_2 = m - m_1$. Observe that $m_1(\xi_2,\xi_3) = m_2(-\xi_2,-\xi_3)$, since $L_0(x_0,\tau,\xi_2,\xi_3,0) = (-1)^m L_0(x_0,-\tau,-\xi_2,-\xi_3,0)$. The circle $\xi_2^2 + \xi_3^2 = 1$ is connected. If a point (ξ_2,ξ_3) is moving to the point $(-\xi_2,-\xi_3)$ along the circle, then the roots of the equation $p(\tau) = 0$ are changing continuously. Moreover, if a root of this equation had a positive imaginary part for some (ξ_2,ξ_3), then it would have it for all (ξ_2,ξ_3) on the whole circle $\xi_2^2 + \xi_3^2 = 1$, since the imaginary part cannot vanish. Therefore, $m_1(\xi_2,\xi_3) = m_1(-\xi_2,-\xi_3) = m_2(-\xi_2,-\xi_3) = m_2(\xi_2,\xi_3)$ and $m = 2m_1$.

Let $f(x) \in L_2(\Omega)$. A function $u(x) \in W_2^m(\Omega)$ is called a *solution* to the equation $Lu = f$, if this equality is fulfilled almost everywhere.

Lemma 6. *Let Ω be a ball: $\sum_{i=1}^n x_i^2 < R^2$, $\sigma(x) = R^2 - |x|^2$. Then for any $\varepsilon > 0$ there is a constant $C(\varepsilon)$ such that the inequality*

$$\int_\Omega \sigma^{2k} |\nabla_k u|^2 \, dx \leq \varepsilon \int_\Omega \sigma^{2m} |\nabla_m u|^2 \, dx + C(\varepsilon) \int_\Omega |u|^2 \, dx$$

holds for all $u(x) \in W_2^m(\Omega)$, $0 \leq k < m$, $|\nabla_k u|^2 = \sum_{|\alpha|=k} |D^\alpha u|^2$.

Proof. Set

$$Q_\delta = \{x \in \mathbf{R}^n : \ |x_i| < \delta/2, \ i = 1, \ldots, n\}.$$

Let us consider the interpolation inequality (Chapter 2, Theorem 66)

$$\int_{Q_1} |\nabla_k u|^2 \, dx \leq \varepsilon_1 \int_{Q_1} |\nabla_m u|^2 \, dx + C(\varepsilon_1) \int_{Q_1} |u(x)|^2 \, dx.$$

Changing variables $\delta x_i = x_i'$, we obtain that

$$\int_{Q_\delta} \delta^{2k} |\nabla_k u|^2 \, dx' \leq \varepsilon_1 \int_{Q_\delta} \delta^{2m} |\nabla_m u|^2 \, dx' + C(\varepsilon_1) \int_{Q_\delta} |u(x)|^2 \, dx'.$$

Denote x' by x again. Then

$$\int_{Q_\delta} \delta^{2k} |\nabla_k u|^2 \, dx \leq \varepsilon_1 \int_{Q_\delta} \delta^{2m} |\nabla_m u|^2 \, dx + C(\varepsilon_1) \int_{Q_\delta} |u(x)|^2 \, dx.$$

This inequality is true for any cube of size δ. Let at first $R = 1$ and

$$\Omega = \cup_1^\infty \Omega_s,$$

where $\Omega_s = \Omega \cap \{x : \ 1 - 2^{-s+1} < |x| < 1 - 2^{-s}\}$. Each domain Ω_s can be covered by closed cubes with nonintersecting inner parts of sizes equal to $(1/\sqrt{n})2^{-s-1}$ and lying in $\Omega_{s+1} \cup \Omega_s \cup \Omega_{s-1}$. To get such a covering one can consider the partition of the space in equal nonintersecting cubes of sizes $(1/\sqrt{n})2^{-s-1}$ and take those that have a nonempty intersection with Ω_s. Writing the above inequality for each cube of the system and summing these inequalities over all cubes, we obtain that

$$\int_{\Omega_s} 2^{(-s-1)2k} |\nabla_k u|^2 \, dx \leq \varepsilon_1 \sqrt{n} \int_{\Omega_{s+1} \cup \Omega_s \cup \Omega_{s-1}} 2^{(-s-1)2m} |\nabla_m u|^2 \, dx$$

$$+ C(\varepsilon_1) \int_{\Omega_{s+1} \cup \Omega_s \cup \Omega_{s-1}} |u(x)|^2 \, dx.$$

Here Ω_0 is an empty set. Observe that in Ω_s the inequalities $2^{-s} \le 1 - |x| \equiv \sigma_1(x) \le 2^{-s+1}$ hold. Therefore

$$(1/4^{2k}) \int_{\Omega_s} \sigma_1^{2k} |\nabla_k u|^2 \, dx \le \varepsilon_1 \sqrt{n} 2^{-2m} \int_{\Omega_{s+1} \cup \Omega_s \cup \Omega_{s-1}} \sigma_1^{2m} |\nabla_m u|^2 \, dx$$
$$+ C(\varepsilon_1) \int_{\Omega_{s+1} \cup \Omega_s \cup \Omega_{s-1}} |u(x)|^2 \, dx.$$

Thus we have

$$\int_\Omega \sigma_1^{2k} |\nabla_k u|^2 \, dx \le \varepsilon \int_\Omega \sigma_1^{2m} |\nabla_m u|^2 \, dx + C(\varepsilon) \int_\Omega |u(x)|^2 \, dx,$$

if $\varepsilon = (\varepsilon_1/\sqrt{n}) 2^{4k-2m}$.

The statement of Lemma 6 for the sphere Ω defined as $\{x : \sum_1^n x_i^2 < R^2\}$ is proved in the same way. Thus

$$\sum_{|\alpha|=k} \int_\Omega \sigma_R^{2k}(x) |D^\alpha u|^2 \, dx \le \varepsilon \int_\Omega \sum_{|\alpha|=m} \sigma_R^{2m}(x) |D^\alpha u|^2 \, dx$$

$$+ C(\varepsilon) \int_\Omega |u(x)|^2 \, dx,$$

where $\sigma_R(x) = R - (\sum_{i=1}^n x_i^2)^{1/2}$. Since for $|x| \le R$ the inequalities

$$R\sigma_R(x) \le \sigma(x) \le 2R\sigma_R(x)$$

hold, we obtain the result. □

Theorem 7. *Let the coefficients $a_\alpha(x)$ of an elliptic operator L be continuous for $|\alpha| = m$ in the closure of a bounded domain Ω and bounded for $|\alpha| < m$. Then for any subdomain Ω' such that $\Omega' \subset\subset \Omega$ there exists a constant C such that*

$$\|u\|_{W_2^m(\Omega')} \le C[\|Lu\|_{L_2(\Omega)} + \|u\|_{L_2(\Omega)}]$$

for all $u \in W_2^m(\Omega)$. The constant C depends only on Ω, Ω' and $a_\alpha(x)$.

Proof. Let $K_\delta^{x_0} \subset \Omega$ be a ball with its center $x_0 \in \overline{\Omega'}$ of a radius δ. Assume at first that $\operatorname{supp} u \subset K_\delta^{x_0}$. Let us show that there is a δ_0 such that for $\delta \le \delta_0$ the inequality

$$\|u\|_{W_2^m(\mathbf{R}^n)} \le C[\|Lu\|_{L_2(\mathbf{R}^n)} + \|u\|_{L_2(\mathbf{R}^n)}]$$

is true. Let us rewrite our equation in the form

$$L_0(x_0, D)u \equiv \sum_{|\alpha|=m} a_\alpha(x_0) D^\alpha u = F,$$

where

$$F = Lu - \sum_{|\alpha|<m} a_\alpha(x) D^\alpha u - \sum_{|\alpha|=m} [a_\alpha(x) - a_\alpha(x_0)] D^\alpha u.$$

Apply the Fourier transform to both sides of this equality. As result we have

$$i^m L_0(x_0, \xi)\tilde{u}(\xi) = \tilde{F}(\xi).$$

The ellipticity implies that

$$|\xi|^m |\tilde{u}(\xi)| \leq C_0 |\tilde{F}(\xi)|.$$

The integration over all $\xi \in \mathbf{R}^n$ gives in virtue of the Parseval equality the inequality

$$\sum_{|\alpha|=m} \|D^\alpha u\|_{L_2(\mathbf{R}^n)} \leq C\|F(x)\|_{L_2(\mathbf{R}^n)}$$

$$\leq C(\|Lu\|_{L_2(\mathbf{R}^n)} + C' \sum_{|\alpha|<m} \|D^\alpha u\|_{L_2(\mathbf{R}^n)}$$

$$+ \sum_{|\alpha|=m} \max_{x \in K_\delta^{x_0}} |a_\alpha(x) - a_\alpha(x_0)| \|D^\alpha u\|_{L_2(\mathbf{R}^n)}).$$

If $\delta \leq \delta_0$ and δ_0 is so small that $|a_\alpha(x) - a_\alpha(x_0)| \leq 1/4C$ for $x \in K_\delta^{x_0}$, then we have:

$$\frac{3}{4} \sum_{|\alpha|=m} \|D^\alpha u\|_{L_2(\mathbf{R}^n)} \leq C\|Lu\|_{L_2(\mathbf{R}^n)} + C_1 \sum_{|\alpha|<m} \|D^\alpha u\|_{L_2(\mathbf{R}^n)}.$$

Let us apply the interpolation inequality

$$\sum_{|\alpha|<m} \|D^\alpha u\|_{L_2(\mathbf{R}^n)} \leq \varepsilon \sum_{|\alpha|=m} \|D^\alpha u\|_{L_2(\mathbf{R}^n)} + C(\varepsilon)\|u\|_{L_2(\mathbf{R}^n)}.$$

Taking $\varepsilon = 1/4C_1$ we obtain that

$$\sum_{|\alpha|=m} \|D^\alpha u\|_{L_2(\mathbf{R}^n)} \leq C_2\|Lu\|_{L_2(\mathbf{R}^n)} + C_2\|u(x)\|_{L_2(\mathbf{R}^n)}.$$

This inequality is proved provided that *supp* $u(x) \subset K_\delta^{x_0}$. If it is not so, then we can take the function equal to $u_1(x) = u(x)(\delta^2 - |x|^2)^m$ for $|x| < \delta$, $u_1 = 0$ for $|x| > \delta$, $\delta < 1$. Obviously, for $|\alpha| \leq m$ we have

$$D^\alpha u_1 = (\delta^2 - |x|^2)^m D^\alpha u + \sum_{|\beta|<|\alpha|} b_\beta(x) D^\beta u, \ |x| < \delta,$$

where $|b_\beta(x)| \leq C(\delta^2 - |x|^2)^{m-|\alpha|+|\beta|}$. It follows that

$$Lu_1 = (\delta^2 - |x|^2)^m Lu + \sum_{|\alpha|<m} A_\alpha(x) D^\alpha u, \ |x| < \delta,$$

where $|A_\alpha(x)| \le C(\delta^2 - |x|^2)^{|\alpha|}$. The function u_1 satisfies the above proved inequality. Therefore,

$$\sum_{|\alpha|=m} \|D^\alpha u_1\|_{L_2(K_\delta^{x_0})} \le C_2 \|Lu_1\|_{L_2(K_\delta^{x_0})} + C_2 \|u_1(x)\|_{L_2(K_\delta^{x_0})}$$

$$\le C_3 \|Lu\|_{L_2(K_\delta^{x_0})} + C_3 \sum_{|\alpha|<m} \|(\delta^2 - |x|^2)^{|\alpha|} D^\alpha u\|_{L_2(K_\delta^{x_0})}.$$

Using the definition of u_1, we obtain that

$$\sum_{|\alpha|=m} \|(\delta^2 - |x|^2)^m D^\alpha u_1\|_{L_2(K_\delta^{x_0})} \le C_4 \|Lu\|_{L_2(K_\delta^{x_0})}$$

$$+ C_4 \sum_{|\alpha|<m} \|(\delta^2 - |x|^2)^{|\alpha|} D^\alpha u\|_{L_2(K_\delta^{x_0})}.$$

Estimating the latter sum with the help of Lemma 6 and choosing small ε, we obtain that

$$\sum_{|\alpha|=m} \|(\delta^2 - |x|^2)^m D^\alpha u\|_{L_2(K_\delta^{x_0})} \le C_5 \|Lu\|_{L_2(K_\delta^{x_0})} + C_5 \|u\|_{L_2(K_\delta^{x_0})}.$$

This gives the inequality

$$\sum_{|\alpha|=m} \|D^\alpha u\|_{L_2(K_{\delta/2}^{x_0})} \le C_6 \|Lu\|_{L_2(K_\delta^{x_0})} + C_6 \|u\|_{L_2(K_\delta^{x_0})}.$$

Therefore

$$\sum_{|\alpha|\le m} \|D^\alpha u\|_{L_2(K_{\delta/2}^{x_0})} \le C_7 \|Lu\|_{L_2(K_\delta^{x_0})} + C_7 \|u\|_{L_2(K_\delta^{x_0})}.$$

Let us cover $\overline{\Omega'}$ by a finite set of balls $K_{\delta/2}^{x_k}$, $k = 1, \dots, N$, such that $K_\delta^{x_k} \subset \Omega$. Applying for each ball $K_{\delta/2}^{x_k}$ the proved inequality and summing over all $K_{\delta/2}^{x_k}$, we obtain the result. □

3.3 Regularity of solutions

In this section we show that the regularity of a solution of an elliptic equation increases when the regularity of the right-hand term increases.

Lemma 8. *Let* $f(x) \in L_2(\mathbf{R}_+^n)$, $\mathbf{R}_+^n = \{x : \ x \in \mathbf{R}^n, \ x_n > 0\}$, *and* $f(x) = 0$ *for* $x \in \mathbf{R}_+^n$, $|x| > 1$. *Consider the equation*

$$Lu = \sum_{|\alpha|=m} a_\alpha D^\alpha u = f,$$

where a_α are constants, L is an elliptic operator, i.e. $\sum_{|\alpha|=m} a_\alpha (i\xi)^\alpha \neq 0$ for $\xi \in \mathbf{R}^n \setminus 0$. Then there exists a solution $u(x) \in W_2^m(\mathbf{R}_+^n)$ such that

$$\|u(x)\|_{W_2^m(\mathbf{R}_+^n)} \leq C\|f(x)\|_{L_2(\mathbf{R}_+^n)},$$

where C is independent of f.

Proof. We define the function f in the whole space $\mathbf{R}^n$, setting $f(x) = 0$ for $|x| > 1$, $x_n < 0$, and $f(x) = \sum_{|\alpha|<m} c_\alpha x^\alpha$ for $|x| < 1$, $x_n < 0$, where the constants c_α are found from the relations

$$\int_{\mathbf{R}^n} f(x) x^\beta \, dx = 0$$

for all β such that $|\beta| < m$. c_α are defined by the relations

$$\sum_{|\alpha|<m} c_\alpha \int_{\substack{x_n<0 \\ |x|<1}} x^\alpha x^\beta \, dx = -\int_{\substack{x_n>0 \\ |x|<1}} f(x) x^\beta \, dx, \ |\beta| < m.$$

The determinant of this system is the Gramian determinant for the system of linearly independent functions x^α and therefore does not vanish. Therefore, c_α exist and

$$|c_\alpha| \leq C \int_{|x|<1} |f(x)| \, dx \leq C_1 \|f\|_{L_2(\mathbf{R}_+^n)}.$$

The constructed function will also be denoted by f. It is defined in $\mathbf{R}^n$ and

$$\|f\|_{L_2(\mathbf{R}^n)} \leq C\|f\|_{L_2(\mathbf{R}_+^n)}.$$

Set

$$v(\xi) = \tilde{f}(\xi) / \sum_{|\alpha|=m} a_\alpha (i\xi)^\alpha, \ \xi \in \mathbf{R}^n \setminus 0,$$

where $\tilde{f}(\xi) = \frac{1}{(2\pi)^n} \int_{\mathbf{R}^n} f(x) e^{-i\xi x} \, dx$. We see that

$$D_\xi^\alpha \tilde{f}(0) = \frac{1}{(2\pi)^n} \int_{\mathbf{R}^n} (ix)^\alpha f(x) \, dx = 0$$

for all α such that $|\alpha| < m$. Since $\tilde{f}(\xi) \in C^\infty(\mathbf{R}^n)$, the Taylor formula implies that

$$|\tilde{f}(\xi)| \leq C_1 |\xi|^m \max_{\substack{|\xi|\leq 1 \\ |\alpha|=m}} |D^\alpha \tilde{f}(\xi)|$$

for $|\xi| \leq 1$. Therefore

$$|\tilde{f}(\xi)| \leq C_2 |\xi|^m \|f(x)\|_{L_2(\mathbf{R}_+^n)} \ \text{ for } \ |\xi| \leq 1.$$

Since $|\sum_{|\alpha|=m} a_\alpha (i\xi)^\alpha| \geq c_1|\xi|^m$, we have $|\xi^\alpha|\cdot|v(\xi)| \leq c_2\|f(x)\|_{L_2(\mathbf{R}^n_+)}$ for $|\xi| \leq 1$ and

$$|\xi^\alpha|\cdot|v(\xi)| \leq c_3|\tilde{f}(\xi)|$$

for $|\xi| > 1$, $|\alpha| \leq m$. Thus

$$\sum_{|\alpha|\leq m} \int_{\mathbf{R}^n} |\xi|^{2|\alpha|}|v(\xi)|^2\,d\xi \leq c_4\|f(x)\|^2_{L_2(\mathbf{R}^n_+)}.$$

Let a function $u(x) \in L_2(\mathbf{R}^n)$ be such that $\tilde{u} = v(\xi)$, i.e. $u(x)$ is the inverse Fourier transform of $v(\xi)$. Then

$$\|u(x)\|_{W_2^m(\mathbf{R}^n_+)} \leq C\|f(x)\|_{L_2(\mathbf{R}^n_+)}.$$

Besides, we have

$$\sum_{|\alpha|=m} a_\alpha(i\xi)^\alpha v(\xi) = \tilde{f}(\xi).$$

Applying the Fourier transform, we see that $Lu = f$. □

Lemma 9. *Let L satisfy the conditions of Lemma 8, $f(x) \in L_2(\mathbf{R}^n_+)$, and $f(x) \equiv 0$ for $x \in \mathbf{R}^n_+$, $|x| > R$. There exists a solution $u(x) \in W_2^m(\mathbf{R}^n_+)$ of the elliptic equation $Lu = f$ such that*

$$\sum_{0\leq|\alpha|\leq m} R^{|\alpha|}\|D^\alpha u\|_{L_2(\mathbf{R}^n_+)} \leq cR^m\|f(x)\|_{L_2(\mathbf{R}^n_+)}.$$

Proof. The proof follows from Lemma 8 after the change of variables $x' = Rx$. □

Theorem 10. *Let $Lu = f(x) \in W_2^k(\Omega)$, $k = 0, 1, 2, \ldots$, $u(x) \in W_2^m(\Omega)$, and let the coefficients of the operator L have continuous in $\overline{\Omega}$ derivatives up to order k. Then for any subdomain Ω' such that $\overline{\Omega'} \subset \Omega$ we have $u(x) \in W_2^{m+k}(\Omega')$ and*

$$\|u(x)\|_{W_2^{m+k}(\Omega')} \leq C[\|f(x)\|_{W_2^k(\Omega)} + \|u(x)\|_{L_2(\Omega)}],$$

where the constant $C =$ const is independent of u, f.

Proof. We will use induction on k. If $k = 0$, the statement follows from Theorem 7. Now assume that the statement is true for some k, $f(x) \in W_2^{k+1}(\Omega)$ and the coefficients of L have in $\overline{\Omega}$ continuous derivatives up to order $k+1$. Let $\Omega' \subset\subset \Omega''$, $\Omega'' \subset\subset \Omega$. The function $v(x) = D^\beta u$, $|\beta| = k$, satisfies the equation $Lv = D^\beta f + F(x)$, where $|F(x)| \leq C\sum_{\substack{1\leq|\gamma|\leq k\\ |\delta|<m+k}} |D^\gamma a_\alpha(x) D^\delta u(x)|$. Therefore

$$Lv = \varphi(x),\ \varphi(x) \in W_2^1(\Omega).$$

Since by the inductive assumption we have $u \in W_2^{m+k}(\Omega'')$ and

$$\|u(x)\|_{W_2^{m+k}(\Omega'')} \leq C[\|f(x)\|_{W_2^k(\Omega)} + \|u(x)\|_{L_2(\Omega)}],$$

we see that $v \in W_2^{m+k}(\Omega'')$, $\varphi \in W_2^1(\Omega'')$ and

$$\|v(x)\|_{W_2^{m+k}(\Omega'')} \le C[\|f(x)\|_{W_2^k(\Omega)} + \|u(x)\|_{L_2(\Omega)}],$$
$$\|\varphi\|_{W_2^1(\Omega'')} \le C[\|f(x)\|_{W_2^{k+1}(\Omega)} + \|u(x)\|_{L_2(\Omega)}].$$

Consider the function

$$v_{j,h} = \frac{1}{h}[v(x_1, \dots, x_{j-1}, x_j + h, x_{j+1}, \dots, x_n) - v(x)],\ x \in \Omega''.$$

If $|h|$ is sufficiently small, then this function satisfies in Ω'' the equation

$$\begin{aligned} Lv_{j,h} &= \sum_{|\alpha|\le m} a_\alpha(x) D^\alpha v_{j,h} \\ &= \varphi_{j,h}(x) - \sum_{|\alpha|\le m} \frac{a_\alpha(x_1, \dots, x_{j-1}, x_j + h, x_{j+1}, \dots, x_n) - a_\alpha(x)}{h} D^\alpha v_{j,h}. \end{aligned}$$

Let us denote the right-hand side by $F_{j,h}(x)$.Theorems 64, 65 of Chapter 2 imply that $F_{j,h} \in L_2(\Omega')$ and

$$\|F_{j,h}\|_{L_2(\Omega')} \le C[\|f(x)\|_{W_2^{k+1}(\Omega)} + \|u(x)\|_{L_2(\Omega)}].$$

The induction assumption yields that $v_{j,h} \in W_2^m(\Omega')$ and

$$\|v_{j,h}\|_{W_2^m(\Omega')} \le C[\|f(x)\|_{W_2^{k+1}(\Omega)} + \|u(x)\|_{L_2(\Omega)}].$$

Therefore, $\partial v/\partial x_j \in W_2^m(\Omega')$ and

$$\|\partial v/\partial x_j\|_{W_2^m(\Omega')} \le C[\|f(x)\|_{W_2^{k+1}(\Omega)} + \|u(x)\|_{L_2(\Omega)}].$$

Since $v = v_\beta = D^\beta u$, we have $u \in W_2^{m+k+1}(\Omega')$ and

$$\begin{aligned} \|u\|_{W_2^{m+k+1}(\Omega')} &\le \sum_{|\beta|=k} \|v_\beta\|_{W_2^{m+1}(\Omega')} \\ &\le \sum_{|\beta|=k} (\sum_{j=1}^{n} \|\partial v_\beta/\partial x_j\|_{W_2^m(\Omega')} + \|v_\beta\|_{L_2(\Omega')}) + \|u(x)\|_{L_2(\Omega)} \\ &\le C[\|f(x)\|_{W_2^{k+1}(\Omega)} + \|u(x)\|_{L_2(\Omega)}]. \end{aligned}$$

Since Ω' is an arbitrary subdomain of Ω, the proof is complete. □

3.4 Boundary problems for elliptic equations

The boundary value problem for an elliptic equation consists in finding a solution in a bounded domain Ω such that on its boundary $\partial\Omega$ the conditions:

$$B_j u = \sum_{|\alpha|\leq m_j} b_{j\alpha}(x) D^\alpha u = \varphi_j(x),\ j = 1,\dots,m/2,\ m_j < m$$

are satisfied. We assume that the order m of the operator L is even. The coefficients $b_{j\alpha}(x)$ are complex-valued measurable functions on $\partial\Omega$. We show that some relations between the operators L and B_j are necessary. These relations are of algebraic nature and are called the *Lopatinsky condition.*

Let $x_0 \in \partial\Omega$, $\vec{n}$ be the unit outer normal vector to $\partial\Omega$ at the point x_0. Consider the polynomial $L(\tau) = L_0(x_0, i\xi + \tau\vec{n})$, where L_0 is a characteristic polynomial of the operator L, $\xi \neq 0$ is a vector orthogonal to $\vec{n}$. Suppose that the algebraic equation $L(\tau) = 0$ has $m/2$ roots with negative real parts: $\tau_1(\xi),\dots,\tau_{m/2}(\xi)$. Let us set $L_+(\tau) = (\tau - \tau_1(\xi))\dots(\tau - \tau_{m/2}(\xi))$. The Lopatinsky condition requires that the polynomials $B_j(x_0, \xi + \tau\vec{n})$, $j = 1,\dots,m/2$ should be linearly independent modulo $L_+(\tau)$, i.e. no linear combination $\sum_{j=1}^{m/2} C_j B_j(x_0, \xi + \tau\vec{n})$ should be a multiple of $L_+(\tau)$, if $\sum_{j=1}^{m/2} |C_j| \neq 0$ and $\xi \neq 0$. It is easy to check that the Lopatinsky condition is invariant with respect to a change of variables $y = g(x)$, if $g \in C^\infty$ and $\det \|\partial g_i/\partial x_j\| \neq 0$.

Theorem 11. *Let the coefficients $a_\alpha(x)$ with $|\alpha| = m$ be continuous in $\overline{\Omega}$ and other coefficients of the operator L be bounded. Let the coefficients $b_{j\alpha}(x)$ with $|\alpha| = m_j$, $j = 1,\dots,m/2$, be continuous and other coefficients of the boundary operators B_j be bounded on $\partial\Omega$. Suppose that the boundary $\partial\Omega$ is a smooth surface, and a function $u(x) \in W_2^m(\Omega)$ satisfies the equation $Lu = f$ almost everywhere in Ω and the boundary conditions*

$$B_j u = \sum_{|\alpha|\leq m_j} b_{j\alpha}(x) D^\alpha u = \varphi_j(x),\ j = 1,\dots,m/2,$$

on an open subset $\Gamma \subset \partial\Omega$. If a subdomain $\Omega_1 \subset \Omega$ is such that $\overline{\partial\Omega \cap \partial\Omega_1} = \Gamma_1 \subset\subset \Gamma$ and the Lopatinsky condition is satisfied, then

$$\|u(x)\|_{W_2^m(\Omega_1)} \leq C[\|f(x)\|_{L_2(\Omega)} + \sum_{j=1}^{m/2} \|\varphi_j\|_{H^{m-m_j-1/2}(\Gamma)} + \|u(x)\|_{L_2(\Omega)}].$$

Proof. Let $x_0 \in \Gamma_1$, $K_\delta^{x_0}$ be the ball with its center at x_0 of a radius δ, and $\Gamma_\delta^{x_0} = K_\delta^{x_0} \cap \partial\Omega \subset \Gamma$ for all $\delta \leq \delta_0 \leq 1$. Assume at first that $K_\delta^{x_0} \cap \partial\Omega$ is lying in the plane $x_n = 0$, $K_\delta^{x_0} \cap \Omega$ is in the half-space $x_n > 0$, and $u(x) \equiv 0$ if $x_n > 0$, $x \notin K_1^{x_0}$. Moreover, assume now that $a_\alpha(x) \equiv a_\alpha(x_0)$ for $|\alpha| = m$, $a_\alpha(x) \equiv 0$ for $|\alpha| < m$, and $b_{j\alpha}(x) \equiv b_{j\alpha}(x_0)$ for $|\alpha| = m_j$, $b_{j\alpha}(x) \equiv 0$ for $|\alpha| < m_j$, $j = 1, \dots, m/2$. By Lemma 8 there exists a function $v(x)$ such that $Lv = f$ in $\mathbf{R}^n_+$ and

$$\sum_{|\alpha| \le m} \|D^\alpha v\|_{L_2(\mathbf{R}^n_+)} \le C \|f\|_{L_2(\mathbf{R}^n_+)}.$$

The function $w = u - v$ should be a solution to the boundary problem

$$Lw = 0 \quad \text{for} \quad x \in \mathbf{R}^n_+, \quad B_j w = \varphi_j - B_j v = \Phi_j, \text{ if } x_n = 0.$$

Observe that

$$\|\Phi_j\|_{H^{m-m_j-1/2}(\Gamma)} \le \|\varphi_j\|_{H^{m-m_j-1/2}(\Gamma)} + C\|f\|_{L_2(\mathbf{R}^n_+)}.$$

For $x_n > 0$ we apply the Fourier transform with respect to the variables $x' = (x_1, \dots, x_{n-1})$. Denote

$$\hat{w}(\xi, x_n) = \int_{\mathbf{R}^{n-1}} w(x', x_n) e^{-i\xi x'} \, dx', \; \xi = (\xi_1, \dots, \xi_{n-1}).$$

For almost all $x_n > 0$ we have

$$|\xi|^{m-1/2} \hat{w} \in L_2(\mathbf{R}^{n-1}).$$

After the Fourier transformation we have

$$L(x_0, i\xi_1, \dots, i\xi_{n-1}, \frac{d}{dx_n}) \hat{w} = 0,$$

$$B_j(x_0, i\xi_1, \dots, i\xi_{n-1}, \frac{d}{dx_n}) \hat{w}|_{x_n=0} = \hat{\Phi}_j(\xi), \; j = 1, \dots, m/2.$$

We have obtained an ordinary differential equation for $\hat{w}(\xi, x_n)$ with the coefficients not dependent on x_n.

Observe that all solutions of the equation $L(x_0, i\xi, d/dx_n) w = 0$ have the form

$$w = \sum_{j=1}^m \int_{\Gamma_1(\xi)} \frac{C_j(\xi) z^{j-1} e^{izx_n}}{L(x_0, i\xi_1, \dots, i\xi_{n-1}, iz)} dz,$$

where $C_j(\xi)$ are arbitrary functions of the variables $\xi = (\xi_1, \dots, \xi_{n-1}) \in \mathbf{R}^{n-1}$ and $\Gamma_1(\xi)$ is a contour in the complex plane, containing the roots of the equation $L(x_0, i\xi_1, \dots, i\xi_{n-1}, iz) = 0$. Moreover, these solutions are either exponentially increasing or exponentially decreasing functions of x_n. However, for almost all ξ we should have $\int_0^\infty |w|^2 \, dx_n < \infty$, and the function w is bounded for almost all ξ and $x_n > 0$. Therefore, we are interested only in the bounded solutions of this equation.

Bounded solutions w_0 of the ordinary differential equation can be written for $\xi \neq 0$ in the form

$$w_0(\xi, x_n) = \sum_{j=1}^{m/2} \int_{\Gamma(\xi)} \frac{C_j(\xi) z^{j-1}}{L^+(\xi, z)} e^{izx_n}\, dz,$$

where $C_j(\xi)$ are arbitrary functions of $\xi \in \mathbf{R}^{n-1}$,

$$L^+(\xi, z) = (z - \tau_1(\xi)) \dots (z - \tau_{m/2}(\xi))$$

and $\tau_j(\xi)$ are the roots of the equation $L(x_0, i\xi_1 \dots i\xi_{n-1}, i\tau) = 0$ with positive imaginary parts. The contour $\Gamma(\xi)$ is constructed in the following way. As Γ_0 one takes an arbitrary closed smooth curve in the half-plane $\operatorname{Im} z > 0$ of the complex plane z, bounding a domain containing the roots $\tau_1(\xi), \dots, \tau_{m/2}(\xi)$ for $|\xi| = 1$, and sets $\Gamma(\xi) = \{z : \frac{z}{|\xi|} \in \Gamma_0\}$. Let us find $C_j(\xi)$ in such a way that the boundary conditions will be satisfied. Then $C_j(\xi)$ should satisfy the equations

$$\sum_{k=1}^{m/2} \int_{\Gamma(\xi)} \frac{C_k(\xi) z^{k-1} B_j(x_0, i\xi, iz)}{L^+(\xi, z)}\, dz = \hat{\Phi}_j(\xi),\ j = 1, \dots, \frac{m}{2}.$$

Let us show that the determinant of this system

$$\det\Big(\int_{\Gamma(\xi)} \frac{z^{s-1} B_j(x_0, i\xi, iz)}{L^+(\xi, z)}\, dz\Big)_{j,s=1,\dots,m/2}$$

does not vanish for $\xi \in \mathbf{R}^{n-1} \setminus 0$. If it vanishes for some ξ^*, then there are constants $A_1, \dots, A_{m/2}$ not all equal to zero and such that

$$\int_{\Gamma(\xi^*)} \frac{z^{k-1} \sum_{j=1}^{m/2} A_j B_j(x_0, i\xi^*, iz)}{L^+(\xi^*, z)}\, dz = 0,\ k = 1, \dots, m/2.$$

This equality means that the polynomial $P(z) = \sum_{j=1}^{m/2} A_j B_j(x_0, i\xi^*, iz)$ is divisible by $L^+(\xi^*, z)$, i.e. each root of the polynomial L^+ is a root of P. To show this, suppose that $L^+(\xi^*, \tau_{m/2}(\xi^*)) = 0$, but $P(\tau_{m/2}(\xi^*)) \neq 0$. Consider the polynomial $Q(z) = (z - \tau_1(\xi^*)) \dots (z - \tau_{m/2-1}(\xi^*)) = z^{m/2-1} + a_1 z^{m/2-2} + \dots + a_{m/2-1}$. We have

$$0 = \int_{\Gamma(\xi^*)} \frac{Q(z)P(z)}{L^+(z)}\, dz = \int_{\Gamma(\xi^*)} \frac{P(z)}{z - \tau_{m/2}(\xi^*)}\, dz = 2\pi i P(\tau_{m/2}(\xi^*)) \neq 0.$$

This contradiction means that the polynomial P is divisible by L^+, i.e. the polynomials $B_j(x_0, i\xi^*, iz)$ are linearly dependent modulo L^+. However, it contradicts the Lopatinsky condition. Thus the determinant of the system, defining the coefficients $C_k(\xi)$, does not vanish and they are defined uniquely.

Writing $C_k(\xi)$ by the Cramer rule, and using the homogeneity of τ in ξ, we obtain that

$$|C_k(\xi)| \leq C \sum_{j=1}^{m/2} |\hat{\Phi}_j(\xi)| \cdot |\xi|^{-(k+m_j-m/2)}.$$

Now we can estimate the function $w_0(\xi, x)$. The latter estimate implies that

$$\sum_{|\alpha|+j=m} \int_{\mathbf{R}^n_+} |\xi^\alpha|^2 |\frac{\partial^j w_0}{\partial x_n^j}|^2 \, dx_n \, d\xi \leq C_1 \sum_{k=1}^{m/2} \int_{\mathbf{R}^{n-1}} |C_k(\xi)|^2 |\xi|^{2m+2k-m-1} \, d\xi$$

$$\leq C_2 \sum_{j=1}^{m/2} \int_{\mathbf{R}^{n-1}} |\hat{\Phi}_j(\xi)|^2 |\xi|^{2m-2m_j-1} \, d\xi.$$

The function w_0 is a solution of the boundary problem on the positive half-line. Such a solution with a finite value of the integral

$$\sum_{|\alpha|+j=m} \int_{\mathbf{R}^n_+} |\xi|^{2\alpha} |\frac{\partial^j w_0}{\partial x_n^j}|^2 \, dx_n \, d\xi \leq \infty$$

is unique for each $\xi \neq 0$. Thus

$$\sum_{|\alpha|=m} \|D^\alpha w_0\|^2_{L_2(\mathbf{R}^n_+)} \leq C \sum_{1}^{m/2} \|\Phi_j\|^2_{H^{m-m_j-1/2}(\mathbf{R}^{n-1})}.$$

Therefore

$$\sum_{|\alpha|=m} \|D^\alpha u\|_{L_2(\mathbf{R}^n_+)} \leq C_1 (\sum_{1}^{m/2} \|\varphi_j\|_{H^{m-m_j-1/2}(\mathbf{R}^{n-1})} + \|f\|_{L_2(\mathbf{R}^n_+)}).$$

Since the support of u is compact, we have

$$\|u\|_{W_2^m(\mathbf{R}^n_+)} \leq C_2 (\sum_{j=1}^{m/2} \|\varphi_j\|_{H^{m-m_j-1/2}(\mathbf{R}^{n-1})} + \|f\|_{L_2(\mathbf{R}^n_+)}).$$

This estimate is valid for solutions of the boundary value problem in $\mathbf{R}^n_+$, vanishing for $|x| \geq \delta$, $(\delta \leq 1)$. The boundary conditions are fulfilled at $x_n = 0$.

Now let the coefficients $a_\alpha(x)$ be continuous for $|\alpha| = m$ and bounded for $|\alpha| < m$, the coefficients $b_{j\alpha}$ have continuous derivatives up to order $m - m_j + 1$ in $\mathbf{R}^n_+$, and the solution u vanish for $|x| > \delta$. We shall show that then the estimate is true for $\delta \leq \delta_0$, if δ_0 is sufficiently small. The considered equation can be rewritten in the form

$$\sum_{|\alpha|=m} a_\alpha(x_0) D^\alpha u = f(x) + \sum_{|\alpha|=m} [a_\alpha(x_0) - a_\alpha(x)] D^\alpha u - \sum_{|\alpha|<m} a_\alpha(x) D^\alpha u = F(x),$$

and the boundary condition in the form

$$\sum_{|\alpha|=m_j} b_{j\alpha}(x_0)D^\alpha u = \varphi_j(x) + \sum_{|\alpha|=m_j}(b_{\alpha j}(x_0) - b_{j\alpha}(x))D^\alpha u - \sum_{|\alpha|<m_j} b_{j\alpha}(x)D^\alpha u = \Phi_j(x)$$

for $x_n = 0$. Set

$$\gamma(\delta) = \sum_{|\alpha|=m} \max_x |a_\alpha(x) - a_\alpha(x_0)| + \sum_{|\alpha|=m_j} \max_x |b_{j\alpha}(x) - b_{j\alpha}(x_0)|$$
$$M = \sum_{|\alpha|\le m} \max_x |a_\alpha(x)| + \sum_{\substack{|\alpha|\le m_j\\ |\beta|\le m-m_j}} \max_x |D^\beta b_{j\alpha}|, \text{ for } x \in K_\delta^{x_0}.$$

The proved inequality implies the estimate

$$\sum_{|\alpha|\le m} \delta^{|\alpha|-m}\|D^\alpha u\|_{L_2(\mathbf{R}_+^n)} \le C(\|f(x)\|_{L_2(\mathbf{R}_+^n)}$$
$$+\sum_{j=1}^{m/2} \|\varphi_j(x)\|_{H^{m-m_j-1/2}(\mathbf{R}^{n-1})} + \gamma \sum_{|\alpha|=m} \|D^\alpha u\|_{L_2(\mathbf{R}_+^n)} + M \sum_{|\alpha|<m} \|D^\alpha u\|_{L_2(\mathbf{R}_+^n)}).$$

Here the value

$$\|(b_{j\alpha}(x_0) - b_{j\alpha}(x))D^\alpha u\|_{H^{m-m_j-1/2}}$$

has been estimated by $\gamma(\delta)\|D^\alpha u\|_{W_2^{m-m_j}(\mathbf{R}_+^n)}$. If δ is sufficiently small, then γ is also small and we obtain our estimate, if $C\gamma < \frac{1}{2}$, $\delta < 1$, $\delta^{-1} > CM$, using the interpolation inequalities as above.

Let us show how to throw away the assumption on the vanishing of u for $|x| \ge \delta$. Let $u(x)$ be a solution to our equation in $K_\delta^{x_0}$, satisfying the boundary conditions on $\Gamma \cap K_\delta^{x_0}$, and δ be so small that the above estimate holds. Let us consider the function $U(x) = u(x)\cdot\theta(x)$, $\theta(x) = \theta_1(x)^m$, $\theta_1(x) = (\delta^2 - |x - x_0|^2)$ for $|x| \le \delta$, $\theta(x) \equiv 0$ for $|x| > \delta$. Observe that $|D^\alpha\theta| \le C|\theta_1|^{m-|\alpha|}$. The function U satisfies the equation

$$\sum_{|\alpha|=m} a_\alpha(x)D^\alpha U = \theta(x)f + \sum_{\substack{|\beta|<m\\ |\alpha|+|\beta|\le m}} \psi_{\alpha\beta}(x)D^\alpha\theta(x)D^\beta u$$

in $\mathbf{R}_+^n$ and the boundary condition

$$\sum_{|\alpha|=m_j} b_{j\alpha}(x)D^\alpha U = \theta(x)\varphi_j + \sum_{\substack{|\beta|<m_j\\ |\alpha|+|\beta|\le m_j}} \psi_{j\alpha\beta}(x)D^\alpha\theta(x)D^\beta u$$

at $x_n = 0$, where $\psi_{\alpha\beta}$, $\psi_{j\alpha\beta}$ are bounded functions. We have

$$\|U\|^2_{W_2^m(K_\delta^{x_0})} \leq C[\|f\|^2_{L_2(\mathbf{R}^n_+)} + \sum_{j=1}^{m/2} \|\varphi_j\|^2_{H^{m-m_j-1/2}(\mathbf{R}^{n-1})} + \sum_{|\alpha|<m} CM \int_{K_\delta^{x_0}} |\theta_1(x)|^{2(m-|\alpha|)} |D^\alpha u|^2\, dx].$$

Substituting the function U on the left-hand side by θu, we obtain

$$\sum_{|\alpha|=m} \int_{K_\delta^{x_0}} \theta_1^{2m} |D^\alpha u|^2\, dx \leq C_1[\|f\|^2_{L_2(\mathbf{R}^n_+)} + \sum_{j=1}^{m/2} \|\varphi_j\|^2_{H^{m-m_j-1/2}(\mathbf{R}^{n-1})} + \sum_{|\alpha|<m} CM \int_{K_\delta^{x_0}} \theta_1^{2(m-|\alpha|)} |D^\alpha u|^2\, dx].$$

Lemma 6 implies that for small $\varepsilon > 0$

$$\sum_{|\alpha|<m} \int_{K_\delta^{x_0}} \theta_1^{2(m-|\alpha|)} |D^\alpha u|^2\, dx \leq \varepsilon \sum_{|\alpha|=m} \int_{K_\delta^{x_0}} \theta_1^{2m} |D^\alpha u|^2\, dx + C(\varepsilon) \|u\|^2_{L_2(K_\delta^{x_0})}.$$

Therefore

$$\sum_{|\alpha|\leq m} \int_{K_{\delta/2}^{x_0}} |D^\alpha u|^2\, dx \leq C[\|f\|^2_{L_2(K_\delta^{x_0})} + \sum_{j=1}^{m/2} \|\varphi_j\|^2_{H^{m-m_j-1/2}(\mathbf{R}^{n-1})} + \|u\|^2_{L_2(K_\delta^{x_0})}].$$

This estimate is proven when $\Omega \cap K_\delta^{x_0}$ is lying in the half-space $\mathbf{R}^n_+$.

Let now x_0 be an arbitrary point of Γ_1. There is a neighborhood U_{x_0} of this point such that for some map $x = g(y)$ the set $\Gamma_1 \cap U_{x_0}$ lies in the plane $y_n = 0$, and the set $\Omega \cap U_{x_0}$ lies in $\mathbf{R}^n_+$. If the neighborhood U_{x_0} is sufficiently small, then the estimate is valid in the variables y and therefore it is valid in the variables x. So we have

$$\sum_{|\alpha|\leq m} \int_{K_{\delta/2}^{x_0}} |D^\alpha u|^2\, dx$$

$$\leq C[\int_{K_\delta^{x_0}} |f(x)|^2\, dx + \sum_{j=1}^{m/2} \|\varphi_j\|^2_{H^{m-m_j-1/2}(\Gamma\cap K_\delta^{x_0})} + \int_{K_\delta^{x_0}} |u(x)|^2\, dx].$$

Let us cover each point $P \in \Gamma_1$ by an open ball $K^P_{\delta_{P/2}}$ so that the latter estimate holds and $K^P_{\delta(P)} \subset \overline{\Omega}$. We can choose a finite subcovering. Let the union of the balls of this subcovering be U. Then

$$\sum_{|\alpha|\leq m} \int_U |D^\alpha u|^2\, dx \leq C[\|f\|^2_{L_2(\Omega)} + \sum_{j=1}^{m/2} \|\varphi_j\|^2_{H^{m-m_j-1/2}(\Gamma)} + \|u\|^2_{L_2(\Omega)}].$$

Set $U' = \Omega_1 \setminus U$. Then by Theorem 10

$$\sum_{|\alpha|\le m} \int_{U'} |D^\alpha u|^2\,dx \le C[\int_\Omega f^2\,dx + \int_\Omega |u(x)|^2\,dx].$$

Since $\Omega_1 = U \cup U'$, we obtain the result. □

3.5 Smoothness of solutions up to boundary

Now we state a theorem on the regularity of solutions in the closed domain.

Theorem 12. *Let $a_\alpha(x)$ and $b_{j\alpha}(x)$ be infinitely differentiable in $\overline{\Omega}$. Let the Lopatinsky condition be satisfied. If $f(x) \in W^k(\Omega)$, $\varphi_j(x) \in H^{m-m_j+k-1/2}(\Gamma)$ and u is a solution to the boundary value problem, then $u(x) \in W^{m+k}(\Omega)$ and the inequality*

$$\|u(x)\|_{W^{m+k}(\Omega)} \le C[\|f(x)\|_{W^k(\Omega)} + \sum_{j=1}^{m/2} \|\varphi_j\|_{H^{m+k-m_j-1/2}(\Gamma)} + \|u(x)\|_{L_2(\Omega)}]$$

is fulfilled with a constant C independent of u.

Proof. Let $x^0 \in \Gamma$ and $K_\delta^{x_0}$ be a ball with its center at x_0 and of radius δ such that there exists a smooth map $x' = g(x)$ of $K_\delta^{x_0} \cap \Omega$ in $\mathbf{R}_+^n$, mapping $\Gamma \cap K_\delta^{x_0}$ in the plane $x_n = 0$. The forms of the equation and the boundary conditions are the same as in the old coordinates x. Let at first $\Omega = g(K_\delta^{x_0})$ and Ω_1 be a subdomain of Ω such that $\partial\Omega \cap \partial\Omega_1 \subset \{x : x_n = 0\}$.

We will use the induction on k. If $k = 0$, then it is proved in Theorem 10. Suppose that it is true for some k and $f(x) \in W_2^{k+1}(\Omega)$, $\varphi(x) \in H^{k+m-m_j+1/2}(\Gamma)$. Consider the function

$$u_h = [\frac{u(x_{i,h}) - u(x)}{h}],\ h \ne 0,$$

where $1 \le i \le n-1$ and $x_{i,h} = (x_1, \dots, x_{i-1}, x_i + h, \dots, x_n)$. This satisfies the equation

$$\frac{1}{h}[\sum_{|\alpha|\le m} (a_\alpha D^\alpha u)(x_{i,h}) - L(x,D)u(x)] = f_h(x),$$

which can be written in the form

$$L(x,D)u_h(x) = -\sum_{|\alpha|\le m} a_{\alpha h} D^\alpha u(x_{i,h}) + f_h(x) = F(h,x).$$

The boundary conditions for u_h have the form:

$$B_j u_h(x) = -\sum_{|\alpha|\le m_j} b_{j\alpha h}(x) D^\alpha u(x_{i,h}) + \varphi_{jh} = \Phi_j(h,x).$$

By the inductive hypothesis we have

$$\|u_h\|_{W_2^{k+m}(\Omega_1)} \leq C[\|F(h,x)\|_{W_2^k(\Omega)} + \sum_{j=1}^{m/2} \|\Phi_j(h,x)\|_{H^{k+m-m_j-1/2}(\Gamma_1)} + \|u\|_{L_2(\Omega_2)}],$$

where the constant C does not depend on h and Ω_2 is a domain such that $\Omega_1 \subset \Omega_2 \subset \Omega$ and $\partial\Omega_1 \cap \{x : x_n = 0\}$ is embedded compactly in an $(n-1)$-dimensional subdomain of the plane $x_n = 0$ belonging to $\partial\Omega_2$, and $\partial\Omega_2 \cap \{x : x_n = 0\}$ is embedded compactly in an $(n-1)$-dimensional subdomain of the plane $x_n = 0$, belonging to $\partial\Omega$. Theorem 64 of Chapter 2 implies that $\partial u/\partial x_i \in W_2^{k+m}(\Omega_1)$, $i = 1, \dots, n-1$. The equation allows us to express $\partial^m u/\partial x_n^m$ in the terms of other derivatives of u of order m and of the derivatives of lower orders. Thus $\partial^m u/\partial x_n^m \in W_2^{k+m}(\Omega_1)$ and the above estimate holds. Choosing a covering of $\partial\Omega_1$ as in Theorem 11, we obtain our statement. □

Corollary 13. *Let $\Omega_1 \subset \Omega$ and either $\Omega_1 \subset\subset \Omega$ or $f \equiv \varphi_j \equiv 0$. Let $u(x) \in W_2^m(\Omega)$ be a solution to the equation in Ω, satisfying the boundary conditions on Γ_1 (if $\Gamma_1 = \overline{\partial\Omega \cap \partial\Omega_1}$ is nonempty). Let the Lopatinsky condition be satisfied. Then*

$$\sup_{\Omega_1} |u(x)| \leq C\|u\|_{L_2(\Omega)}.$$

Proof. It follows from Theorem 12 and Theorem 69 of Chapter 2, since $C(\Omega_1) \subset W_2^{m+k}(\Omega)$ for $m + k > 2n$. □

The Lopatinsky condition was introduced in [Lop]. A similar condition for $m = 2$ there is also in [Sha]. The theory of general boundary value problems for elliptic equations of any order was developed in the papers Schechter [Sche1], Agmon-Douglis-Nirenberg [ADN], and Peetre [Pee]. The complete statement of the theory can be found in [Höl]. A different presentation of the theory of elliptic boundary value problems can be found in the books of Morrey [Mo] and Nečas [Ne]. See also the book of Bers-John-Schechter [BJS] and the article of Agranovich and Vishik [AgV].

Chapter 4

Spectral Properties of Elliptic Operators

4.1 Variational principle

Let Ω be a bounded domain in $\mathbf{R}^n$ and Γ its boundary. Consider the integrals

$$D[\varphi] = \int_\Omega \sum_{j=1}^n (\frac{\partial \varphi}{\partial x_j})^2 \, dx, \; H[\varphi] = \int_\Omega \varphi(x)^2 \, dx$$

for real-valued functions φ from $H^1(\Omega)$.

Theorem 1. *A function $u(x)$ giving the minimal value λ to the functional $D[\varphi]$ in the set K of functions $\varphi \in H_0^1(\Omega)$ such that $H[\varphi] = 1$ is an eigenfunction of the Laplace operator, i.e. $-\Delta u = \lambda u$ in Ω.*

Proof. It is clear that the function u also gives the minimal value to the functional $F[\varphi] = D[\varphi]/H[\varphi]$ in the space $H_0^1(\Omega)$. Therefore, for any $\varphi \in H_0^1(\Omega)$ we have $F(u + \varepsilon\varphi) \geq 0$ for $\varepsilon > 0$, and $\frac{d}{d\varepsilon} F[u + \varepsilon\varphi]|_{\varepsilon=0} = 0$. Calculating the derivative, we obtain the equality

$$\int_\Omega [\sum_{j=1}^n \frac{\partial u}{\partial x_j} \frac{\partial \varphi}{\partial x_j} - \lambda u \varphi] \, dx = 0,$$

where $\lambda = D[u]/H[u]$. This equality means that $u(x)$ is a generalized solution of the Dirichlet problem. Theorem 10 of Chapter 3 implies that $u \in C^\infty(\Omega)$. After integrating by parts we see that

$$\int_\Omega [\Delta u + \lambda u] \cdot \varphi(x) \, dx = 0.$$

Since this is true for all $\varphi \in C_0^\infty(\Omega)$, the proof is complete. □

Let us show that the minimal value discussed in Theorem 1 is always attained.

Theorem 2. *Let K be the set of functions $\varphi \in H_0^1(\Omega)$ such that $H[\varphi] = 1$. There exists a function $u \in K$ such that $D[u] \le D[v]$ for all $v \in K$.*

Proof. Let $m = \inf_{\varphi \in K} D[\varphi]$. Consider a sequence $\{\varphi_k\} \subset K$ such that $D[\varphi_k] \to m$. Direct calculation shows that

$$D[\frac{\varphi_k - \varphi_l}{2}] + D[\frac{\varphi_k + \varphi_l}{2}] = \frac{1}{2}D[\varphi_k] + \frac{1}{2}D[\varphi_l].$$

Let $\varepsilon > 0$ be an arbitrary number, and k and l be so large that $D[\varphi_k] \le m + \varepsilon,\ D[\varphi_l] \le m + \varepsilon$. Observe that

$$D[\frac{\varphi_k + \varphi_l}{2}] \ge mH[\frac{\varphi_k + \varphi_l}{2}].$$

The sequence $\{D\varphi_k\}$ has a limit and therefore is bounded. Thus the sequence $\{\varphi_k\}$ is bounded in $H_0^1(\Omega)$ and compact in $L_2(\Omega)$. Choosing a subsequence, we can assume that it is converging in $L_2(\Omega)$. From the obvious equality

$$H[\frac{\varphi_k + \varphi_l}{2}] = \frac{1}{2}H[\varphi_k] + \frac{1}{2}H[\varphi_l] - H[\frac{\varphi_k - \varphi_l}{2}] = 1 - H[\frac{\varphi_k - \varphi_l}{2}]$$

it follows that $H[\frac{\varphi_k + \varphi_l}{2}] \ge 1 - \frac{\varepsilon}{m}$ for big k and l, so that $D[\frac{\varphi_k + \varphi_l}{2}] \ge m - \varepsilon$. Therefore

$$D[\frac{\varphi_k - \varphi_l}{2}] \le \frac{1}{2}(m + \varepsilon) + \frac{1}{2}(m + \varepsilon) - (m - \varepsilon) = 2\varepsilon,$$

if k and l are large enough. This means that the sequence $\{\varphi_k\}$ is a Cauchy sequence in $H_0^1(\Omega)$ and therefore it converges to a function $u \in H_0^1(\Omega)$. Since $D[\varphi_k] \to D[u]$, the proof is complete. □

The variational approach allows us to also find other eigenfunctions. Suppose that functions $u_j \in H_0^1(\Omega)$ are given for $j = 1, \dots, k-1$ such that

$$H[u_j] = 1, \int_\Omega u_j u_i \, dx = 0,\ i = 1, \dots, k-1, \int_\Omega (\sum_{l=1}^n \frac{\partial u_j}{\partial x_l}\frac{\partial \varphi}{\partial x_l} - \lambda_j u_j \varphi)\, dx = 0$$

for all $\varphi \in H_0^1(\Omega)$. Consider the linear space $L(k)$ of functions from $H_0^1(\Omega)$ satisfying the conditions

$$\int_\Omega v \cdot u_j \, dx = 0,\ j = 1, \dots, k-1.$$

Theorem 3. *Let $K(k)$ be the set of functions $\varphi \in L(k)$ such that $H[\varphi] = 1$. This set is closed and contains a function u, giving the minimal value λ_k to the Dirichlet integral. This function satisfies in Ω the equation $-\Delta u_k = -\lambda_k u_k$.*

Proof. If $v_m \in L(k)$ and $v_m \to v$ in $H_0^1(\Omega)$, then by the Friedrichs inequality $\|v_m - v\|_{L_2(\Omega)} \le C\|v_m - v\|_{H_0^1(\Omega)}$ and $v_m \to v$ in $L_2(\Omega)$. Thus $\int v \cdot u_j \, dx = 0$

for $j = 1, \dots, k-1$ and the limit function belongs to $L(k)$, i.e. $L(k)$ is closed. Therefore, $K(k)$ is closed, too.

Consider a sequence $\{v_j\}$ of functions from $L(k)$ such that

$$H[v_j] = 1, \lim_{j\to\infty} D[v_j] = \lambda_k.$$

The sequence $\{v_j\}$ is bounded in $H_0^1(\Omega)$ and therefore is compact in $L_2(\Omega)$. Thus we can suppose that $v_j \to u$ in $L_2(\Omega)$. The equality

$$H[\frac{v_j+v_i}{2}] = \frac{1}{2}H[v_j] + \frac{1}{2}H[v_i] - H[\frac{v_j-v_i}{2}] = 1 - H[\frac{v_j-v_i}{2}]$$

implies that for big i and j

$$H[\frac{v_j+v_i}{2}] > 1 - \varepsilon/\lambda_k.$$

Since for all $\varphi \in L(k)$ we have $D[\varphi] \geq \lambda_k H[\varphi]$ and $(v_i+v_j)/2 \in L(k)$, it follows that

$$D[\frac{v_i+v_j}{2}] \geq \lambda_k H[\frac{v_i+v_j}{2}] > \lambda_k - \varepsilon.$$

The relations

$$D[\frac{v_i-v_j}{2}] = \frac{1}{2}D[v_i] + \frac{1}{2}D[v_j] - D[\frac{v_i+v_j}{2}], D[v_i] < \lambda_k + \varepsilon, \ D[v_j] < \lambda_k + \varepsilon$$

yield the inequality

$$D[\frac{v_i-v_j}{2}] < \frac{1}{2}(\lambda_k+\varepsilon) + \frac{1}{2}(\lambda_k+\varepsilon) - (\lambda_k-\varepsilon) = 2\varepsilon,$$

i.e. the sequence $\{v_j\}$ is converging in $H_0^1(\Omega)$ to a function u_k. Since $L(k)$ is closed, $u_k \in L(k)$. Moreover, $H[u_k] = 1, \ D[u_k] = \lambda_k$. If $\varphi \in L(k)$, then

$$\frac{d}{d\varepsilon}\frac{D(u_k+\varepsilon\varphi)}{H(u_k+\varepsilon\varphi)} = 0 \ \text{ at } \ \varepsilon = 0,$$

because $\varepsilon = 0$ is the minimum point of this function. Calculating the derivative, we see that

$$\int_\Omega (\sum_{i=1}^n \frac{\partial u_k}{\partial x_i}\frac{\partial \varphi}{\partial x_i} \cdot H[u_k] - u_k \cdot \varphi \cdot D[u_k])\, dx = 0.$$

Since $D[u_k] = \lambda_k H[u_k]$, we have

$$\int_\Omega (\Delta u_k(x) + \lambda_k u_k(x))\varphi(x)\, dx = 0$$

for all $\varphi \in L(k)$. This means that $\Delta u_k(x) + \lambda_k u_k(x) = \sum_{j=1}^{k-1} C_j u_j(x)$ with some constants C_j. Multiplying this equality by $u_i(x)$ with $i \in [1,k]$ and integrating over Ω, we obtain that

$$C_i \int_\Omega u_j(x)^2\, dx = \int_\Omega (\Delta u_k(x) + \lambda_k u_k(x)) u_i(x)\, dx$$

$$= \int_\Omega u_k(x) \Delta u_i(x)\, dx = \lambda_i \int_\Omega u_k(x) u_i(x)\, dx = 0.$$

Therefore, $C_i = 0$ for $i = 1, \dots, k-1$ and $-\Delta u_k = \lambda_k u_k$ in Ω. □

Let us show that the number of the eigenvalues is infinite.

Theorem 4. *There exists a nondecreasing infinite sequence of numbers $\{\lambda_k\}$ and a sequence of functions $\{u_k\} \in H_0^1(\Omega)$ such that*

$$\int (\sum_{j=1}^n \frac{\partial u_k}{\partial x_j} \frac{\partial \varphi}{\partial x_j} - \lambda_k u_k \varphi)\, dx = 0$$

for all $\varphi \in H_0^1(\Omega)$ and $\int_\Omega u_k u_l\, dx = 0$ for $k \neq l$, $\int_\Omega u_k^2(x)\, dx = 1$,

$$\int_\Omega \sum_{i=1}^n \frac{\partial u_k}{\partial x_i} \frac{\partial u_l}{\partial x_i}\, dx = 0 \text{ for } k \neq l, \quad \int_\Omega \sum_{i=1}^n (\frac{\partial u_k}{\partial x_i})^2\, dx = \lambda_k.$$

Moreover, $\lim_{k\to\infty} \lambda_k = +\infty$.

Proof. Theorem 2 implies the existence of a number λ_1 and of a function u_1 such that

$$\int (\sum_{j=1}^n \frac{\partial u_1}{\partial x_j} \frac{\partial \varphi}{\partial x_j} - \lambda_1 u_1 \varphi)\, dx = 0 \quad \text{for} \quad \varphi \in H_0^1(\Omega),$$

and $H[u_1] = 1$, $D[u_1] = \lambda_1$. Consider the one-dimensional subspace S_1, generated by the function u_1.

Theorem 3 implies the existence of a number λ_2 and of a function u_2 such that

$$\Delta u_2 + \lambda_2 u_2 = 0 \text{ in } \Omega; \; H[u_2] = 1, \; D[u_2] = \lambda_2, \; \int u_1 u_2\, dx = 0.$$

Of course, then $\lambda_2 \geq \lambda_1$. Considering the subspace S_2 generated by the functions u_1, u_2, by Theorem 3 we can find λ_3 and u_3 and so on. We can continue this process and find the infinite sequences $\{\lambda_j\}$ and $\{u_j\}$. We have

$$\lambda_1 \leq \lambda_2 \leq \cdots \leq \lambda_k \leq \cdots.$$

The set of the functions u_j is orthonormed in $L_2(\Omega)$.

Assume now that $\lambda_j \leq C$ for all $j = 1, 2, \dots$. Then $D[u_k] \leq C$ and $H(u_k) = 1$. Therefore

$$\|u_k\|^2_{H_1(\Omega)} \leq D[u_k] + H[u_k] \leq C + 1.$$

This allows us to extract from $\{u_k\}$ a subsequence $\{u_{k_i}\}$ converging in $L_2(\Omega)$ and such that $H[u_{k_i} - u_{k_j}] \to 0$. However this is impossible, since $H[u_{k_i} - u_{k_j}] = H[u_{k_i}] - 2\int u_{k_i}u_{k_j}\,dx + H[u_{k_j}] = 2$. Thus the sequence $\{\lambda_k\}$ cannot be bounded and $\lim_{k\to\infty} \lambda_k \to +\infty$. □

Now we will consider the *Courant principle*, which allows us to immediately define the kth eigenvalue and the kth eigenfunction without finding the preceding eigenfunctions.

Theorem 5 (Courant). *Let $v_1(x), \dots, v_{k-1}(x)$ be a set of functions continuous in Ω. Let $\mu(v_1, \dots, v_{k-1})$ be the minimal value of $D[\varphi]$ in the class $L(k)$ of functions from $H_1(\Omega)$, satisfying the conditions $H[\varphi] = 1$, $\int_\Omega v_j(x)\varphi(x)\,dx = 0$, $j = 1, \dots, k-1$. Then the kth eigenvalue λ_k is equal to $\sup_{v_1,\dots,v_{k-1}} \mu(v_1, \dots, v_{k-1})$.*

Proof. By Theorem 4 the number λ_k coincides with $\mu(v_1, \dots, v_{k-1})$, if one takes as v_j the jth eigenfunction for $j = 1, \dots, k-1$. Therefore it suffices to verify the inequality $\mu(v_1, \dots, v_{k-1}) \le \lambda_k$ for an arbitrary set of functions $v_1, \dots, v_{k-1}$. In turn to see this, it suffices to find a function w_k such that $H[w_k] = 1$, $\int v_j w_k\,dx = 0$ for $j = 1, \dots, k-1$ and $D[w_k] \le \lambda_k$. Let us look for w_k in the form $w_k = \sum_{j=1}^k \alpha_j \varphi_j$, where $\varphi_1, \dots, \varphi_k$ are the k first eigenfunctions of the problem and $H[\varphi_j] = 1$. Since $\int \varphi_i \varphi_j\,dx = 0$ for $i \ne j$, we have $H[w_k] = \sum_{j=1}^k \alpha_j^2 = 1$. The conditions $\int_\Omega w_k v_i\,dx = 0$ give the system of equations

$$\sum_{j=1}^k \alpha_j \int \varphi_j v_i(x)\,dx = 0,\ i = 1, \dots, k-1.$$

This is a linear system of $k-1$ equations for k unknown values $\alpha_1, \dots, \alpha_k$. Such a system always has a solution $\alpha_1^0, \dots, \alpha_k^0$ such that $\sum_{j=1}^k {\alpha_j^0}^2 = 1$. The mutual orthogonality of φ_j implies that

$$D[\sum_{j=1}^k \alpha_j^0 \varphi_j] = \sum_{j=1}^k {\alpha_j^0}^2 D[\varphi_j] = \sum_{j=1}^k {\alpha_j^0}^2 \lambda_j.$$

However $\lambda_j \le \lambda_k$ for $j = 1, \dots, k-1$. Therefore, $D[\sum_{j=1}^k \alpha_j^0 \varphi_j] \le \lambda_k \sum_{j=1}^k {\alpha_j^0}^2 = \lambda_k$. □

As a simple corollary of this theorem we have the following.

Theorem 6. *Let $\alpha > 0$ and let there exist a set of linearly independent functions $\varphi_1, \dots, \varphi_k$ from $H_0^1(\Omega)$ such that*

$$H[\varphi_j] = 1,\ D[\varphi_j] < \alpha,\ j = 1, \dots, k.$$

Then the number N of the eigenvalues λ_j such that $\lambda_j < \alpha$ is not less than k.

Inversely, let there exist a set of linearly independent functions $\varphi_1, \dots, \varphi_k$ from $H_0^1(\Omega)$ such that $H[\varphi_j] = 1$ and $D[u] \ge \alpha$, if $u \in H^1(\Omega)$ and $\int u\varphi_j\,dx = 0$

for all $j = 1, \dots, k$. Then the number N of the eigenvalues λ_j such that $\lambda_j < \alpha$ does not exceed k.

Now let

$$Lu = \sum_{\substack{|\alpha|\le m\\ |\beta|\le m}} D^\alpha(a_{\alpha\beta}(x)D^\beta u)$$

be a linear symmetric elliptic operator of order $2m$ and

$$(Lu,u) = \int_\Omega \sum_{\substack{|\alpha|\le m\\ |\beta|\le m}} a_{\alpha\beta}(x)D^\alpha u \cdot \overline{D^\beta u(x)}\,dx \ge c_0 \sum_{|\alpha|=m} \int_\Omega |D^\alpha u|^2\,dx$$

for all $u \in C_0^\infty(\Omega)$, $c_0 = \text{const} > 0$.

Set

$$H[u] = \int_\Omega |u|^2\,dx,\ D[u] = \operatorname{Re}(Lu,u).$$

Theorems 1–6 proved above are true for the eigenfunctions of the Dirichlet problem:

$$Lu = \lambda u \text{ in } \Omega,$$
$$u = 0,\ \frac{\partial u}{\partial \nu} = 0, \dots, \frac{\partial^{m-1} u}{\partial \nu^{m-1}} = 0 \text{ on } \Gamma.$$

These can be proved by a simple modification of the corresponding proofs above. We leave this to the readers as a simple exercise.

Let us show that the first eigenvalue of the Dirichlet problem for the Laplace operator is simple and the first eigenfunction is positive in Ω. To prove this we need the following lemma.

Lemma 7. *Let u be a nontrivial nonnegative solution of the equation*

$$-\Delta u = \lambda u \text{ in } \Omega$$

for some $\lambda > 0$. Then $u(x)$ does not vanish in Ω.

Proof. Suppose that u vanishes at $x_0 \in \Omega$. Let B be a ball of radius R lying in Ω, having x_0 on its boundary and such that $u(x) > 0$ in B.

The radius R can be assumed to be so small that $2\sqrt{\lambda+1}R < \pi$. Let the center of B coincide with the origin and $k = \sqrt{\lambda+1}$. The function $v(x) = u(x)/\cos\ kr$ satisfies in B the equation

$$Lv \equiv \Delta v - 2k\tan(kr)\frac{\partial v}{\partial r} - v(\lambda + \frac{k(n-1)}{r}\tan(kr)) = 0.$$

Moreover, $v(x) > 0$ in B and $v(x_0) = 0$. Then $\nabla v(x_0) = 0$. Let Q be the set of points of B whose distance from x_0 is less than $R/4$. Set $w(x) = e^{-\gamma r} - e^{-\gamma R}$, where $\gamma > 0$. Then $w(x_0) = 0$, $w(x) > 0$ in Q, and

$$Lw = e^{-\gamma r}[\gamma^2 - \frac{\gamma(n-1)}{r} + 2k\gamma\cdot\tan(kr) - (1 - e^{-\gamma(R-r)})(\lambda + \frac{k(n-1)}{r}\tan(kr))].$$

Since $\frac{3}{4}R < r < R$ in Q, we can enlarge γ and get the inequality $Lw > 0$ in Q.

The function $z = v - \varepsilon w$ is positive on the boundary of Q outside the point x_0, if $\varepsilon > 0$ is sufficiently small. Since $L(v - \varepsilon w) < 0$, the function z cannot have a negative minimal value in Q. So $z > 0$ in Q and $\partial z(x_0)/\partial r \leq 0$. However, then

$$\frac{\partial v(x_0)}{\partial r} \leq \varepsilon \frac{\partial w(x_0)}{\partial r} = -\varepsilon\gamma e^{-\gamma R} < 0.$$

But this inequality contradicts the equality $\nabla v(x_0) = 0$, obtained above. □

Theorem 8. *The eigenfunction of the Dirichlet problem, corresponding to the first eigenvalue, has a constant sign in Ω. This function is unique up to a constant factor.*

Proof. Let $\lambda_1 = \inf D(u)/H(u)$ and u_0 be a function giving the minimal value. Since $|u_1| \in H_0^1(\Omega)$, if $u_1 \in H_0^1(\Omega)$ and $D(|u_0|) = D(u_0)$, $H(|u_0|) = H(u_0)$, the function $|u_0|$ is an eigenfunction corresponding to the eigenvalue $\lambda = \lambda_1$. By Lemma 7 this function does not vanish anywhere in Ω.

Now let v be another eigenfunction corresponding to λ_1. Substituting v by the function $v_1 = v + ku_0$ we can have the function v_1 be orthogonal to u_0, i.e.

$$\int v_1(x)u_0(x)\,dx = 0.$$

Since $u_0(x) > 0$ in Ω, the function $v_1(x)$ changes its sign in Ω. The equality $D(|v_1|)/H(|v_1|) = \lambda_1$ implies that the function $|v_1|$ is also an eigenfunction. Since $|v_1|$ vanishes at an inner point of Ω, Lemma 7 implies that $v_1 \equiv 0$. □

Theorem 9. *Let $\Omega' \subset \Omega$ and $\lambda_1(\Omega)$, $\lambda_1(\Omega')$ be the first eigenvalues of the Dirichlet problem for the domains Ω and Ω', respectively. If* $\operatorname{meas}(\Omega \setminus \Omega') > 0$, *then $\lambda_1(\Omega) < \lambda_1(\Omega')$.*

Proof. The variational principle implies that $\lambda_1(\Omega) \leq \lambda_1(\Omega')$. Suppose that $\lambda_1(\Omega) = \lambda_1(\Omega') = \lambda$. Then the eigenfunction of the Dirichlet problem in the domain Ω' with the eigenvalue λ expanded by zero values on $\Omega \setminus \Omega'$ is an eigenfunction in Ω. However, it vanishes at some points of Ω, contrary to Lemma 7. □

4.2 The spectrum of a self-adjoint operator

Here we consider some properties of unbounded operators. We recall some notions introduced in Sections 1.8–1.10, slightly modifying them.

Let H be a Hilbert space and A a linear operator with a dense domain $D(A)$. The set of all Ax with $x \in D(A)$ is denoted by $R(A)$ or $\operatorname{im} A$.

Let D^* be a linear manifold in H such that $v \in D^*$ if there is an element $w \in H$ such that

$$(w, u) = (v, Au) \text{ for all } u \in D(A).$$

It is clear that there exists at most one such element w: if $(w_1, u) = (w_2, u)$ for all $u \in D(A)$ then $w_1 = w_2$ because $D(A)$ is dense in H. Therefore, one can set $w = A^*v$ and $D(A^*) = D^*$. The operator A^* is called *adjoint* to A. Its domain can be not dense in H, so that $(A^*)^*$ is not always defined. Moreover, if it is defined, it can be different from A.

An operator A is called *self-adjoint* if $A^* = A$.

If $A^* = A$, then $(Au, v) = (u, Av)$ for all $u, v \in D(A)$. However, this equality is not yet sufficient for self-adjointness, since $D(A^*)$ can be wider than $D(A)$. If $(Au, v) = (u, Av)$ for all $u, v \in D(A)$, the operator A is called *symmetric.*

An operator B is *an extension* of A if $D(A) \subset D(B)$ and $A = B$ on $D(A)$. If an operator A has a unique self-adjoint extension, it is said to be *essentially self-adjoint.*

Theorem 10. *If a linear operator A with a dense domain is invertible, then A^* is also invertible and $(A^*)^{-1} = (A^{-1})^*$.*

Proof. Since A^{-1} is bounded, the operator $(A^{-1})^*$ is also bounded. If $g \in D(A^*)$ and $h \in H$, then

$$((A^{-1})^*A^*g, h) = (A^*g, A^{-1}h) = (g, AA^{-1}h) = (g, h).$$

Therefore, $(A^{-1})^*A^*g = g$.

On the other hand, if $f \in H, g \in D(A)$, then

$$(A^*(A^{-1})^*f, g) = ((A^{-1})^*f, Ag) = (f, A^{-1}Ag) = (f, g)$$

and by the definition of the operator A^* we have

$$(A^{-1})^*f \in D(A^*) \text{ and } A^*(A^{-1})^*f = f$$

for $f \in H$. Thus $(A^*)^{-1}$ exists and is equal to $(A^{-1})^*$. □

Theorem 11. *If an operator A is self-adjoint, bounded, and invertible, then A^{-1} is self-adjoint.*

Proof. It is easy to see that the range of A is dense in H. Indeed, if it is not so, then there exists an element $y \in E$, $y \neq 0$, such that $(Ax, y) = 0$ for all $x \in E$. Therefore, $y \in D_{A^*}$, $(x, A^*y) = 0$, and $Ay = A^*y = 0$. But this is impossible, since A is invertible. Since the range of A is dense in H, we have

$$(A^{-1})^* = (A^*)^{-1} = A^{-1}.$$

Indeed, the domain of A^{-1} is dense in H and the operator $(A^{-1})^*$ is well defined. Namely, $x \in D_{(A^{-1})^*}$, if $(x, A^{-1}y) = (x^*, y)$ for some x^* and all $y \in D_{A^{-1}}$. Then by definition, $x^* = (A^{-1})^*y$. This formula is true for all $x \in D_{(A^{-1})^*}$, $y \in D_A$. Put $A^{-1}y = z$. Then $(x, z) = (x^*, Az)$ for all $x \in D_{(A^{-1})^*}$, $z \in D_A$. This means that $x = A^*x^*$, i.e. $x^* = (A^*)^{-1}x$. Therefore, $(A^*)^{-1} = (A^{-1})^*$ on the domain of $(A^{-1})^*$. Now let x belong to the domain of $(A^*)^{-1}$, which is dense in H, i.e. $x = A^*v$. Then

$$(x, y) = (v, Ay)$$

for all $y \in H$. In particular, $(A^*v, A^{-1}y) = (v, y)$, if y belongs to $D_{A^{-1}}$.

We can see that $A^*v \in D_{(A^{-1})^*}$, i.e. $x \in D_{(A^{-1})^*}$. Thus $(A^*)^{-1} = (A^{-1})^*$. Since $A^* = A$, the proof is complete. □

A resolvent set $\rho(A)$ of an operator A is a set of complex numbers λ for which the operator $(\lambda I - A)^{-1}$ is defined and bounded. The set $\mathbf{C} \setminus \rho(A)$ is called the *spectrum* of the operator A and is denoted as $\sigma(A)$. If the map $\lambda I - A$ is not one-to-one, then λ belongs to the *point spectrum* $\sigma_p(A)$. Otherwise, if this map is one-to-one and $\text{im}(\lambda I - A)$ is dense in H, but does not coincide with H, we say that λ belongs to the *continuous spectrum* $\sigma_c(A)$.

One also often uses the notion "*essential spectrum*". It consists of the points $\lambda \in \sigma(A)$ that are not isolated eigenvalues of a finite multiplicity. It is easy to see that λ belongs to the essential spectrum $\sigma_{\text{ess}}(A)$ if and only if there exists an infinite sequence $\{v_j\}$ of mutually orthogonal unit vectors such that $\|Av_j - \lambda v_j\| \to 0$ as $j \to \infty$.

A linear operator A is *closed* if the conditions $x_n \in D(A)$, $x_n \to x$, $Ax_n \to y$ imply that $x \in D(A)$ and $Ax = y$.

Theorem 12. *A linear closed symmetric operator in a Hilbert space H is self-adjoint if and only if its spectrum is lying on the real axis.*

Proof. Let $\lambda = \alpha + i\beta$, α and β be real, and $\beta \neq 0$. Since the operator A is symmetric, we have

$$\begin{aligned}
\|(\lambda I - A)\varphi\|^2 &= ((\alpha + i\beta)\varphi - A\varphi, (\alpha + i\beta)\varphi - A\varphi) \\
&= \|\alpha\varphi - A\varphi\|^2 + i\beta(\varphi, (\alpha + i\beta)\varphi - A\varphi) - i\beta(\alpha\varphi - A\varphi, \varphi) \\
&= \|\alpha\varphi - A\varphi\|^2 + \beta^2\|\varphi\|^2 \geq \beta^2\|\varphi\|^2.
\end{aligned}$$

Therefore, the set $R(\lambda I - A)$ is closed.

If $A^* = A$, then $\ker(\lambda I - A) = \emptyset$ for $\lambda = \alpha + i\beta$ with $\beta \neq 0$, i.e. $\sigma(A) \subset \mathbf{R}$.

On the other hand, let $\sigma(A) \subset \mathbf{R}$. Then for $\beta \neq 0$ the operator $\lambda I - A$, $\lambda = \alpha + i\beta$, is invertible. If $\psi \in D(A^*)$, then there is an element $\varphi \in D(A)$ such that

$$A\varphi - \lambda\varphi = A^*\psi - \lambda\psi,$$

i.e. $A^*(\psi - \varphi) = \lambda(\psi - \varphi)$. By Theorem 10 the operator $A^* - \lambda I$ is invertible, so that $\psi = \varphi$ and $\psi \in D(A)$. Therefore, $D(A^*) = D(A)$, i.e. $A^* = A$. □

Theorem 13. *Let A be a linear symmetric operator in a Hilbert space H.*

If $D(A) = H$ or $R(A) = H$, then A is self-adjoint.

If $D(A)$ is dense in H and $R(A) = H$, then there exists an inverse operator A^{-1}, which is symmetric.

Proof. If $D(A) = H$, then from $D(A) \subset D(A^*)$ it follows that $D(A^*) = H$ and therefore $A = A^*$.

If $Au = 0$, then $(u, Av) = (Au, v) = 0$ for $v \in D(A)$. Therefore, the element u is orthogonal to $R(A)$. If $R(A) = H$, then $u = 0$, i.e. the operator A is invertible. Theorem 10 implies that A^{*-1} exists and is equal to $(A^{-1})^*$. Since $D(A) \subset D(A^*)$, we have $D(A^{-1}) \subset D(A^{*-1}) = D((A^{-1})^*)$, and thus the operator A^{-1} is symmetric.

If $R(A) = H$, then $D(A^{-1}) = H$ and A^{-1} is self-adjoint, i.e. $(A^{-1})^* = A^{-1}$. But then $A^{-1} = A^{*-1}$, i.e. $A = A^*$. □

Theorem 14. *If A is a linear closed symmetric operator in a Hilbert space H, then its spectrum can be one of the following four sets in* $\mathbf{C}$*:*

(a) *The whole complex plane;*
(b) *The closure of the upper half-plane;*
(c) *The closure of the lower half-plane;*
(d) *A subset of the real axis.*

Proof. Let $\lambda = \alpha + i\beta$, α and β be real, and $\beta \neq 0$. Since the operator A is symmetric, we have

$$\begin{aligned}\|(\lambda I - A)\varphi\|^2 &= ((\alpha + i\beta)\varphi - A\varphi, (\alpha + i\beta)\varphi - A\varphi)\\ &= \|\alpha\varphi - A\varphi\|^2 + i\beta(\varphi, (\alpha + i\beta)\varphi - A\varphi) - i\beta(\alpha\varphi - A\varphi, \varphi)\\ &= \|\alpha\varphi - A\varphi\|^2 + \beta^2\|\varphi\|^2 \geq \beta^2\|\varphi\|^2\end{aligned}$$

if $\varphi \in D(A)$. So $\operatorname{im}(\lambda I - A)$ is a closed subset in H. Let us show that its orthogonal complement is $\ker(\overline{\lambda} I - A^*)$.

Indeed, if $A^*w = \overline{\lambda}w$ and $u = \lambda v - Av$, then

$$(u, w) = (\lambda v - Av, w) = (v, \overline{\lambda}w) - (v, A^*w) = 0.$$

Inversely, if $(u, w) = 0$ for all elements u of the form $u = \lambda v - Av$, $v \in D(A)$, then

$$0 = (u, w) = (\lambda v - Av, w) = (v, \overline{\lambda}w) - (v, A^*w)$$

and since $D(A)$ is dense in H, it follows that $A^*w = \overline{\lambda}w$.

Let us verify now that the dimensions of the subspaces $\ker(\lambda I - A^*)$ and $\ker((\lambda + \mu)I - A^*)$ coincide for small $|\mu|$. Indeed, let $u \in \ker((\lambda + \mu)I - A^*)$, $v \in \ker(\lambda I - A^*)$, and $\|u\| = \|v\| = 1$, $(u, v) = 0$. Then as was proved above, $u \in R(\overline{\lambda} I - A)$, so that $u = \overline{\lambda}w - Aw$ for some $w \in D(A)$. Therefore

$$0 = (((\lambda + \mu)I - A^*)u, w) = (u, (\overline{\lambda} I - A)w) + \mu(u, w) = \|u\|^2 + \mu(u, w),$$

which is impossible for $|\mu| < |\beta|$, if $\|w\| \leq \|u\|/|\beta|$. Therefore, for small $|\mu|$ the space $\ker((\lambda + \mu)I - A^*)$ does not contain elements orthogonal to $\ker(\lambda I - A^*)$. Thus

$$\dim\ker((\lambda + \mu)I - A^*) \leq \dim\ker(\lambda I - A^*).$$

Similar arguments show that

$$\dim \ker(\lambda I - A^*) \leq \dim \ker((\lambda + \mu)I - A^*),$$

so that these dimensions are equal if $|\mu| < |\beta|$. Therefore, $\dim \ker(\lambda I - A^*)$ is locally constant in the upper half-plane and in the lower half-plane.

However, for $\operatorname{Im} \lambda \neq 0$ the operator $\lambda I - A$ has a bounded inverse operator defined everywhere if $\ker(\lambda I - A^*) = \emptyset$. Thus the upper half-plane of $\mathbf{C}$ is always lying either in the spectrum of the operator A or in the set of its regular points and the same is true for the lower half-plane. □

Corollary 15. *If A is a closed symmetric operator, then the codimension of the subspace $R(\lambda I - A)$ is constant for λ in the upper (or in the lower) half-plane.*

Corollary 16. *A closed symmetric operator A is self-adjoint if and only if $\ker(\lambda I - A^*) = \emptyset$ for $\operatorname{Re} \lambda \neq 0$ (or, which is the same, if $R(\lambda I - A) = H$ for $\operatorname{Re} \lambda \neq 0$).*

4.3 The Friedrichs extension

A symmetric operator A is called *semi-bounded from below* if

$$(Au, u) \geq m\|u\|^2 \quad \text{for all} \quad u \in D(A).$$

If $m = 0$, then A is called *positive.* If $m > 0$, then it is *strictly positive.*

Theorem 17 (Neumann). *A symmetric operator A semi-bounded from below is extendible to a self-adjoint operator $\tilde{A}$ semi-bounded with the same constant m. The equality $R(\tilde{A}) = H$ is valid if and only if the operator A is strictly positive.*

Proof. Since A can be replaced by $A - (m-1)I$, we can without loss of generality suppose that $(Au, u) \geq \|u\|^2$. Consider the scalar product $[u, v] = (Au, v)$ in $D(A)$. It is clear that $[u, v] = \overline{[v, u]}$, $[u, u] > 0$, if $u \neq 0$. Denote by H_A the completion of $D(A)$ with respect to the norm $\|u\|_A = [u, u]^{1/2}$.

Since $[u, u] \geq \|u\|^2$, the space H_A can be identified with a subspace in H. Let $H_1 = D(A^*) \cap H_A$. It is evident that $D(A) \subset H_1 \subset D(A^*)$. Let $\tilde{A}$ be the restriction of A^* on H_1.

Verify that $\tilde{A}$ is a self-adjoint operator. To this end by Theorem 13 it suffices to show that $R(\tilde{A}) = H$ and $\tilde{A}$ is symmetric.

Let $v \in H$. Since $|(u, v)| \leq \|u\| \cdot \|v\| \leq \|v\| \cdot \|u\|_A$ for $u \in D(A)$, the form (u, v) is a continuous functional defined on a dense subset of H_A. Therefore, by the Riesz theorem there is an element $v^* \in H_A$ such that

$$(u, v) = [u, v^*] = (Au, v^*), \quad u \in D(A).$$

Thus $v^* \in D(A^*) \cap H_A$ and $A^* v^* = \tilde{A} v^* = v$. This means that $R(\tilde{A}) = H$.

Let us show now that $\tilde{A}$ is symmetric and $(\tilde{A}u, u) \geq (u, u)$. If $u, v \in D(\tilde{A}) = H_A \cap D(A^*)$, then there are sequences $\{u_n\}, \{v_m\} \in D(A)$ such that $\|u - u_n\|_A \to 0$, $\|v - v_m\|_A \to 0$. So

$$\lim_{m,n\to\infty} (Au_n, v_m) = \lim_{n\to\infty} (Au_n, v) = \lim_{n\to\infty} (u_n, \tilde{A}v) = (u, \tilde{A}v).$$

On the other hand,

$$\lim_{m,n\to\infty} (Au_n, v_m) = \lim_{m,n\to\infty} (u_n, Av_m) = \lim_{m\to\infty} (u, Av_m) = \lim_{m\to\infty} (\tilde{A}u, v_m) = (\tilde{A}u, v).$$

Therefore, $(\tilde{A}u, v) = (u, \tilde{A}v)$, $u, v \in D(\tilde{A})$, i.e. A is symmetric.

Analogously,

$$(\tilde{A}u, u) = \lim_{n\to\infty} (\tilde{A}u, u_n) = \lim_{n\to\infty} (u, Au_n) = \lim_{n\to\infty} [u, u_n] = [u, u] \geq \|u\|^2. \qquad \square$$

The Neumann theorem gives one of the possible extensions of an operator A. There exist other self-adjoint extensions of A. The constructed extension is called a *"rigid" Friedrichs extension.*

The question remains: Does any symmetric operator have a self-adjoint extension? In general, the answer is negative. However, for a differential symmetric operator a self-adjoint extension does always exist.

4.4 Examples of linear unbounded operators

We present below several examples.

Example 18. Consider on $[0, 1]$ the operator d/dt defined on the set of absolutely continuous functions. Its domain is dense in $L_2([0, 1])$. The sequence $f_m(t) = e^{imt}/m$ is converging to 0 in $L_2([0, 1])$, and $\|(d/dt)f_m(t)\|_{L_2(0,1)} = 1$. Therefore, this operator is unbounded.

Example 19. Consider the differential expression

$$L(D) = \sum_{|\alpha|=2m} a_\alpha D^\alpha$$

such that a_α are real constants. Suppose that the differential expression $L^*(D) = \sum_{|\alpha|=2m}(-1)^{|\alpha|} a_\alpha D^\alpha$ coincides with L. L defines an operator on $C_0^\infty(\Omega)$, where Ω is an arbitrary domain in $\mathbf{R}^n$, which can coincide with the whole $\mathbf{R}^n$. We do not assume that the boundary $\partial\Omega$ is smooth. L defines a linear unbounded operator $L_2(\Omega) \to L_2(\Omega)$ with its domain dense in $L_2(\Omega)$. The Green formula implies that this operator is symmetric, i.e. $(Lu, v)_{L_2(\Omega)} = (u, L^*v)_{L_2(\Omega)}$ for any $u \in C_0^\infty(\Omega)$, $v \in C_0^\infty(\Omega)$.

The symmetric differential operator L with constant coefficients has a closed symmetric extension. Let L_0 denote the closure of the constructed operator. It is called the *minimal* operator, generated by the differential expression L. The operator adjoint to L_0 is called the *maximal* operator, generated by the differential expression L. It is easy to see that $L_0 = L_0^{**}$.

Let now an operator L be *elliptic*, i.e. the polynomial $L(i\lambda)$ does not vanish in $\mathbf{R}^n \setminus 0$, and let Ω be a bounded domain in $\mathbf{R}^n$. Then the Parseval formula implies that the inequality

$$(Lu, u) \geq 0, \; u \in C_0^\infty(\Omega)$$

holds (perhaps after substituting L for $-L$). Moreover

$$(Lu, u) \geq b(u, u), \; \forall u \in C_0^\infty(\Omega), \; b = \text{const} > 0$$

and even

$$(Lu, u) \geq b_1(u, u)_{W_2^m(\Omega)}, \; \forall u \in C_0^\infty(\Omega), \; b_1 = \text{const} > 0.$$

It is easy to see that the domain of the operator L_0 coincides with $W_{2,0}^{2m}(\Omega)$. (We assume that the domain Ω is bounded, otherwise this is not always true.) The domain of the operator L_0^* is the set of functions $u(x)$ such that $u(x) \in L_2(\Omega)$, $Lu \in L_2(\Omega)$, and $u \in W_2^{2m}(\Omega')$ for any subdomain Ω' such that $\overline{\Omega'} \subset \Omega$.

The minimal operator L_0 has self-adjoint extensions. We have considered above one of them, called the *Friedrichs extension.* This extension has been defined as the operator $u \mapsto L_F u$ with its domain consisting of the functions from $W_{2,0}^m(\Omega)$ such that $Lu \in L_2(\Omega)$.

If the coefficients a_α of the operator L are smooth functions of x, then the operator L_0 and its extension can be defined analogously.

For an operator with nonsmooth coefficients, the Friedrichs extension is defined by using the bilinear form

$$B[u, v] = \int_\Omega \sum_{\substack{|\alpha| \leq m \\ |\beta| \leq m}} a_{\alpha\beta}(x) D^\alpha u \overline{D^\beta v} \, dx.$$

Let the coefficients $a_{\alpha\beta}(x)$ be bounded measurable functions such that $B[u, v] = \overline{B[v, u]}$, $B[u, u] \geq \gamma \|u\|^2_{W_{2,0}^m(\Omega)}$ for any $u(x) \in W_{2,0}^m(\Omega)$, where $\gamma > 0$ is a constant independent of u. Let D_B be a set of functions $u \in W_{2,0}^m(\Omega)$ such that there is a function $f(x) \in L_2(\Omega)$, for which $B[u, v] = \int_\Omega f\overline{v} \, dx$ when $v \in W_{2,0}^m(\Omega)$. Define the operator L_F on D_B, setting $L_F u = f$. This operator is self-adjoint. In the case of smooth coefficients $a_{\alpha\beta}(x)$ it coincides with the Friedrichs extension of the operator generated by the differential expression $L = \sum_{|\alpha| \leq m, |\beta| \leq m} (-1)^{|\beta|} D^\beta a_{\alpha\beta}(x) D^\alpha$. The embedding theorems imply that the operator L^{-1} is compact as an operator $L_2(\Omega) \to L_2(\Omega)$ and therefore its spectrum is discrete. Moreover, this spectrum is positive.

Example 20. Let $\Omega = \mathbf{R}^n$, and $L = \Delta$ be the Laplace operator. Then *the operator L_0 is self-adjoint.* To prove this we have to show that the domain of the maximal operator coincides with D_{L_0}. Let $u \in D_{L_{\max}}$, i.e. $u \in L_2(\mathbf{R}^n)$, $\Delta u \in L_2(\mathbf{R}^n)$. Applying the Fourier transform we see that $u \in W_2^2(\mathbf{R}^n)$. Let $\theta(x) \equiv 1$ for $|x| < 1$, $\theta(x) \equiv 0$ for $|x| > 2$, $\theta(x) \in C^\infty(\mathbf{R}^n)$. Set $u_k = \theta(x/k)u(x)$. It is obvious that $\Delta u_k \to \Delta u$ in $L_2(\mathbf{R}^n)$ as $k \to \infty$ and $u_k \to u$ in $L_2(\mathbf{R}^n)$. Each function $u_k \in D_{L_0}$. Since the operator L_0 is closed, we see that $u(x) \in D_{L_0}$.

Example 21. Let $\Omega = \mathbf{R}^n$, L be a differential expression with constant coefficients such that $L(D) = L^*(D)$, where L^* is the formally adjoint operator, and $V(x)$ be a bounded real function. Then *the minimal operator generated by* $L + V$ *is self-adjoint.* Indeed, let u belong to the domain of the maximal operator. Then $u \in L_2(\mathbf{R}^n)$, $(L+V)u \in L_2(\mathbf{R}^n)$, and therefore $Lu \in L_2(\mathbf{R}^n)$. We can show, as was done, in Example 20, that there is a sequence of functions $u_k(x)$ from $W_{2,0}^m(\mathbf{R}^n)$, converging in $L_2(\mathbf{R}^n)$ to u and such that the sequence $\{Lu_k\}$ converges to Lu. Then the sequence $\{Lu_k + V(x)u_k\}$ converges in $L_2(\mathbf{R}^n)$ to $Lu + V(x)u$, i.e. the domains of the maximal and the minimal operators coincide.

Example 22. The Sturm-Liouville operator

$$Lu = -(p(x)y')' + q(x)y, \ 0 < x < \infty,$$

where the functions $p(x)$, $p'(x)$, $q(x)$ are continuous and bounded, $p(x) \geq p_0 > 0$, $q(x) \geq q_0 = \text{const}$, considered on the set of functions of $W_2^2([0,1])$ such that $y(0) = 0$, $y(1) = 0$, is semi-bounded.

Example 23. Let $\Omega \subset \mathbf{R}^n$ be a bounded domain. Consider the operator $u \mapsto \Delta u = \sum_{i=1}^n \partial^2 u/\partial x_i^2$ in the space $C_0^\infty(\Omega)$. This is an unbounded operator: $L_2(\Omega) \to L_2(\Omega)$. Denote it as Δ_0. Let us show that Δ_0 *has a closure.* Let $u_k(x) \in C_0^\infty(\Omega)$, $u_k \to 0$, $\Delta u_k \to g$ in $L_2(\Omega)$. We shall show that then $g = 0$.

Let $v(x) \in W_{2,0}^1(\Omega)$ be a generalized solution to the equation $\Delta v = g$. Then $(u_k - v)$ is a generalized solution from $W_{2,0}^1(\Omega)$ to the equation $\Delta(u_k - v) = \Delta u_k - g = f_k$. The definition of generalized solution implies that

$$\int_\Omega \sum_{i=1}^n \frac{\partial \psi}{\partial x_i} \frac{\partial (u_k - v)}{\partial x_i} \, dx = -\int_\Omega f_k \psi \, dx, \quad \text{for } \psi \in W_{2,0}^1(\Omega).$$

Setting $\psi = u_k - v$, we obtain that

$$\begin{aligned}\int_\Omega |\nabla(u_k - v)|^2 \, dx &\leq \int_\Omega |f_k| \, |u_k - v| \, dx \\ &\leq \Big(\int_\Omega |f_k|^2 \, dx\Big)^{1/2} \Big(\int_\Omega |u_k - v|^2 \, dx\Big)^{1/2}.\end{aligned}$$

Using the Friedrichs inequality we obtain that

$$\int_\Omega |\nabla(u_k - v)|^2 \, dx \leq C \int_\Omega |f_k|^2 \, dx \to 0$$

as $k \to \infty$. Estimating $\int_\Omega |u_k - v|^2 \, dx$ with the help of the Friedrichs inequality, we see that $u_k \to v$ in $L_2(\Omega)$. Therefore, $v = 0$ and thus $g = 0$.

Let us denote the closure of Δ_0 as $\Delta_{\min}$. It is a closed symmetric operator and its domain is the set $W_2^2(\Omega)\cap W_{2,0}^1(\Omega)$. Indeed, if $u \in D_{\Delta_{\min}}$, then there exists a sequence $u_k \in C_0^\infty(\Omega)$ such that $\|u-u_k\|_{L_2} \to 0$, $\|\Delta u-\Delta u_k\|_{L_2} \to 0$ as $k \to \infty$. Therefore, the integral

$$\int_\Omega |\nabla(u-u_k)|^2\,dx = -\int_\Omega (\overline{u}-\overline{u_k})(\Delta u-\Delta u_k)\,dx$$

tends to 0. Thus $u \in W_{2,0}^1(\Omega)$. Integrating by parts we see that

$$\int_{\mathbf{R}^n} \sum_{|\alpha|=2} |D^\alpha(u_k-u_{k+p})|^2\,dx = \int_{\mathbf{R}^n} |\Delta(u_k-u_{k+p})|^2\,dx \to 0$$

as $k \to \infty$. Here the functions u_k are defined outside Ω by zero values. Therefore, the sequence u_k is compact in $W_2^2(\Omega)$ and its limit is $u \in W_2^2(\Omega)$.

Let now $\Delta_{\max}$ be the Laplace operator, defined on the set of functions u such that $u \in L_2(\Omega)$, $\Delta u \in L_2(\Omega)$, where Δ is calculated as a sum of generalized derivatives. We will show that $\Delta^*_{\max} = \Delta_{\max}$. Let $u \in D_{\Delta_{\min}}$, $v \in D_{\Delta_{\max}}$. Then $\int_\Omega(\Delta u)v\,dx = \lim_{k\to\infty}\int(\Delta u_k)v\,dx$, $u_k \in C_0^\infty$. Therefore, $\int_\Omega(\Delta u)v\,dx = \lim_{k\to\infty}\int u_k\Delta v\,dx$, where Δv is the sum of the generalized derivatives of $v(x)$ of second order, so that $\Delta v \in L_2(\Omega)$. Since $u_k \to u$ in $L_2(\Omega)$, we have

$$\int_\Omega (\Delta u)v\,dx = \int_\Omega u(\Delta v)\,dx,$$

i.e. $D_{\Delta_{\max}} \subset D_{\Delta^*_{\min}}$.

Now let $v \in D_{\Delta^*_{\min}}$. This means that for all $u \in D_{\Delta_{\min}}$

$$\int_\Omega v\Delta u\,dx = \int_\Omega wu\,dx,\ w \in L_2(\Omega).$$

Let $v_1 \in W_{2,0}^1(\Omega)$ be a generalized solution to the equation $\Delta v_1 = w$. Then

$$\int_\Omega v_1\Delta u\,dx = -\int_\Omega \sum_{i=1}^n \frac{\partial v_1}{\partial x_i}\frac{\partial u}{\partial x_i}\,dx.$$

This equality follows from the Green formula for $u \in C_0^\infty(\Omega)$ and holds for $u \in D_{\Delta_{\min}}$ by the definition of $\Delta_{\min}$. Similarly, the definition of generalized solution implies that

$$-\int_\Omega \sum_{i=1}^n \frac{\partial v_1}{\partial x_i}\frac{\partial u}{\partial x_i}\,dx = \int_\Omega wu\,dx.$$

Thus $\int(v-v_1)\Delta u\,dx = 0$ for $u \in C_0^\infty(\Omega)$ and $\Delta(v-v_1) = 0$ in Ω, i.e. $\Delta v = w \in L_2(\Omega)$.

By the Friedrichs inequality the operator $-\Delta_{\min}$ is positively defined. Thus it has self-adjoint extensions. We can describe one of them, corresponding to the Dirichlet conditions. The domain of this operator is the set of u such that $u \in W_{2,0}^1(\Omega)$, $u \in W_2^2(\Omega')$, for any $\Omega' \subset\subset \Omega$ and $\Delta u \in L_2(\Omega)$.

Example 24. Let us consider the spectrum of the operator $-\Delta$ in the case $\Omega = \mathbf{R}^n$. This operator is positive, since for $u \in C_0^\infty(\mathbf{R}^n)$ we have

$$-\int_\Omega u\Delta u\, dx = \int_\Omega |\nabla u|^2\, dx \geq 0.$$

If $\lambda \notin \overline{\mathbf{R}}_+$, then the operator $-\Delta - \lambda I$ is invertible:

$$(-\Delta - \lambda I)^{-1} f = (2\pi)^{-n} \int \tilde{f}(\xi)(|\xi|^2 - \lambda)^{-1} e^{ix\xi}\, d\xi.$$

Let us show that each point $\lambda > 0$ belongs to the essential spectrum of the operator $(-\Delta)$. For this we will construct a noncompact in $L_2(\mathbf{R}^n)$ sequence of functions $u_k(x)$ from $C_0^\infty(\mathbf{R}^n)$ such that $\|\Delta u_k - \lambda u_k\|_{L_2(\mathbf{R}^n)} \to 0$. Set for this $u_k = e^{i\sqrt{\lambda}x_1}\theta_k(x - x_k)\cdot \varepsilon_k$, where $\theta_k \equiv 1$ for $|x| < k$, $\theta_k \equiv 0$ for $|x| > 2k$, $|\nabla\theta_k| \leq Ck^{-1}$, $|\Delta\theta_k| \leq Ck^{-2}$, $\theta_k \in C_0^\infty(\mathbf{R}^n)$, and x_k are chosen in such a way that $\operatorname{supp}\theta_k$ and $\operatorname{supp}\theta_i$, $i < k$, have no common points, $|x_k| > k^2$. The constant ε_k is such that $\|u_k\|_{L_2(\mathbf{R}^n)} = 1$, so that $\varepsilon_k = O(k^{-n/2})$. The sequence u_k is not compact in $L_2(\mathbf{R}^n)$. However

$$\|\Delta u_k - \lambda u_k\|^2_{L_2(\mathbf{R}^n)} \leq Ck^{-2}\varepsilon_k^2 \cdot k^n.$$

Since the sequence $\varepsilon_k^2 k^n$ is bounded, we see that

$$\|\Delta u_k - \lambda u_k\|_{L_2(\mathbf{R}^n)} \to 0$$

as $k \to \infty$. Thus the semi-axis $\lambda \geq 0$ consists of points of the essential spectrum. Observe that a point $\lambda \geq 0$ cannot belong to the discrete spectrum, since applying the Fourier transform to the equation $-\Delta u = \lambda u$, $u \in L_2(\mathbf{R}^n)$, we see readily that $u = 0$.

4.5 Self-adjointness of the Schrödinger operator

Now we discuss some conditions for the Schrödinger operator to be self-adjoint.

Theorem 25. *Let A be a self-adjoint operator and B a symmetric one such that $D(B) \supset D(A)$ and*

$$\|Bu\| \leq q\|Au\| + C\|u\|, \quad u \in D(A),\ q < 1.$$

Then the operator $A + B$ is essentially auto-adjoint.

Proof. Let $\beta > 0$. Since

$$\|(A + i\beta I)u\|^2 = \|Au\|^2 + \beta^2\|u\|^2, \quad u \in D(A),$$

we get, substituting $u = (A + i\beta I)^{-1}v$, that

$$\|v\|^2 = \|A(A + i\beta I)^{-1}v\|^2 + \beta^2\|(A + i\beta I)^{-1}v\|^2$$

and hence

$$\|A(A + i\beta I)^{-1}\| \leq 1, \quad \|(A + i\beta I)^{-1}\| \leq \beta^{-1}.$$

Since by condition

$$\|B(A+i\beta I)^{-1}v\| \le q\|A(A+i\beta I)^{-1}v\| + C\|(A+i\beta I)^{-1}v\|,$$

we obtain that

$$\|B(A+i\beta I)^{-1}v\| \le (q+C\beta^{-1})\|v\|.$$

If β is so big that $q+C\beta^{-1} < 1$, then the point -1 is regular for the operator $Q = B(A+i\beta I)^{-1}$. Thus $R(I+Q) = H$. The equality

$$(I+Q)(A+i\beta I)u = (A+B+i\beta I)u, \quad u \in D(A),$$

implies that $R(A+B+i\beta I) = H$. Similarly one can verify that $R(A+B-i\beta I) = H$. Corollary 16 implies that the operator $A+B$ is self-adjoint. □

Theorem 26. *Let $L_{\min}$ be the minimal operator for $L = -\Delta$ in $L_2(\mathbf{R}^n)$ and $L_{\max}$ be its maximal operator, so that $D(L_{\min}) = C_0^\infty(\mathbf{R}^n), D(L_{\max}) = H^2(\mathbf{R}^n)$. The operator $L_{\max}$ is self-adjoint. The operator $L_{\min}$ is essentially self-adjoint and its self-adjoint extension is $L_{\max}$.*

Proof. The self-adjointness of the operator $L_{\max}$ follows from its representation in the form $\mathcal{F}^{-1}|\xi|^2\mathcal{F}$, where $\mathcal{F}$ is the Fourier transform.

Show that $L^*_{\min} = L_{\max}$. Let $u \in D(L^*_{\min})$. Then

$$(-\Delta v, u) = (L_{\min}v, u) = (v, L^*_{\min}u), \quad v \in C_0^\infty(\mathbf{R}^n).$$

Since $L^*_{\min}u \in L_2(\mathbf{R}^n)$, we see that $-\Delta u \in L_2(\mathbf{R}^n)$ in the sense of distribution theory. So $u \in D(L_{\max})$ and $L^*_{\min}u = -\Delta u = L_{\max}u$. On the other hand, if $u \in D(L_{\max})$, then $-\Delta u \in L_2(\mathbf{R}^n)$ and for all $v \in C_0^\infty(\mathbf{R}^n)$ we have $(-\Delta v, u) = (v, -\Delta u)$. Therefore, $u \in D(L^*_{\min})$ and $L^*_{\min}u = -\Delta u$. □

Theorem 27. *If $V(x) \to 0$ as $|x| \to \infty$, then the essential spectrum of the Schrödinger operator $L = -\Delta + V(x)$ coincides with $[0,\infty)$.*

Proof. As in Example 24 we set $u_k(x) = \theta_k(x)e^{i\sqrt{k}x_1}\varepsilon_k$.
Since

$$\int |V(x)u_k(x)|^2\,dx \le \max_{|x|>k^2} |V(x)|^2\,dx \to 0 \quad \text{and} \quad \|(-\Delta - \lambda I)u_k\|_{L_2(\mathbf{R}^n)} \to 0$$

we see that each $\lambda \ge 0$ belongs to the essential spectrum. □

Theorem 28. *If $V(x)$ is real, $V(x) \ge a$, then the operator $L = -\Delta + V(x)$ with its domain $D_L = C_0^\infty(\mathbf{R}^n)$ is essentially self-adjoint.*

Proof. It is evident that the operator L is symmetric and

$$(Lu, u) = \int (|\nabla u|^2 + V(x)|u|^2)\,dx \ge a\|u\|^2.$$

Therefore, the statement follows from Theorem 17. □

Theorem 29. *If $n \geq 3$, $V(x)$, is real and $V(x) \geq a - \dfrac{4}{(n-2)^2|x|^2}$, where a is a constant, then the operator $L = -\Delta + V(x)$ with its domain $D_L = C_0^\infty(\mathbf{R}^n)$ is essentially self-adjoint.*

Proof. As in Theorem 28 it suffices to prove the boundedness of L from below. We have

$$(Lu, u) = \int (|\nabla u|^2 + V(x)|u|^2)\,dx \geq \int (|\nabla u|^2 - \frac{4|u|^2}{(n-2)^2|x|^2})\,dx + a\|u\|^2,$$

so that in virtue of the Hardy inequality we have $(Lu, u) \geq a\|u\|^2$. □

Theorem 30. *If a measurable locally bounded function $V(x)$ is such that*

$$\liminf_{|x|\to\infty} V(x) \geq a,$$

then the operator $L = -\Delta + V(x)$ is semi-bounded from below and has a discrete spectrum on $(-\infty, a)$, so that for any $\varepsilon > 0$ the spectrum of L on $(-\infty, a-\varepsilon)$ consists of a finite number of eigenvalues of finite multiplicities.

We will use in the proof the following Glazman's lemma.

Lemma 31. *If $\liminf_{|x|\to\infty} V(x) \geq a$ and $u \in D_L$, then $(Lu, u) = \int(|\nabla u|^2 + V(x)|u|^2)\,dx < \infty$.*

Proof. Since $V(x) \geq C$, we can substitute L by $L - (C-1)I$ and assume that $V(x) \geq 1$ so that $\int V(x)|u|^2\,dx \geq \int |u(x)|^2\,dx$. Introduce in $D(L)$ the norm $\|u\|_\Gamma = (\int(|u|^2 + |Lu|^2)\,dx)^{1/2}$. If $u \in C_0^\infty(\mathbf{R}^n)$, then

$$(Lu, u) = \int (|\nabla u|^2 + V(x)|u|^2)\,dx.$$

The convergence of the sequence $\{u_k\}$ in the graph norm $\|\cdot\|_\Gamma$ implies its convergence in $H_1(\mathbf{R}^n)$ and in the space L_2 with the weight function $V(x)$. Therefore, for the limit function $u(x)$ the integral $\int(|\nabla u|^2 + V(x)|u|^2)\,dx$ has a finite value and is equal to (Lu, u). □

Proof of Theorem 30. In virtue of the variational principle it suffices to prove that for $\varepsilon > 0$ the dimension of the subspace S of functions u of D_L, such that

$$(Lu, u) \leq (a - \varepsilon)\|u\|^2,$$

is finite. By Lemma 31 this inequality is equivalent to the following one:

$$\int (|\nabla u|^2 + (V(x) - a + \varepsilon)|u|^2)\,dx \leq 0, \ u \in S.$$

Let $R > 0$ be such that $V(x) \geq a - \varepsilon/2$ for $|x| \geq R$. Let $V(x) \geq m$ everywhere. Then

$$\int_{|x|\leq R} |\nabla u|^2\,dx + \int_{|x|>R} [|\nabla u|^2 + \frac{\varepsilon}{2}|u|^2]\,dx \leq C\int_{|x|\leq R} |u|^2\,dx, \ u \in S,$$

if $C \geq 0$ and $C \geq a - m - \varepsilon$. Let B be the operator of restriction of functions from S on the ball $K_R = \{x : |x| \leq R\}$. This operator is continuous in $L_2(\mathbf{R}^n)$ and injective in virtue of the latter estimate. So it is sufficient to prove that the subspace BS has a finite dimension. However, by the same estimate we have $\|u\|_{H^1(K_R)} \leq \|u\|_{L_2(K_R)}$, $u \in BS$. Therefore, the unit ball in the space $BS \cap L_2(K_R)$ is compact and so the subspace S is finite dimensional. □

Example 32. Let us consider the differential expression $P(x, D) = (-\Delta)^m$. The corresponding minimal operator is a positively defined self-adjoint operator $L_2(\mathbf{R}^n) \to L_2(\mathbf{R}^n)$. This operator is positive, therefore, all points $\lambda < 0$ are regular. Let us show that all points $\lambda \geq 0$ belong to the continuous spectrum. Let $\lambda > 0$ and $u_\varepsilon(x) = e^{i\xi x_1}\theta_\varepsilon(x)$, where $\xi^{2m} = \lambda$, $\xi > 0$, $\theta_\varepsilon(x) = \theta(\varepsilon x)$, $\theta(x) \in C_0^\infty(\mathbf{R}^n)$, $\theta(x) \equiv 1$ for $|x| < 1$, and $\theta(x) \equiv 0$ for $|x| > 2$. It is easy to see that

$$\|Pu_\varepsilon - \lambda u_\varepsilon\|^2_{L_2(\mathbf{R}^n)} \leq \varepsilon^2 \int_{1/\varepsilon<|x|<2/\varepsilon} C\,dx = \varepsilon^{2-n}$$

and $\|u_\varepsilon\|^2_{L_2(\mathbf{R}^n)} \geq \varepsilon^{-n}$. Thus $\|(P - \lambda)u_\varepsilon\|^2/\|u_\varepsilon\|^2 \to 0$ as $\varepsilon \to 0$. The point $\lambda = 0$ also belongs to the continuous spectrum as a limit point. Using the Fourier transform one can check that the points $\lambda \geq 0$ are not eigenvalues.

Example 33. Consider the differential expression

$$P(x, D) = (-\Delta)^m + Q(x),$$

where $Q(x)$ is a bounded function, $Q(x) \to 0$ as $|x| \to \infty$. The corresponding minimal operator is a bounded from below self-adjoint operator. If $Q(x) \geq 0$, then this operator is positively defined.

Let us show that if $Q(x) \leq 0$, $Q(x) \not\equiv 0$, $n \leq 2m$, then the minimal operator P_0 is not positively defined. Take for that a function $\psi(t) \in C^\infty(\mathbf{R}^1)$, such that $1 \geq \psi(t) \geq 0$, $\psi(t) = 1$ for $t \leq 0$, $\psi(t) \equiv 0$ for $t > \ln 2$. Then

$$(P_0\psi_\varepsilon, \psi_\varepsilon)_{L_2} = \int_{\mathbf{R}^n} P(\psi_\varepsilon)\psi_\varepsilon\,dx,$$

where $\psi_\varepsilon = \psi(\ln \varepsilon|x|)$, $\varepsilon > 0$. We have

$$\begin{aligned}
(P\psi_\varepsilon, \psi_\varepsilon)_{L_2} &\leq C \int_{1/\varepsilon<|x|<2/\varepsilon} \sum_{|\alpha|=m} |D^\alpha\psi_\varepsilon|^2\,dx + \int_{\mathbf{R}^n} \psi_\varepsilon^2 Q(x)\,dx \\
&\leq C_1 \int_{1/\varepsilon<|x|<2/\varepsilon} (\varepsilon^2/|x|^{2m})\,dx + \int_{|x|<2/\varepsilon} \psi_\varepsilon^2 Q(x)\,dx \\
&\leq C_2\varepsilon^2 \int_{1/\varepsilon<|x|<2/\varepsilon} |x|^{n-2m-1} d|x| - C_0,
\end{aligned}$$

where $C_0 = -\int_{|x|<2/\varepsilon} Q(x)\,dx > 0$. If ε is small, then C_0 is larger than a positive constant and $(P\psi_\varepsilon, \psi_\varepsilon) < 0$ for small positive ε. Thus the operator P is not positive.

Chapter 5

The Sturm-Liouville Problem

5.1 Elementary properties

Let us consider the equation

$$-(p(x)y'(x))' + q(x)y(x) = \lambda\rho(x)y(x)$$

on the segment $0 \leq x \leq l$, assuming that the real-valued functions p, p', q, ρ are continuous on this segment and

$$p(x) \geq p_0 > 0, \ \rho(x) \geq \rho_0 > 0, \ q(x) \geq 0.$$

The *Sturm-Liouville problem* consists in finding values of λ for which this equation has a nontrivial solution, satisfying the boundary conditions

$$\alpha y(0) + \beta y'(0) = 0, \ \gamma y(l) + \delta y'(l) = 0,$$

where $\alpha, \beta, \gamma, \delta$ are real numbers such that $\alpha^2 + \beta^2 \neq 0$, $\gamma^2 + \delta^2 \neq 0$. The corresponding nontrivial solution $y(x)$ is called an *eigenfunction* and λ is an *eigenvalue.*

Example 1. The eigenvalues λ of the problem

$$-y'' = \lambda y, \ 0 \leq x \leq \pi, \ y(0) = y(\pi) = 0,$$

are equal to k^2, $k = 1, 2, \ldots$, and the corresponding eigenfunctions are $y_k = A_k \sin kx$, $A_k \neq 0$.

Proposition 2. *If y and z are two solutions to the Sturm-Liouville problem with the same value of λ, then they are linearly dependent, i.e. there are two constants C_1, C_2 such that $C_1y(x) + C_2z(x) = 0$, $|C_1| + |C_2| \neq 0$.*

Proof. Since $\alpha y(0) + \beta y'(0) = 0$, $\alpha z(0) + \beta z'(0) = 0$, there are two constants C_1, C_2 such that $|C_1| + |C_2| \neq 0$, $C_1y(0) + C_2z(0) = 0$, and $C_1y'(0) + C_2z'(0) = 0$. By the uniqueness theorem the function $u(x) = C_1y(x) + C_2z(x)$ vanishes in $[0, 1]$. Therefore, the functions y and z are linearly dependent. □

Proposition 3. *If y and z are two solutions of the Sturm-Liouville problem with different values of λ, then*

$$\int_0^l \rho(x)y(x)z(x)\,dx = 0.$$

Proof. The functions y and z satisfy the equations

$$-(p(x)y'(x))' + q(x)y(x) = \lambda\rho(x)y(x),$$
$$-(p(x)z'(x))' + q(x)z(x) = \mu\rho(x)z(x).$$

Multiplying the first equation by $z(x)$ and the second by $y(x)$ and subtracting, we get the equality

$$-(p(x)y'(x))'z(x) + (p(x)z'(x))'y(x) = (\lambda - \mu)\rho(x)y(x)z(x).$$

We integrate over $(0, l)$. We then have

$$\int_0^l (p(x)y'(x))'z(x)\,dx = p(x)y'(x)z(x)|_0^l - \int_0^l p(x)y'(x)z'(x)\,dx,$$
$$\int_0^l (p(x)z'(x))'y(x)\,dx = p(x)y(x)z'(x)|_0^l - \int_0^l p(x)y'(x)z'(x)\,dx.$$

In virtue of the boundary conditions

$$p(l)y'(l)z(l) - p(l)y(l)z'(l) = 0,\ p(0)y'(0)z(0) - p(0)y(0)z'(0) = 0.$$

Therefore

$$\int_0^l [(p(x)y'(x))'z(x) - (p(x)z'(x))'y(x)]\,dx = 0$$

and $(\lambda - \mu)\int_0^l \rho(x)y(x)z(x)\,dx = 0$. Since $\lambda - \mu \neq 0$, the proof is complete. □

Corollary 4. *The Sturm-Liouville problem has at most a countable set of eigenvalues.*

Proof. The Hilbert space $L_2(\rho, [0, l])$ with the scalar product

$$(y, z) = \int_0^l \rho(x)y(x)\overline{z}(x)\,dx$$

is separable. Therefore it contains at most a countable set of mutually orthogonal vectors. □

Proposition 5. *If $\alpha\beta \le 0$, $\gamma\delta \ge 0$, then all eigenvalues of the Sturm-Liouville problem are nonnegative. Moreover, if $|\alpha| + |\gamma| \ne 0$, then they are positive.*

Proof. Multiply both sides of the equality

$$-(p(x)y'(x))' + q(x)y(x) = \lambda\rho(x)y(x)$$

by $y(x)$ and integrate the result over $[0, l]$. Since

$$\int_0^l (p(x)y'(x))'y(x)\,dx = p(l)y'(l)y(l) - p(0)y'(0)y(0) - \int_0^l p(x)y'^2(x)\,dx,$$

we obtain the equality

$$-p(l)y'(l)y(l)+p(0)y'(0)y(0)+\int_0^l [p(x)y'(x)^2+q(x)y(x)^2]\,dx = \lambda\int_0^l \rho(x)y(x)^2\,dx.$$

If $\beta = 0$, then $\alpha \ne 0$ and $p(0)y(0)y'(0) = 0$. If $\beta \ne 0$, then $y'(0) = -\alpha y(0)/\beta$ and $p(0)y'(0)y(0) = -\alpha p(0)y(0)^2/\beta \ge 0$. Similarly, if $\delta = 0$, then $\gamma \ne 0$ and $p(l)y(l)y'(l) = 0$, and for $\delta = 0$ we have $p(l)y(l)y'(l) = -p(l)\gamma y(l)^2/\delta$. Therefore

$$Ay(l)^2 + By(0)^2 + \int_0^l [p(x)y'(x)^2 + q(x)y(x)^2]\,dx = \lambda\int_0^l \rho(x)y(x)^2\,dx.$$

The left-hand part of this equality is nonnegative, since $A = 0$ or $A = \gamma p(l)/\delta \ge 0$, $B = 0$ or $B = -\alpha p(0)/\beta \ge 0$. Therefore, $\lambda \ge 0$.

If $\lambda = 0$, then $y'(x) \equiv 0$ and therefore $y \equiv const$. If in addition $A \ne 0$ or $B \ne 0$, then $y(x) \equiv 0$. □

Proposition 6. *Let $\alpha\beta \le 0$, $\gamma\delta \ge 0$, and $|\alpha| + |\gamma| \ne 0$. Let y_1 be a solution of the equation*

$$-(p(x)y'(x))' + q(x)y(x) = 0$$

for $0 < x < l$, satisfying the conditions

$$\alpha y_1(0) + \beta y_1'(0) = 0, \quad |y_1(0)| + |y_1'(0)| \ne 0.$$

Let y_2 be a solution of the same equation satisfying the conditions

$$\gamma y_2(l) + \delta y_2'(l) = 0, \quad |y_2(l)| + |y_2'(l)| \ne 0.$$

Then the functions y_1 and y_2 are linearly independent.

Proof. If this is not the case, then we have $\gamma y_1(l) + \delta y_1'(l) = 0$. Multiplying both parts of the equality

$$-(p(x)y_1'(x))' + q(x)y_1(x) = 0$$

by $y_1(x)$ and integrating over $(0, l)$, we obtain

$$\int_0^l p(x)y_1'(x)^2\,dx + \int_0^l q(x)y_1(x)^2\,dx + p(0)y_1(0)y_1'(0) - p(l)y_1(l)y_1'(l) = 0.$$

Repeating the arguments of the proof of Proposition 5, one can easily show that $p(0)y_1'(0)y_1(0) \geq 0$, $p(l)y_1'(l)y_1(l) \leq 0$. Therefore, a sum of nonnegative terms is equal to zero. Then all terms vanish and thus $y_1(x) \equiv \text{const}$. But then $\alpha = \gamma = 0$, which contradicts our assumption. □

Theorem 7 (Liouville). *Let y_1 and y_2 be solutions of the equation*

$$-(p(x)y'(x))' + q(x)y(x) = 0$$

and $W(x) = y_1(x)y_2'(x) - y_1'(x)y_2(x)$ their Wronskian. Then $p(x)W(x) = const.$

Proof. The derivative $W'(x)$ is equal to

$$y_1(x)y_2''(x) - y_1''(x)y_2(x) = -\frac{p'(x)}{p(x)}(y_1(x)y_2'(x) - y_1'(x)y_2(x)),$$

i.e. $W'(x) = -p'(x)W(x)/p(x)$. This means that $(W(x)p(x))' = 0$. □

Definition 8. Let $|\alpha| + |\gamma| \neq 0$. The Green function of the Sturm-Liouville problem is a function $G(x, \xi)$, continuous for $0 \leq x \leq l$, $0 \leq \xi \leq l$ such that

$$G(x, \xi) = Cy_1(x)y_2(\xi) \text{ for } 0 \leq x \leq \xi,$$

$$G(x, \xi) = Cy_2(x)y_1(\xi) \text{ for } 0 \leq \xi \leq x,$$

where y_1 and y_2 are the functions defined in Proposition 6 and the constant C is such that $Cp(x)[y_2'(x)y_1(x) - y_2(x)y_1'(x)] = -1$. The latter condition is admissible by Theorem 7.

Proposition 9. *Let the operator*

$$Ly(x) = -(p(x)y')' + q(x)y(x)$$

be defined on the space $D(L)$ of twice differentiable functions, satisfying the boundary conditions

$$\alpha y(0) + \beta y'(0) = 0, \ \gamma y(l) + \delta y'(l) = 0.$$

Then the integral operator

$$Af(x) = \int_0^l G(x, \xi)f(\xi)\, d\xi,$$

where $G(x, \xi)$ is the Green function, is inverse to L, i.e.

$$LAf = f \text{ for } f \in C[0, l], \quad ALu = u \text{ for } u \in D(L).$$

Proof. Let $f \in C[0,l]$. Then

$$\begin{aligned}
LAf &= -\frac{d}{dx}[p(x)\int_0^x G'_x(x,\xi)f(\xi)\,d\xi + p(x)\int_x^l G'_x(x,\xi)f(\xi)\,d\xi] \\
&\quad + q(x)\int_0^l G(x,\xi)f(\xi)\,d\xi \\
&= C\int_0^x [-\frac{d}{dx}(p(x)y_1(\xi)y_2'(x)) + q(x)y_1(\xi)y_2(x)]f(\xi)\,d\xi \\
&\quad + C\int_x^l [-\frac{d}{dx}(p(x)y_2(\xi)y_1'(x)) + q(x)y_2(\xi)y_1(x)]f(\xi)\,d\xi \\
&\quad + p(x)[G'_x(x,x+0) - G'_x(x,x-0)]f(x).
\end{aligned}$$

The integrals on the right-hand side vanish since $Ly_1 = 0, Ly_2 = 0$. Moreover,

$$p(x)[G'_x(x,x+0) - G'_x(x,x-0)] = Cp(x)[y_1'(x)y_2(x) - y_1(x)y_2'(x)] = 1,$$

so that $LAf(x) = f(x)$. On the other hand, let $u \in C^2(0,l)$ satisfy the boundary conditions. Then

$$\begin{aligned}
ALu(x) &= \int_0^l G(x,\xi)Lu(\xi)\,d\xi \\
&= C\int_0^x y_1(\xi)y_2(x)[-(p(\xi)u'(\xi))' + q(\xi)u(\xi)]\,d\xi \\
&\quad + C\int_x^l y_2(\xi)y_1(x)[-(p(\xi)u'(\xi))' + q(\xi)u(\xi)]\,d\xi.
\end{aligned}$$

Since $Ly_1(\xi) = 0,\ Ly_2(\xi) = 0$, after integrating by parts we obtain that

$$\begin{aligned}
ALu(x) &= -Cy_1(\xi)y_2(x)p(\xi)u'(\xi)|_0^x + Cy_1'(\xi)y_2(x)p(\xi)u(\xi)|_0^x \\
&\quad - Cy_1(x)y_2(\xi)p(\xi)u'(\xi)|_x^l + Cy_2'(\xi)y_1(x)p(\xi)u(\xi)|_x^l \\
&= -Cy_1(x)y_2(x)p(x)u'(x) + Cy_1(0)y_2(x)p(0)u'(0) \\
&\quad + Cy_1'(x)y_2(x)p(x)u(x) - Cy_1'(0)y_2(x)p(0)y(0) \\
&\quad - Cy_1(x)y_2(l)p(l)y'(l) + Cy_1(x)y_2(x)p(x)y'(x) \\
&\quad + Cy_1(x)y_2'(l)p(l)u(l) - Cy_1(x)y_2'(x)p(x)u(x) \\
&= -Cp(x)W(x)u(x) + Cp(0)y_2(x)[y_1(0)u'(0) - u(0)y_1'(0)] \\
&\quad - Cp(l)y_1(x)[u'(l)y_2(l) - u(l)y_2'(l)] = u(x),
\end{aligned}$$

since

$$y_1(0)u'(0) - u(0)y_1'(0) = 0,\ u'(l)y_2(l) - u(l)y_2'(l) = 0. \qquad \square$$

The definition implies the symmetry of the Green function:

$$G(x,\xi) = G(\xi,x).$$

Therefore, the operator A is symmetric. Proposition 9 implies that an eigenfunction φ of the Sturm-Liouville problem, corresponding to the eigenvalue $\lambda \neq 0$, is an eigenfunction of the operator A with the eigenvalue $1/\lambda$.

The operator A is an integral operator with a symmetric continuous kernel. From Theorem 37 of Chapter 1 it follows that it is compact in $L_2(0,l)$. In virtue of Theorem 34 of Chapter 1 the set of eigenvalues of such an operator can have only the limit point $\lambda = 0$. Therefore, if $\lambda_1 < \lambda_2 < \ldots$ is the sequence of the eigenvalues of the Sturm-Liouville problem, then $\lambda_j \to \infty$ as $j \to \infty$. By the same Theorem 34 the set of the eigenfunctions of the Sturm-Liouville problem is an orthogonal base in the space $L_2(0,l)$.

The Sturm-Liouville problem can be reduced to a problem of finding an extremum of some functional. This functional can be chosen in such a way that the Euler-Lagrange equation for it coincides with the Sturm-Liouville equation on the class of functions satisfying the corresponding boundary conditions.

Definition 10. $\mathcal{H}$ is the subspace of functions of $H^1(0,l)$, such that $y(0) = 0$ if $\beta = 0$, and $y(l) = 0$ if $\delta = 0$.

Let

$$F[y] = \int_0^l [p(x)y'(x)^2 + q(x)y(x)^2]\, dx + k_1 y^2(0) + k_2 y^2(l),$$

where $k_1 = -p(0)\alpha/\beta$ if $\beta \neq 0$, and $k_1 = 0$ if $\beta = 0$; $k_2 = p(l)\gamma/\delta$ if $\delta \neq 0$, and $k_2 = 0$ if $\delta = 0$. We will suppose that $\alpha\beta \leq 0$ and $\gamma\delta \geq 0$, so that $k_1 \geq 0$ and $k_2 \geq 0$.

The condition $F[y] = 0$ for $y \in \mathcal{H}$ implies that $y = 0$ on the segment $(0,l)$. It is obvious that the values of F in the space $\mathcal{H}$ are bounded from below, since $F[u] \geq 0$.

Let us look for the minimal value of the functional F in the class Γ. The latter is defined as the set of the functions y from $\mathcal{H}$ such that $\int \rho(x)y(x)^2\, dx = 1$.

Proposition 11. *Let $m = \inf_\Gamma F[y]$. Then there exists a function $y \in \Gamma$, such that $F[y] = m$.*

Proof. Observe that for $y \in \Gamma$ one has

$$F[y] \geq p_0 \int_0^l y'(x)^2\, dx + k_1 y(0)^2 \geq c_0 \|y\|^2_{H^1(0,l)},$$

where c_0 is a positive constant. (Remember that $y(0) = 0$ if $k_1 = 0$.) Therefore, one can take $F[y]$ as the square of a norm of y in $\mathcal{H}$.

Let $\{y_k\}$ be a minimizing sequence in Γ, i.e. $F[y_k] \to m$ as $k \to \infty$. Then $\|y_k\|^2_{\mathcal{H}} \leq m + 1$ for all big enough values of k. In virtue of Theorem 20 of Chapter 1 and Theorem 46 of Chapter 2 this bounded sequence contains a weakly in $\mathcal{H}$ converging subsequence, which also converges in the space $L_2(0,l)$ to a function y. Thus the limit function y is an element of $\mathcal{H}$, and $\int_0^l \rho(x)y(x)^2\, dx = 1$. Let us prove that this subsequence converges in $\mathcal{H}$. Since the functional F is quadratic, the following identity

$$F[\frac{y_k - y_l}{2}] + F[\frac{y_k + y_l}{2}] = \frac{1}{2}F[y_k] + \frac{1}{2}F[y_l]$$

is valid. Let $\varepsilon > 0$ and k and l be so big that for y_k, y_l from the subsequence one has $F[y_k] \leq m + \varepsilon,\ F[y_l] \leq m + \varepsilon$, and

$$\int_0^l \rho(x)[\frac{y_k(x) - y_l(x)}{2}]^2\, dx \leq \varepsilon^2.$$

Then

$$\begin{aligned}\int_0^l \rho(x)[\frac{y_k(x) + y_l(x)}{2}]^2\, dx &= \int_0^l \rho(x)[y_k(x) + \frac{y_k(x) - y_l(x)}{2}]^2\, dx\\ \geq (1-\varepsilon)\int_0^l \rho(x) y_k(x)^2\, dx &- \frac{1}{\varepsilon}\int_0^l \rho(x)[\frac{y_k(x) - y_l(x)}{2}]^2\, dx\\ \geq (1-\varepsilon) - \varepsilon = 1 - 2\varepsilon.\end{aligned}$$

Therefore,

$$\begin{aligned} & F[\frac{y_k(x) + y_l(x)}{2}] \geq m(1-2\varepsilon)\\ \text{and}\quad & F[\frac{y_k(x) - y_l(x)}{2}] \leq m + \varepsilon - m(1-2\varepsilon) = \varepsilon(1+2m).\end{aligned}$$

This means that the subsequence $\{y_k\}$ converges in $\mathcal{H}$. Since it converges in $\mathcal{H}$ weakly to y, its limit in $\mathcal{H}$ is equal to y, too. We have then that $F[y] = m$ because the functional F is continuous in $\mathcal{H}$. □

Proposition 12. *Let $y \in \Gamma$ and $F[y] = m = \inf_{z\in\Gamma} F[z]$. Then*

$$-(p(x)y')' + q(x)y(x) = \lambda\rho(x)y(x),$$

where $\lambda = m$ is the minimal eigenvalue of the Sturm-Liouville problem.

Proof. It is evident that

$$m = \inf_{z\in\mathcal{H}, z\neq 0} \frac{F[z]}{\int_0^1 \rho(x)z(x)^2\, dx}.$$

Let u be an element of $\mathcal{H}$. Consider the function of $t \in \mathbf{R}$

$$g(t) = \frac{F[y+tu]}{\int_0^1 \rho(x)(y(x)+tu(x))^2\, dx}.$$

We have $g(t) \geq g(0) = m$. Therefore, $g'(0) = 0$. This means that for all $u \in \mathcal{H}$

$$\int_0^l [p(x)y'(x)u'(x) + q(x)y(x)u(x)]\, dx + k_1 y(0)u(0) + k_2 y(l)u(l)$$
$$= \lambda \int_0^l \rho(x)y(x)u(x)\, dx,$$

where $\lambda = \text{const}$. In particular, setting $u = y$, we obtain that $\lambda = m$. Therefore,

$$\int_0^l [p(x)y'(x)u'(x) + q(x)y(x)u(x) - m\rho(x)y(x)u(x)]\, dx$$
$$+ k_1 y(0)u(0) + k_2 y(l)u(l) = 0.$$

Since this equality is valid for all $u \in C_0^\infty(0,l)$, it implies the existence of a generalized derivative of the function $p(x)y'(x)$ such that

$$-(p(x)y'(x))' + q(x)y(x) - m\rho(x)y(x) = 0.$$

The continuity of the function y implies then that y'' is continuous in $(0,l)$. Besides, after integration by parts we see that

$$p(x)y'(x)u(x)|_{x=0}^{x=l} + k_1 y(0)u(0) + k_2 y(l)u(l) = 0$$

or

$$[p(l)y'(l) + k_2 y(l)]u(l) - [p(0)y'(0) - k_1 y(0)]u(0) = 0.$$

If $k_2 = 0$, then by our construction $y(l) = 0$. If $k_2 \neq 0$, then a value of $u(l)$ can be arbitrary and we can take $u(0) = 0$. Hence $p(l)y'(l) + k_2 y(l) = 0$, i.e. $\gamma y(l) + \delta y'(l) = 0$. Similarly, if $k_1 \neq 0$, then u can be chosen so that $u(l) = 0, u(0) \neq 0$, and therefore, $p(0)y'(0) - k_1 y(0) = 0$, which implies that $\alpha y(0) + \beta y'(0) = 0$. Thus y is a solution of the Sturm-Liouville problem with the eigenvalue $\lambda = m$. Since for any solution z of the Sturm-Liouville problem the equality

$$\int_0^l [p(x)z'(x)^2 + q(x)z(x)^2]\, dx + k_1 z^2(0) + k_2 z^2(l) = \lambda \int_0^1 \rho(x)z(x)^2\, dx$$

holds, we have the relation $\lambda \geq m$, i.e. m is the minimal eigenvalue. □

Proposition 13. *Let $y \in \Gamma$ and $F[y] = m = \inf_\Gamma F[y]$. Then $y(x) \geq 0$ or $y(x) \leq 0$ for all $x \in [0,l]$. Moreover, $y(x) \neq 0$ for $x \in (0,l)$.*

Proof. We have

$$\int_0^l \rho(x)y(x)^2\, dx = 1, \quad \int_0^l [p(x)y'(x)^2 + q(x)y(x)^2]\, dx + k_1 y^2(0) + k_2 y^2(l) = m.$$

The function $z(x) = |y(x)|$ satisfies these conditions, too, since $z^2 = y^2$, and the function y' has a finite number of zeroes on $(0,l)$. Proposition 12 implies that z is a solution of the Sturm-Liouville problem with $\lambda = m$. But then it follows from Proposition 2 that $y(x) = cz(x)$ and since $z(x) \geq 0$ the proof is complete.

Since the function y does not change its sign, it cannot vanish in $(0, l)$, since otherwise the relation $y(x_0) = 0$ implies that $y'(x_0) = 0$, and by the uniqueness theorem $y(x) \equiv 0$ on $(0, l)$. □

One can use the variational principle for finding other eigenfunctions. Let $\mathcal{H}_1$ be the subset in $\mathcal{H}$ of functions y satisfying the condition

$$\int_0^l \rho(x)y(x)y_0(x)\,dx = 0,$$

where y_0 is the above-found eigenfunction with the eigenvalue m. Let Γ_1 be the subset in $\mathcal{H}_\infty$ of the functions y satisfying the condition

$$\int_0^l \rho(x)y(x)^2\,dx = 1.$$

Let

$$m_1 = \inf_{\Gamma_1}[\int_0^l [p(x)y'(x)^2 + q(x)y(x)^2]\,dx + k_1y^2(0) + k_2y^2(l)].$$

As in the proof of Proposition 11 one can verify that the minimal value m_1 is attained on a function $y_1 \in \Gamma_1$.

Let us show that y_1 is a solution of the Sturm-Liouville problem with $\lambda = m_1$. Let u be an element of $\mathcal{H}_1$. Consider the function of $t \in \mathbf{R}$

$$g(t) = \frac{F[y_1 + tu]}{\int_0^1 \rho(x)(y_1(x) + tu(x))^2\,dx}.$$

It is obvious that $g(t) \geq g(0) = m_1$, i.e. the function g has a minimal value at $t = 0$. Therefore, $g'(0) = 0$, i.e.

$$\int_0^1 [-(p(x)y_1'(x))' + q(x)y_1(x) - \mu\rho(x)y_1(x)]u(x)\,dx = 0$$

for all $u \in \mathcal{H}_1$. Therefore

$$-(p(x)y_1'(x))' + q(x)y_1(x) - \mu\rho(x)y_1(x) = C\rho(x)y_0(x).$$

Multiplying this equality by $y_0(x)$ and integrating the result over $(0, l)$, we see that

$$C\int_0^1 \rho y_0^2\,dx = (Ly_1, y_0) = (y_1, Ly_0) = m\int_0^1 \rho y_0 y_1\,dx = 0.$$

Therefore, $C = 0$ and $F[y_1] = \mu \int \rho(x)y_1(x)^2\,dx$, i.e. $\mu = m_1$. □

Let us show that m_1 is the nearest to m eigenvalue. It is clear that $m_1 \geq m$, since the minimal value over a subspace is bigger than the minimal value over a whole space. Moreover $m_1 \neq m$, since by Proposition 2 in the case when $m_1 = m$, the functions y_0 and y_1 are linearly dependent, which contradicts the condition $\int_0^l \rho(x)y_0(x)y_1(x)\,dx = 0$. Therefore, $m_1 > m$. An eigenvalue λ such that $m < \lambda < m_1$ cannot exist, since for the corresponding eigenfunction z we would then have that $F[z] = \lambda < m_1$ and $\int_0^l \rho(x)y_0(x)z(x)\,dx = 0$, contrary to the definition of m_1.

The same arguments show that there exists an infinite sequence $y_0(x)$, $y_1(x)$, ... of eigenfunctions of the Sturm-Liouville problem such that

$$\int_0^l \rho(x)y_j(x)y_k(x)\,dx = 0 \quad \text{for} \quad k = 0,\dots,j-1$$

and the corresponding eigenvalues are such that $0 < m = \lambda_0 < \lambda_1 < \cdots$. Moreover, the function y_j gives the minimal value to the functional F on the class of functions u from Γ, satisfying the conditions $\int_0^l \rho(x)y_k(x)u(x)\,dx = 0$ for $k = 0,\dots,j-1$.

Theorem 14 (Courant). *Let n be a natural number and L_n be an n-dimensional subspace in $\mathcal{H}$. Let Γ_n be the set of functions u of Γ, such that*

$$\int_0^l \rho(x)u(x)v(x)\,dx = 0 \ \text{ for all } \ v \in L_n.$$

Let at last

$$\mu_n = \sup_{L_n} \inf_{u\in\Gamma_n} F[u].$$

Then μ_n is the $(n+1)$-th eigenvalue of the Sturm-Liouville problem, i.e. $\mu_n = \lambda_n$.

Proof. Let L_n be the subspace generated by the eigenfunctions $y_0, y_1, \dots, y_{n-1}$. Then μ_n is the $(n+1)$-th eigenvalue of the Sturm-Liouville problem, as has been shown above. Thus it suffices to prove that $\mu_n \leq \lambda_n$ for any subspace L_n. In turn, for that it is sufficient to find a function $u \in \Gamma$, such that $F[u] \leq \lambda_n$ and

$$\int_0^l \rho(x)u(x)\varphi_j(x)\,dx = 0,\ j = 1,\dots,n,$$

where the functions $\varphi_1(x),\dots,\varphi_n(x)$ form a base in L_n.

Let us look for a function u in the form $u(x) = \sum_{j=0}^n \alpha_j y_j(x)$. It is obvious that the boundary conditions are satisfied for any α_j. The orthogonality conditions give the system of equations

$$\sum_{j=0}^n \alpha_j \int_0^l \rho(x)y_j(x)\varphi_k(x)\,dx = 0,\ k = 1,\dots,n.$$

This system of n equations for $n+1$ unknown values of $\alpha_0,\dots,\alpha_n$ always has a nontrivial solution, which can be normed in such a way that

$$\int_0^l \rho(x)u(x)^2\,dx = \sum_{k=0}^n \alpha_k^2 = 1,$$

provided that $\int_0^l \rho(x)y_j^2(x)\,dx = 1,\ j = 0,1,\dots,n$. At last, we have

$$\begin{aligned} F[u] &= \int_0^l [p(x)(\sum_{k=0}^n \alpha_k y_k'(x))^2 + q(x)(\sum_{k=0}^n \alpha_k y_k(x))^2]\,dx + k_1 u(0)^2 + k_2 u(l)^2 \\ &= \sum_{j,k=0}^n \alpha_j \alpha_k \int_0^l L y_k(x)\cdot y_j(x)\,dx = \sum_{j,k=0}^n \alpha_j \alpha_k \lambda_k \int_0^l \rho(x) y_k(x) y_j(x)\,dx \\ &= \sum_{k=0}^n \lambda_k \alpha_k^2. \end{aligned}$$

Since $\lambda_k \le \lambda_n$ for $k = 0,1,\dots,n$, it follows that $F[u] \le \lambda_n \sum_{k=0}^n \alpha_k^2 = \lambda_n$. The proof is complete. □

We will show that the eigenfunction y_n has exactly n zeroes in $(0,l)$. The following Sturm theorem serves as the basis for the proof.

Theorem 15 (Sturm). *Let there be functions u and v that do not vanish identically on $(0,l)$ and satisfy the equations*

$$-(p(x)u')' + q_1(x)u = 0,\ -(p(x)v')' + q_2(x)v = 0,$$

where $p(x) > 0$ and $q_1(x) \ge q_2(x)$ everywhere in $(0,l)$. Let x_1 and x_2 be two zeroes of u. Then there is at least one zero x_0 of the function v such that $x_1 < x_0 \le x_2$.

Proof. Multiplying the first equation by v and the second by u and subtracting we get the equality

$$(pu'v - puv')' = (q_1 - q_2)uv.$$

Let $u(x) > 0$ in (x_1, x_2). Integrating the obtained equality over (x_1, x_2), we see that

$$(pu'v - puv')_{x=x_1}^{x=x_2} = \int_{x_1}^{x_2} (q_1(x) - q_2(x))u(x)v(x)\,dx.$$

Assume that $v(x) \ne 0$ for $x_1 < x \le x_2$. We can suppose that $v(x) > 0$, when $x_1 < x \le x_2$. Then the latter integral is positive. Since by the uniqueness theorem $u'(x_1) \ne 0$ and $u'(x_2) \ne 0$, we have $u'(x_1) > 0,\ u'(x_2) < 0$. Therefore,

$$(pu'v - puv')_{x=x_1}^{x=x_2} = p(x_2)u'(x_2)v(x_2) - p(x_1)u'(x_1)v(x_1) < 0,$$

which is impossible. □

We can also prove the following more general theorem.

Theorem 15A (Sturm). *Let there be functions u and v that do not vanish identically on $(0,l)$ and satisfy the equations*

$$-(p_1(x)u')' + q_1(x)u = 0,\ -(p_2(x)v')' + q_2(x)v = 0,$$

where $p_1(x) \geq p_2(x) > 0$ and $q_1(x) \geq q_2(x)$ everywhere in $(0,l)$. Let x_1 and x_2 be two zeroes of u. Then there is at least one zero x_0 of the function v such that $x_1 < x_0 \leq x_2$.

Proof. Let $u(x) > 0$ in (x_1, x_2). Assume that $v(x) \neq 0$ for $x_1 < x \leq x_2$. We can suppose that $u(x) > 0$ and $v(x) > 0$, when $x_1 < x \leq x_2$. Multiplying the first equation by u and the second by u^2/v and subtracting, we get the equality

$$-(p_1 u')'u + (p_2 v')'u^2/v + (q_1 - q_2)u^2 = 0.$$

Integrating the obtained equality over (x_1, x_2), we get that

$$\int_{x_1}^{x_2} (q_1(x) - q_2(x))u(x)^2\,dx + \int_{x_1}^{x_2} ((p_2 v')'u^2/v - (p_1 u')'u)\,dx = 0.$$

Integrating by parts and using the fact that $u(x_1) = u(x_2) = 0$, we can see that

$$\int_{x_1}^{x_2} (q_1(x) - q_2(x))u(x)^2\,dx + \int_{x_1}^{x_2} (p_1(x) - p_2(x))u'(x)^2\,dx + \int_{x_1}^{x_2} p_2(u'^2 - v'(u^2/v)')\,dx = 0.$$

Remark that the two first integrals are positive. On the other hand,

$$u'^2 - v'(u^2/v)' = u'^2 - 2uu'v'/v + u^2v'^2/v^2 = (u' - uv'/v)^2,$$

so that the latter integral is positive, too, unless $u(x) = Cv(x)$. Thus all three integrals must vanish, and this is possible only if $q_1 = q_2$, $p_1 = p_2$, $u = Cv$ on $[x_1, x_2]$. □

Theorem 16. *Let the functions u and v not vanish identically on $(0,l)$ and satisfy the equations*

$$-(p(x)u')' + q_1(x)u = 0, \quad -(p(x)v')' + q_2(x)v = 0,$$

where $p(x) > 0$, $q_1(x) > q_2(x)$ everywhere in $(0,l)$. Let

$$\alpha u(0) + \beta u'(0) = 0, \quad \alpha v(0) + \beta v'(0) = 0,$$

where $\beta \neq 0$, $\beta\alpha \leq 0$. If $u(x_0) = 0$ and $0 < x_0 < l$, then the function v has a zero in the interval $(0, x_0)$.

Proof. Without loss of generality we can assume that x_0 is the least positive zero of u, and $u > 0$ in $(0, x_0)$. Suppose that the statement is not valid and $v(x) \neq 0$ for $x \in (0, x_0)$. Then we can assume that $v(x) > 0$ for these x. Multiplying the first equation by v and the second by u and subtracting, we get the equality

$$(pu'v - puv')' = (q_1 - q_2)uv.$$

Integrating the obtained equality over $(0, x_0)$, we see that

$$(pu'v - puv')|_{x=0}^{x=x_0} = \int_0^{x_0} (q_1(x) - q_2(x))u(x)v(x)\,dx.$$

The latter integral is positive. By the uniqueness theorem, $u'(x_0) \neq 0$ and therefore $u'(x_0) < 0$. Therefore, $(pu'v - puv')_{x=0}^{x=x_0} = p(x_0)u'(x_0)v(x_0) \leq 0$, which is impossible. □

Theorem 17. *Let the functions u and v not vanish identically on $(0, l)$ and satisfy the equations*

$$-(p(x)u')' + q_1(x)u = 0, \ -(p(x)v')' + q_2(x)v = 0,$$

where $p(x) > 0$, $q_1(x) > q_2(x)$ everywhere in $(0, l)$. Let

$$\gamma u(l) + \delta u'(l) = 0, \ \gamma v(l) + \delta v'(l) = 0,$$

where $\delta \neq 0$, $\delta\gamma \geq 0$. If $u(x_0) = 0$ and $0 < x_0 < l$, then the function v has a zero in the interval (x_0, l).

Proof. Without loss of generality we can assume that $u > 0$ in (x_0, l). Suppose that the statement is not valid and $v(x) \neq 0$ for $x \in (x_0, l)$. Then we can assume that $v(x) > 0$ for these x. Multiplying the first equation by v and the second by u and subtracting we get the equality

$$(pu'v - puv')' = (q_1 - q_2)uv.$$

Integrating the obtained equality over (x_0, l), we see that

$$(pu'v - puv')|_{x=x_0}^{x=l} = \int_{x_0}^{l} (q_1(x) - q_2(x))u(x)v(x)\,dx.$$

The latter integral is positive. By the uniqueness theorem, $u'(x_0) \neq 0$ and therefore $u'(x_0) > 0$, $u'(l) \leq 0$. Thus

$$(pu'v - puv')_{x=x_0}^{x=l} = -p(x_0)u'(x_0)v(x_0) \leq 0,$$

which is impossible. □

Theorem 18 (The second comparison theorem). *Let the functions u and v not vanish identically on $(0, l)$ and satisfy the equations*

$$-(p(x)u')' + q_1(x)u = 0, \ -(p(x)v')' + q_2(x)v = 0,$$

where $p(x) > 0$, $q_1(x) > q_2(x)$ everywhere in $(0, l)$. Let

$$u(0) = v(0) = a, \ u'(0) = v'(0) = b.$$

If the function u has k zeroes for $0 < x < l$, then the number of zeroes of v is not less than k and the jth zero of the function v is less than the jth zero of u.

Proof. If x_1 is the first zero of the function u, then, by Theorem 16, the function v has a zero x_1' such that $0 < x_1' < x_1$. Let x_j, x_j' be the jth zeroes of the functions u and v, respectively. Let $x_j' \leq x_j$. Let us show that $x_{j+1}' \leq x_{j+1}$.

Indeed, if it is not so, then $x_{j+1}' > x_{j+1}$ and the function v does not vanish for $x_j < x \leq x_{j+1}$, which contradicts Theorem 15. □

Theorem 19. *Let $0 < \lambda_0 < \lambda_1 < \cdots$ be the sequence of the eigenvalues of the Sturm-Liouville problem and $y_n \neq 0$ be a solution of this problem with $\lambda = \lambda_n$. Then the function y_n has exactly n zeroes in $(0, l)$.*

Proof. The function y_n satisfies the equation

$$(p(x)y_n'(x))' + [\lambda_n \rho(x) - q(x)]y_n(x) = 0.$$

By Proposition 13 $y_0(x) \neq 0$ in $(0, l)$. Since $\int_0^l \rho(x) y_0(x) y_1(x)\, dx = 0$, and the function ρy_0 has a constant sign, the function y_1 has to change its sign and therefore it has at least one zero in $(0, l)$.

Let us denote by $\varphi(x, \lambda)$ the solution of the equation

$$(p(x)y'(x))' + [\lambda \rho(x) - q(x)]y(x) = 0,$$

satisfying the following initial conditions:

$$\begin{aligned} &\text{if } \beta \neq 0 \text{ then} \quad \varphi(0, \lambda) = 1, \quad \varphi'(0, \lambda) = -\alpha/\beta; \\ &\text{if } \beta = 0 \text{ then} \quad \varphi(0, \lambda) = 0, \quad \varphi'(0, \lambda) = 1. \end{aligned}$$

The zeroes of the function $\varphi(x, \lambda)$ are simple by the uniqueness theorem and thus the function changes its sign at each of them. Since $\varphi(x, \lambda)$ is continuous in λ, the zeroes also depend on λ continuously and by Theorem 15 the zeroes are moving to the left as λ grows.

If $\delta = 0$, i.e. $y(l) = 0$, the statement has been proved: if λ increases, all zeroes of $\varphi(x, \lambda)$ move to the left and the appearance of a new zero at the point $x = l$ when $\lambda = \mu$ means that μ is an eigenvalue. Therefore, the function $y_n(x) = \varphi(x, \lambda_n)$ has exactly n zeroes on $(0, l)$.

Now let $\delta \neq 0$. As above, the function y_0 does not change its sign, and the function y_1, being orthogonal to y_0, has at least one zero. If it had two zeroes, then the function $(\varphi(x, \lambda_1)\exp(\gamma x/\delta))'$ would vanish at some point between them. As λ is decreasing, the zeroes of $\varphi(x, \lambda)$ are moving to the right. Therefore we have $\gamma\varphi(l, \lambda) + \delta\varphi'(l, \lambda) = 0$ for a value of $\lambda \in (\lambda_0, \lambda_1)$. Since this is impossible, the function y_1 has exactly one zero in $(0, l)$.

Suppose that the function y_j has j zeroes in $(0, l)$. Let us show that y_{j+1} has $j+1$ zeroes. Indeed, the number of zeroes is not less than j by the Sturm theorem. This number cannot be more than $j + 1$, since as above in this case there is an eigenvalue $\lambda \in (\lambda_j, \lambda_{j+1})$. Finally, this number is not equal to j by Theorems 16 and 17. □

Lemma 20 (Gronwall). *If u and v are two continuous functions with positive values and for $t \geq 0$ the inequality*

$$u(t) \leq a + \int_0^t u(s)v(s)ds$$

holds with a positive constant a, then

$$u(t) \leq ae^{\int_0^t v(s)\,ds}.$$

Proof. Since

$$\frac{u(t)v(t)}{a + \int_0^t u(s)v(s)\,ds} \leq v(t),$$

integrating with respect to t, we get

$$\ln[a + \int_0^t u(s)v(s)\,ds] - \ln a \leq \int_0^t v(s)\,ds$$

and

$$u(t) \leq a + \int_0^t u(s)v(s)\,ds \leq ae^{\int_0^t v(s)\,ds}.$$ □

Theorem 21. *If all solutions of the equation*

$$y''(t) + a(t)y(t) = 0$$

are bounded for $0 \leq t < \infty$ and $\int_0^\infty |b(t)|\,dt < \infty$, then all solutions of the equation

$$z''(t) + (a(t) + b(t))z(t) = 0$$

are bounded. Moreover, each solution of the second equation is represented in the form

$$z(t) = [C_1 + \alpha(t)]y_1(t) + [C_2 + \beta(t)]y_2(t),$$

where y_1, y_2 are two linearly independent solutions to the first equation; C_1, C_2 are constants, and the functions α and β tend to zero as $t \to \infty$.

Proof. Let $y_1(t), y_2(t)$ be two linearly independent solutions of the equation $y''(t)+a(t)y(t) = 0$ satisfying the conditions

$$y_1(0) = 0,\ y_1'(0) = 1;\ y_2(0) = 1,\ y_2'(0) = 0.$$

By the Liouville theorem we have $W(t) = y_1'(t)y_2(t) - y_2'(t)y_1(t) = 1$. It is easy to see that the function

$$y(t) = C_1y_1(t) + C_2y_2(t) + \int_0^t [y_1(t)y_2(s) - y_1(s)y_2(t)]f(s)\,ds$$

satisfies the equation

$$y''(t) + a(t)y(t) = f(t)$$

and the initial conditions

$$y(0) = C_1y_1(0) + C_2y_2(0) = C_2,\ y'(0) = C_1y_1'(0) + C_2y_2'(0) = C_1.$$

Therefore, every solution $z(t)$ of the second equation is represented in the form

$$z(t) = C_1 y_1(t) + C_2 y_2(t) - \int_0^t [y_1(t)y_2(s) - y_1(s)y_2(t)]b(s)z(s)\,ds.$$

The solutions z_1, z_2, satisfying the initial conditions

$$z_1(0) = 0,\ z_1'(0) = 1;\ z_2(0) = 1,\ z_2'(0) = 0,$$

are represented in the form

$$z_1(t) = y_1(t) - \int_0^t [y_1(t)y_2(s) - y_1(s)y_2(t)]b(s)z_1(s)\,ds,$$
$$z_2(t) = y_2(t) - \int_0^t [y_1(t)y_2(s) - y_1(s)y_2(t)]b(s)z_2(s)\,ds.$$

Since $|y_1(t)| \le C,\ |y_2(t)| \le C$, we see that

$$|z_1(t)| + |z_2(t)| \le 2C + 2C^2 \int_0^t |b(s)|(|z_1(s)| + |z_2(s)|)\,ds.$$

Lemma 20 implies that

$$|z_1(t)| + |z_2(t)| \le 2Ce^{2C^2 \int_0^\infty |b(s)|\,ds} \le C_3.$$

Therefore, the integrals $\int_0^\infty b(s)y_j(s)z_k(s)\,ds$ with $j, k = 1, 2$ converge and

$$\int_0^t b(s)y_j(s)z_k(s)\,ds = a_{jk} + \alpha_{jk}(t),$$

where $\alpha_{jk}(t) \to 0$ as $t \to \infty$. Thus

$$z_1(t) = y_1(t)[1 - a_{21} - \alpha_{21}(t)] + y_2(t)[a_{11} + \alpha_{11}(t)],$$
$$z_2(t) = y_2(t)[1 + a_{12} + \alpha_{12}(t)] - y_1(t)[a_{22} + \alpha_{22}(t)].$$

Therefore, each solution of the second equation has the form

$$\begin{aligned} z(t) &= C_1 z_1(t) + C_2 z_2(t) \\ &= y_1(t)[C_1 - C_1 a_{21} - C_2 a_{22} - C_1\alpha_{21}(t) - C_2\alpha_{22}(t)] \\ &\quad + y_2(t)[C_2 + C_2 a_{12} + C_1 a_{11} + C_1\alpha_{11}(t) + C_2\alpha_{12}(t)] \\ &= y_1(t)[C_1' + \alpha(t)] + y_2(t)[C_2' + \beta(t)], \end{aligned}$$

where C_1', C_2' are constants and $\alpha(t) \to 0,\ \beta(t) \to 0$ as $t \to \infty$. □

Consider now a special case of the Sturm-Liouville problem:

$$y'' + V(x)y + \lambda y = 0,\ 0 \le x \le 1,\ y(0) = 0,\ y(1) = 0,$$

assuming that $V(x) \le 0$. It is easy to see that $\lambda > 0$.

Theorem 22. *Let the eigenfunctions y_n of the latter problem be normed in such a way that $\int_0^1 y_n^2(x)\,dx = 1$. Then*

$$\lambda_n = n^2\pi^2 + 2a\pi + \gamma_n,\ y_n(x) = \sqrt{2}\sin n\pi x + h_n(x),$$

where as $n \to \infty$

$$\gamma_n = O(\frac{1}{n}),\ h_n(x) = O(\frac{1}{n}),\ h_n'(x) = O(1),\ h_n''(x) = O(n).$$

Proof. Let $\varphi(x,\lambda)$ be the solution of this equation, satisfying the initial conditions:

$$\varphi(0,\lambda) = 0,\ \varphi'(0,\lambda) = 1.$$

Then the eigenvalues of the Sturm-Liouville problem are defined by the equation

$$\varphi(1,\lambda) = 0.$$

Set $\lambda = \mu^2$. Then the function $\varphi(x,\lambda)$ satisfies the equation

$$\varphi(x,\lambda) = \frac{1}{\mu}\sin\mu x + \frac{1}{\mu}\int_0^x \sin\mu(t-x)V(t)\varphi(t,\lambda)\,dt.$$

Since by Proposition 9 $\lambda_n \to \infty$ as $n \to \infty$, applying Lemma 20, we can see that $\varphi(x,\lambda) \to 0$ as $\lambda \to \infty$. The eigenvalues are defined by the equation

$$\sin\mu + \int_0^1 \sin\mu(t-1)V(t)\varphi(t,\lambda)\,dt = 0,$$

$$\text{or}\quad \sin\mu\,[1 - \int_0^1 \cos\mu t\, V(t)\varphi(t,\lambda)\,dt] + \cos\mu\int_0^1 \sin\mu t\, V(t)\varphi(t,\lambda)\,dt = 0.$$

Since $\varphi(x,\lambda) \to 0$ as $\mu \to \infty$, this equation implies that $\sin\mu \approx 0$ as $\mu \to \infty$, i.e. $\mu = n\pi + \alpha_n$, where $\alpha_n \to 0$ as $n \to \infty$.

Moreover, $\sin\mu = (-1)^n \sin\alpha_n,\ \cos\mu = (-1)^n\cos\alpha_n$, so that

$$\sin\alpha_n[1 - \int_0^1 \cos\mu t\, V(t)\varphi(t,\lambda)\,dt] + \cos\alpha_n\int_0^1 \sin\mu t\, V(t)\varphi(t,\lambda)\,dt = 0,$$

and once again applying Lemma 20, we see that $\varphi(x,\lambda) = \frac{\sin n\pi x}{n\pi} + O(\frac{1}{n^2})$. Hence it follows that

$$\begin{aligned} n\pi\tan\alpha_n &= -\frac{\int_0^1 V(t)\sin\mu t\,\sin n\pi t\,dt + O(1/n)}{1 - \int_0^1 V(t)\sin n\pi t\,\cos\mu t\,dt/\pi n + O(1/n^2)} \\ &= -\frac{1}{2}\int_0^1 V(t)\,dt + \frac{1}{2}\int_0^1 V(t)\cos 2\pi n t\,dt + o(1). \end{aligned}$$

Therefore, $\alpha_n = \frac{a}{n} + \beta_n$, where

$$a = -\frac{1}{2\pi}\int_0^1 V(t)\,dt,\ \beta_n = \frac{1}{2\pi n}\int_0^1 V(t)\cos 2\pi nt\,dt + o(\frac{1}{n}) = o(\frac{1}{n}),$$

since $\int_0^1 V(t)\cos 2\pi nt\,dt \to 0$, as $n \to \infty$. Therefore

$$\lambda_n = (n\pi + \alpha_n)^2 = n^2\pi^2 + 2a\pi + o(1)$$

and if the functions $y_n(x) = \varphi(x, \lambda_n)$ are normed in such a way that $\int_0^1 y_n^2(x)\,dx = 1$, then $y_n(x) = \sqrt{2}\sin n\pi x + h_n(x)$, where

$$h_n(x) = O(\frac{1}{n}),\ h_n'(x) = O(1),\ h_n''(x) = O(n)$$

as $n \to \infty$. □

5.2 On the first eigenvalue of a Sturm-Liouville problem

Let us consider the Sturm-Liouville problem

$$y''(x) + \lambda q(x)y(x) = 0$$

on the segment $0 \le x \le 1$, with the boundary conditions

$$y(0) = 0,\ y(1) = 0.$$

We will study the dependence of the first eigenvalue λ_1 on the potential q.

Denote by R_β the set of measurable on $(0,1)$ functions q with nonnegative values such that

$$\int_0^1 q(x)^\beta\,dx = 1,$$

where β is a real number, $\beta \neq 0$. The variational principle implies that the first eigenvalue λ_1 can be found as

$$\lambda_1 = \inf_{y \in C_0^\infty(0,1)} \frac{\int_0^1 y'(x)^2\,dx}{\int_0^1 q(x)y(x)^2\,dx}.$$

We will estimate the values

$$m_\beta = \inf_{q \in R_\beta} \lambda_1,\ M_\beta = \sup_{q \in R_\beta} \lambda_1.$$

Set

$$L[q,y] = \frac{\int_0^1 y'(x)^2\,dx}{\int_0^1 q(x)y(x)^2\,dx},\ G[y] = \frac{\int_0^1 y'(x)^2\,dx}{(\int_0^1 y(x)^p\,dx)^{2/p}},\quad \text{where}\quad p = \frac{2\beta}{\beta - 1}.$$

The main result of this section is the following.

Theorem 23. *If* $\beta > 1$, *then*

$$m_\beta = \frac{(\beta-1)^{1+1/\beta}}{\beta(2\beta-1)^{1/\beta}} B^2(\frac{1}{2}, \frac{1}{2} - \frac{1}{2\beta}),$$

and $M_\beta = \infty$, *where* B *is the Euler beta function:*

$$B(a,b) = \int_0^1 x^{a-1}(1-x)^{b-1}\, dx.$$

There exist functions $u(x) \in H_0^1(0,1)$ *and* $q(x) \in R_\beta$ *such that*

$$\inf_y L[q,y] = L[q,u] = m_\beta.$$

If $\beta = 1$, *then* $M_1 = \infty$ *and* $m_1 = 4$.
If $0 < \beta < 1/2$, *then*

$$M_\beta = \frac{(1-\beta)^{1+1/\beta}}{\beta(1-2\beta)^{1/\beta}} B^2(\frac{1}{2}, \frac{1}{2\beta}),$$

and $m_\beta = 0$. *There exist functions* $u(x) \in H_0^1(0,1)$ *and* $q(x) \in R_\beta$ *such that* $\inf_y L[q,y] = L[q,u] = M_\beta$.
If $\beta < 0$, *then*

$$M_\beta = -\frac{(1-\beta)^{1+1/\beta}}{\beta(1-2\beta)^{1/\beta}} B^2(\frac{1}{2}, \frac{1}{2} - \frac{1}{2\beta}),$$

and $m_\beta = 0$. *There exist functions* $u(x) \in H_0^1(0,1)$ *and* $q(x) \in R_\beta$ *such that* $\inf_y L[q,y] = L[q,u] = M_\beta$.
If $1/2 \leq \beta < 1$, *then* $M_\beta = \infty$ *and* $m_\beta = 0$.

Proof. 1. If $\beta > 1$, then we have by the Hölder inequality

$$\int_0^1 qy(x)^2\, dx \leq (\int_0^1 q(x)^\beta\, dx)^{1/\beta} (\int_0^1 |y(x)|^p\, dx)^{2/p} = (\int_0^1 |y(x)|^p\, dx)^{2/p},$$

where $p = 2\beta/(\beta-1)$, for any $y \in H_0^1(0,1)$. Therefore, $\lambda_1 \geq m$, where $m = \inf_y G[y]$ in the space $H_0^1(0,1)$. Remark that the homogeneity allows us to assume that $\int_0^1 |y(x)|^p\, dx = 1$. Let $\{y_k\}$ be a sequence of functions of this class, such that

$$\int_0^1 |y_k(x)|^p\, dx = 1, \quad \int_0^1 y_k'(x)^2\, dx \to m.$$

This sequence is bounded in $H_0^1(0,1)$, therefore it is weakly compact in this space and compact in $C[0,1]$. We will assume that this sequence is converging uniformly and weakly in $H_0^1(0,1)$ to a function u. Then

$$\int_0^1 |u(x)|^p\,dx = 1, \quad \int_0^1 u'(x)^2\,dx \le \lim_{k\to\infty} \int_0^1 y_k'(x)^2\,dx = m,$$

and therefore $\int_0^1 u'(x)^2\,dx = m$. Since $G[y]$ has a minimal value at $y = u$, we have

$$\frac{d}{dt}G[u+tz] = 0 \quad \text{at} \quad t = 0$$

for an arbitrary function z of the class $H_0^1(0,1)$. This means that

$$\int_0^1 u'(x)z'(x)\,dx - m\int_0^1 |u(x)|^{p-2}u(x)z(x)\,dx = 0$$

for all $z \in H_0^1(0,1)$. This equality yields that the function u' has a generalized derivative, equal to $-m|u|^{p-2}u$, i.e.

$$u'' + m|u|^{p-2}u = 0$$

almost everywhere in $(0,1)$. Since u is a continuous function, we have $u'' \in C[0,1]$, so the equation is true in the classical sense.

Since $G[|y|] = G[y]$ for all y, we can assume that $y_k(x) \ge 0$ and thus $u(x) \ge 0$. Then by the uniqueness theorem for the Cauchy problem $u(x) > 0$ in $(0,1)$. Multiplying both sides of the equation

$$u''(x) + mu(x)^{p-1} = 0$$

by $2u'$ and integrating over $(0,x)$, we obtain that

$$u'(x)^2 + \frac{2m}{p}u(x)^p = C.$$

Integrating the both sides of this equality over (0,1) and taking into account that

$$\int_0^1 u'(x)^2\,dx = m, \quad \int_0^1 u(x)^p\,dx = 1,$$

we obtain $m(1+2/p) = C$.

Let b be a point at which the function u has a maximal value M. Since $u'' = -mu^{p-1} < 0$, such a point exists and is unique. If $b \ne 1/2$, then we can assume that $b < 1/2$, since $u(x)$ can be replaced by $u(1-x)$. The function $u_1(x) = u(2b-x)$ satisfies the same equation on $(b, 2b)$ as u and $u(b) = u_1(b) = M$, $u'(b) = u_1'(b) = 0$. Therefore, these functions coincide and $u(2b) = u(0) = 0$, i.e. $b = 1/2$. Since

$$u'(x) = \sqrt{C - \frac{2m}{p}u(x)^p}$$

for $0 \le x \le \frac{1}{2}$, we have

$$\int_0^M \frac{dy}{\sqrt{C - 2my^p/p}} = \frac{1}{2}.$$

Since $u'(1/2) = 0$, we have

$$M = u(\frac{1}{2}) = \left(\frac{pC}{2m}\right)^{1/p} = (1 + \frac{p}{2})^{1/p}.$$

Changing the variable of integration $y = Mt$, we obtain the equality

$$M^{1-p/2}\sqrt{\frac{p}{2m}}\int_0^1 \frac{dt}{\sqrt{1-t^p}} = \frac{1}{2}.$$

Observe that

$$\int_0^1 \frac{dt}{\sqrt{1-t^p}} = \frac{1}{p}B(\frac{1}{p}, \frac{1}{2}), \quad \text{so that} \quad M^{p/2-1} = \sqrt{\frac{2}{pm}}B(\frac{1}{p}, \frac{1}{2}).$$

The obtained relations allows us to find

$$m = C(\beta) = \frac{(\beta-1)^{1+1/\beta}}{\beta(2\beta-1)^{1/\beta}}B^2(\frac{1}{2}, \frac{1}{2} - \frac{1}{2\beta}).$$

2. Let now $\beta = 1$. Since $\int_0^1 q(x)y(x)^2\,dx \le \max y(x)^2$, we have

$$\lambda_1 \ge m = \inf_{y \in H_0^1(0,1)} \frac{\int_0^1 y'(x)^2\,dx}{\max y(x)^2}.$$

The value of m can be found according to the following lemma.

Lemma 24. *If $y \in H_0^1(0,1)$, then*

$$\max_{x\in(0,1)} y(x)^2 \le \frac{1}{4}\int_0^1 y'(x)^2\,dx.$$

The equality is attained on the function $y_0(x) = 1/2 - |x - 1/2|$.

Proof. Without loss of generality we can assume that $y(x) \ge 0$ for $0 < x < 1$. Let $M \equiv \max y(x)^2 = y(b)^2$. Then by the Hölder inequality

$$y(b)^2 \le b\int_0^b y'(x)^2\,dx, \; y(b)^2 \le (1-b)\int_b^1 y'(x)^2\,dx,$$

hence

$$(\frac{1}{b} + \frac{1}{1-b})y(b)^2 \le \int_0^1 y'(x)^2\,dx.$$

Since $1/b + 1/(1-b) \ge 4$ for $0 < b < 1$, the proof of the lemma is complete. □

Lemma 24 implies that $\lambda_1 \geq 4$ if $\beta = 1$.

3. Let $\beta < 0$. Set

$$q(x) = \begin{cases} (1-\varepsilon)^{1/\beta}\varepsilon^{-1/\beta}, & \text{if } 0 < x < \varepsilon, \\ (1-\varepsilon)^{-1/\beta}\varepsilon^{1/\beta}, & \text{if } \varepsilon < x < 1, \end{cases}$$

where $\varepsilon > 0$ is a small number. Let $y_0(x) = 1/2 - |x - 1/2|$. Then

$$\lambda_1 \leq \frac{1}{\int_0^1 q(x) y_0(x)^2\, dx} \leq C\varepsilon^{-1/\beta}.$$

Therefore, λ_1 can be arbitrarily small.

4. Let $0 < \beta < 1$. Set

$$q(x) = \begin{cases} (2\varepsilon)^{-1/\beta}, & \text{if } |x - 1/2| < \varepsilon, \\ 0, & \text{if } |x - 1/2| > \varepsilon, \end{cases}$$

where $\varepsilon > 0$ is a small number. Let y_0 be a smooth function, vanishing at the points $x = 0$ and $x = 1$, which is equal to 1 in $(1/3, 2/3)$. Then

$$\lambda_1 \leq \frac{C}{\int_{1/2-\varepsilon}^{1/2+\varepsilon} (2\varepsilon)^{-1/\beta}\, dx} = C_1 \varepsilon^{1/\beta - 1} \to 0$$

as $\varepsilon \to 0$. Therefore λ_1 can be arbitrarily small.

5. Let $\beta > 1/2$. Set

$$q(x) = \begin{cases} \varepsilon^{-1/\beta}, & \text{if } 0 < x < \varepsilon, \\ 0, & \text{if } \varepsilon < x < 1, \end{cases}$$

where $\varepsilon > 0$ is a small number. Then

$$\int_0^1 q(x) y(x)^2\, dx = \varepsilon^{-1/\beta} \int_0^\varepsilon y(x)^2\, dx \leq \varepsilon^{2-1/\beta} \int_0^\varepsilon y'(x)^2\, dx$$

and hence $\lambda_1 \geq \varepsilon^{1/\beta - 2}$. Therefore, in this case $M_\beta = \infty$.

6. If $\beta = 1/2$, we can set $q(x) = \varepsilon^2 / 4x^{\varepsilon - 2}$, so that $\int_0^1 q(x)^{1/2}\, dx = 1$. Then

$$\int_0^1 q(x) y(x)^2\, dx = \frac{\varepsilon^2}{4} \int_0^1 x^{\varepsilon-2} y(x)^2\, dx \leq C_1 \varepsilon^2 \int_0^1 y'(x)^2\, dx.$$

Therefore, $\lambda_1 \geq C_1^{-1} \varepsilon^{-2}$ and $M_{1/2} = \infty$.

7. Let $0 < \beta < 1/2$. Then by the Hölder inequality

$$\int_0^1 q(x)^\beta\, dx \leq \left(\int_0^1 q(x) y(x)^2\, dx\right)^\beta \left(\int_0^1 y(x)^p\, dx\right)^{1-\beta},$$

where $p = 2\beta/(\beta-1)$ so that $0 > p > -2$. Therefore, $L[q, y] \le G[y]$. Set $y_0(x) = x^\gamma$ for $0 < x < 1/2$ and $y_0(x) = (1-x)^\gamma$ for $1/2 < x < 1$, where $-1/p > \gamma > 1/2$ so that $y_0 \in H_0^1(0,1)$. Then the integral $\int_0^1 y_0(x)^p\,dx$ is converging and thus $\lambda_1 \le C_1$. Let $m = \inf_{y\in H_0^1(0,1)} G[y]$. Consider a minimizing sequence $\{y_k\}$ such that

$$\int_0^1 y_k(x)^p\,dx = 1, \ \int_0^1 y_k'(x)^2\,dx \to m.$$

There exists a subsequence $\{y_{n_k}\}$ uniformly converging to a function $u \in H_0^1(0,1)$. By the Fatou theorem, we have $\int_0^1 u(x)^p\,dx \le 1$ and $\int_0^1 u'(x)^2\,dx \le m$. Therefore, $G[u] \le m$ and since m is the minimal possible value of G, we have $G[u] = m$. Since $G[|y|] = G[y]$ for all y, we can assume that $y_k(x) \ge 0$ and thus $u(x) \ge 0$. The function u satisfies the same equation as in step 1, i.e. the equation

$$u'' + m|u|^{p-2}u = 0,$$

almost everywhere in $(0,1)$. Since u is a continuous function, we see that the equation is true in the classical sense in each interval where $u \ne 0$. If $u(x_0) = 0$, $u(x_1) = 0$, $u(x) > 0$ for $x_0 < x < x_1$ and $0 \le x_0 < x_1 \le 1$, $x_1 - x_0 = \kappa < 1$, then we can consider the function $v(x) = u(x_0 + \kappa x)$ and since

$$v''(x) + m\kappa^2 v^{p-1} = 0, \ v(0) = 0, \ v(1) = 0,$$

we see that $G[v] = m\kappa^2 < m$, which is impossible. So $u(x) > 0$ in $(0,1)$. Furthermore, the equation

$$u'(x)^2 + \frac{2m}{p}u(x)^p = C$$

holds for $0 < x < 1$ with $C = m(1 + 2/p) < 0$.

Let b be a point at which the function u has a maximal value M. Since $u'' = -mu^{p-1} < 0$, such a point exists and is unique. If $b \ne 1/2$, then we can assume that $b < 1/2$, since $u(x)$ can be replaced by $u(1-x)$. The function $u_1(x) = u(2b-x)$ satisfies the same equation on $(b, 2b)$ as u and $u(b) = u_1(b) = M$, $u'(b) = u_1'(b) = 0$. Therefore, by the uniqueness theorem for the Cauchy problem these functions coincide and $u(2b) = u(0) = 0$, i.e. $b = 1/2$. We have

$$\int_0^M \frac{dy}{\sqrt{C - 2my^p/p}} = \frac{1}{2}.$$

Since $u'(1/2) = 0$, we have

$$M = u(1/2) = (\frac{pC}{2m})^{1/p} = (1 + p/2)^{1/p}.$$

Changing the variable of integration $y = Mt$, we obtain the equality

$$M^{1-p/2}\sqrt{\frac{-p}{2m}}\int_0^1 \frac{dt}{\sqrt{t^p - 1}} = \frac{1}{2}.$$

Note that

$$\int_0^1 \frac{dt}{\sqrt{t^p-1}} = -\frac{1}{p}B(\frac{1}{2},\frac{1}{2}-\frac{1}{p}), \quad \text{so that} \quad m = C(\beta) = \frac{(1-\beta)^{1+1/\beta}}{\beta(1-2\beta)^{1/\beta}}B^2(\frac{1}{2},\frac{1}{2\beta}).$$

Since $\inf_y L[q,y] \le \inf_y G[y]$ and $L[u^{p-2},u] = C(\beta)$, we see that $M_\beta = C(\beta)$.

8. If $\beta < 0$, then by the Hölder inequality

$$\int_0^1 y(x)^{2\beta/(\beta-1)}\,dx \le (\int_0^1 q(x)y(x)^2\,dx)^{\beta/(\beta-1)}(\int_0^1 q(x)^\beta\,dx)^{1/(1-\beta)}.$$

Therefore, $L[q,y] \le G[y]$. Set $y_0(x) = |x-1/2| - 1/2$. Then

$$\lambda_1 \le \frac{1}{(\int_0^1(|x-1/2|-1/2)^{2\beta/(\beta-1)}\,dx)^{(\beta-1)/\beta}}$$

and so $M_\beta < \infty$. Consider a minimizing sequence $\{y_k\}$ such that

$$\int_0^1 y_k(x)^p\,dx = 1, \quad \int_0^1 y_k'(x)^2\,dx \to m.$$

There exists a subsequence $\{y_{n_k}\}$ uniformly converging to $u \in H_0^1(0,1)$. Since $p > 0$, the sequence $\{y_{n_k}^p\}$ converges uniformly to u^p. Therefore, $\int_0^1 u(x)^p\,dx \le 1$, and $\int_0^1 u'(x)^2\,dx \le m$. Therefore, $G[u] \le m$ and since m is the minimal possible value of G, we have $G[u] = m$. The function u satisfies the same equation as in step 1, and $C(\beta)$ is defined by the same formula as for $\beta > 1$. Since $\inf_y L[q,y] \le \inf_y G[y] = C(\beta)$, we see that $M_\beta = C(\beta)$. □

5.3 On other estimates of the first eigenvalue

Let us consider the Sturm-Liouville problem

$$(p(x)y')' + \lambda y(x) = 0$$

on the segment $0 \le x \le 1$, with the boundary conditions

$$y(0) = 0, \; y(1) = 0.$$

We will study the dependence of the first eigenvalue λ_1 on the function p. Let us denote by R_α the set of nonnegative measurable functions p on $[0,1]$ such that

$$\int_0^1 p(x)^\alpha\,dx = 1,$$

where α is a real number, $\alpha \neq 0$. Set

$$L[p,y] = \frac{\int_0^1 p(x)y'(x)^2\,dx}{\int_0^1 y(x)^2\,dx}, \quad G[y] = \frac{(\int_0^1 y'(x)^r\,dx)^{2/r}}{\int_0^1 y(x)^2\,dx}, \quad r = \frac{2\alpha}{\alpha - 1}.$$

The main result of this section is the following.

Theorem 25. *Let*

$$M_\alpha = \sup_{p\in R_\alpha} \lambda_1, \; m_\alpha = \inf_{p\in R_\alpha} \lambda_1,$$

$$C(\alpha) = \frac{2\alpha+1}{\alpha}\Big(\frac{\alpha+1}{2\alpha+1}\Big)^{1-1/\alpha} B^2\Big(\frac{1}{2}, \frac{1}{2} + \frac{1}{2\alpha}\Big), \text{ for } \alpha < -1 \text{ and for } \alpha > 0;$$

$$C(\alpha) = -4\frac{2\alpha+1}{\alpha}\Big(\frac{1+\alpha}{1+2\alpha}\Big)^{1-1/\alpha}\Big(\int_0^\infty \frac{dt}{(1+t^2)^{1/2-1/2\alpha}}\Big)^2 \text{ if } -1/2 < \alpha < 0.$$

If $\alpha > -1/2$, $\alpha \neq 0$, then $M_\alpha = C(\alpha)$ and $m_\alpha = 0$. There exist functions $p \in R_\alpha$, $z \in H_0^1(0,1)$, such that $z'(x)^2 = p(x)^{\alpha-1}$ and

$$\inf_y L[p,y] = L[p,z] = C(\alpha).$$

If $\alpha \leq -1$, then $m_\alpha = C(\alpha)$ and $M_\alpha = \infty$. There exist functions $p \in R_\alpha$, $z \in H_0^1(0,1)$, such that $z'(x)^2 = p(x)^{\alpha-1}$ and

$$\inf_y L[p,y] = L[p,z] = C(\alpha).$$

If $-1 < \alpha \leq -1/2$, then $M_\alpha = \infty$ and $m_\alpha = 0$.

The proof of Theorem 25 is based on the variational principle, according to which $\lambda_1 = \inf_{y\in H_0^1(0,1)} L[p,y]$.

Proof. 1. If $\alpha > 0$, then we can take a function y vanishing in $[0,1/2]$ and such that $\int_0^1 y^2\,dx = 1$. Since the function p can have arbitrarily small values in $[1/2,1]$, the value of λ_1 cannot be bounded from below by a positive constant.

2. Let $0 > \alpha > -1$. Let us show that in this case λ_1 cannot be bounded below by a positive constant.

Set the following:

$$y(x) = \begin{cases} x/\varepsilon, & \text{if } 0 < x < \varepsilon, \\ 1, & \text{if } \varepsilon < x < 1-\varepsilon, \\ (1-x)/\varepsilon, & \text{if } 1-\varepsilon < x < 1, \end{cases}$$

$$p(x) = \begin{cases} \delta, & \text{if } 0 < x < \varepsilon \text{ or } 1-\varepsilon < x < 1, \\ \varepsilon^{-1}, & \text{if } \varepsilon < x < 1-\varepsilon, \end{cases}$$

where δ is a number such that

$$\int_0^1 p(x)^\alpha\,dx = 2\varepsilon\delta^\alpha + (1-2\varepsilon)\varepsilon^{-\alpha} = 1,$$

so that $\delta \le C_1\varepsilon^{-1/\alpha}$. It is evident that

$$\int_0^1 y(x)^2\,dx = 1 - 2\varepsilon + 2\varepsilon/3 = 1 - 4\varepsilon/3.$$

On the other hand,

$$\int_0^1 p(x)y'(x)^2\,dx = 2\delta/\varepsilon \le 2C_1\varepsilon^{-1-1/\alpha}.$$

Therefore

$$\lambda_1 \le \frac{\int_0^1 p(x)y'(x)^2\,dx}{\int_0^1 y(x)^2\,dx} \le C\varepsilon^{-1-1/\alpha},$$

and since $-1 - 1/\alpha > 0$, the value of λ_1 can be arbitrarily small.

3. Let $\alpha \le -1$. Then $1 \le r < 2$ and by the Hölder inequality

$$\begin{aligned}\int_0^1 y'(x)^r\,dx &= \int_0^1 p(x)^{r/2}y'(x)^r \cdot p(x)^{-r/2}\,dx\\ &\le \Big(\int_0^1 p(x)y'(x)^2\,dx\Big)^{r/2}\Big(\int_0^1 p(x)^\alpha\,dx\Big)^{1/(1-\alpha)}.\end{aligned}$$

Therefore, for any admissible p we have $L[p,y] \ge G[y]$. Let $m = \inf_y G[y]$. Since $y(x) = \int_0^x y'(t)\,dt$, we have

$$\int_0^1 y(x)^2\,dx \le \int_0^1 \Big(\int_0^x |y'(t)|\,dt\Big)^2\,dx \le \Big(\int_0^1 |y'(t)|^r\,dt\Big)^{2/r}$$

and thus $m \ge 1$.

Consider a sequence $\{y_k\}$ such that $\int_0^1 |y_k'(x)|^r\,dx = 1$ and $\int_0^1 |y_k(x)|^2\,dx \to 1/m$. There exists a subsequence $\{y_{n_k}\}$ converging uniformly to a function $z \in H_0^1(0,1)$ such that $\int_0^1 |z'(x)|^r\,dx = 1$, $\int_0^1 |z(x)|^2\,dx = 1/m$. The function z satisfies the Euler-Lagrange equation

$$(|z'(x)|^{r-2}z'(x))' + mz(x) = 0.$$

Multiplying by z' and integrating we obtain

$$\frac{r-1}{r}|z'(x)|^r + \frac{m}{2}z^2 = C.$$

Integrating the last equality over $(0,1)$ we see that $C = 3/2 - 1/r > 0$. The function z is even with respect to $x = 1/2$, increasing in $(0,1/2)$ from 0 to M and decreasing in $(1/2,1)$. Since $z'(1/2) = 0$, we have

$$mM^2 = 2C = 3 - 2/r.$$

We have

$$\int_0^M \frac{dz}{(C - mz^2/2)^{1/r}} = \frac{1}{2}\left(\frac{r}{r-1}\right)^{1/r}.$$

Substituting $z = My$, we see that

$$\int_0^1 \frac{dy}{(1-y^2)^{1/r}} = \frac{1}{2}\left(\frac{r}{r-1}\right)^{1/r}\frac{C^{1/r}}{M}.$$

Observe that

$$\int_0^1 \frac{dy}{(1-y^2)^{1/r}} = \frac{1}{2}B(\frac{1}{2}, 1-\frac{1}{r}).$$

Therefore

$$M^{-1} = \left(\frac{r-1}{r}\right)^{1/r} C^{-1/r} B(\frac{1}{2}, 1-\frac{1}{r})$$

and $m = 2CM^{-2} = C(\alpha)$. Since $L[p, y] \geq C(\alpha)$ and $L[z'^{2/(\alpha-1)}, z] = C(\alpha)$, it follows that $m_\alpha = C(\alpha)$.

4. If $\alpha > 1$, then by the Hölder inequality

$$\int_0^1 p(x)y'(x)^2\,dx \leq (\int_0^1 p(x)^\alpha\,dx)^{1/\alpha}(\int_0^1 y'(x)^r\,dx)^{2/r},$$

where $r = 2\alpha/(\alpha-1) > 2$. Set $u(x) = 1/2 - |x - 1/2|$. Then

$$\lambda_1 \leq \frac{(\int_0^1 u'(x)^r)^{2/r}}{\int_0^1 u(x)^2\,dx} = C.$$

We can repeat the same arguments as above in step 3 to find the optimal functions p and z. Moreover, then $M = C(\alpha)$ and if $\alpha = 1$, then

$$\lambda_1 \leq \frac{\max u'(x)^2}{\int_0^1 u(x)^2\,dx} = 12 = \lim_{\alpha\to 1} C(\alpha).$$

5. Let $0 < \alpha < 1$ and p be a function of the set R_α. Set $u'(x)^2 = p(x)^{\alpha-1}$ and construct the function u in such a way that $u(0) = u(1) = 0$; it increases monotonically on $(0, b)$ and decreases monotonically on $(b, 1)$, where $b \in (0, 1)$. Let $M = \max u(x)^2$. It is evident that

$$\int_0^1 u'(x)^r\,dx = \int_0^1 p(x)^\alpha\,dx = 1.$$

Let $b \geq 1/2$. The measure of points $x \in (0, b)$ such that $u'(x)^r \geq 4$ is less than $1/4$. Therefore the supplementary set E on $(0, b)$ has the measure $\geq 1/4$ and at the points x of this set we have $u'(x) \geq 4^{1/r}$, because $r < 0$. Set now $z(0) = 0$, $z'(x) = 4^{1/r}$ at the points of E and $z'(x) = 0$ at other points of $(0, b)$. Then

$$\int_0^1 u(x)^2\, dx \geq \int_0^r z(x)^2\, dx \geq \int_0^{1/4} 4^{2/r} x^2\, dx = c_0$$

and therefore

$$M_\alpha \leq \sup_{p \in R_\alpha} \frac{\int_0^1 p(x) u'(x)^2\, dx}{\int_0^1 u(x)^2\, dx} \leq \frac{1}{c_0}.$$

To prove the existence of the optimal functions p and z we need some lemmas.

Lemma 26. *Let r be a real number, $r < 2/3$, $r \neq 0$. Then the function*

$$F(x, y) = x^{2/r} y^{3-2/r} + (1-x)^{2/r}(1-y)^{3-2/r}, \ 0 \leq x \leq 1, \ 0 \leq y \leq 1$$

has a minimal value $F_{\min} = 1/4$ at the point $(1/2, 1/2)$.

Proof. Observe that for the two exponents $2/r$ and $3 - 2/r$ one is always positive and the other is negative. For definiteness let $r < 0$. Then for $y \neq 0, 1$

$$\lim_{x \to +0} F(x, y) = +\infty, \quad \lim_{x \to 1-0} F(x, y) = +\infty.$$

If $y = 0$, then $F(x, 0) = (1-x)^{2/r} \geq 1$ and for $y = 1$ we have $F(x, 1) = x^{2/r} \geq 1$. Therefore, the values of F on the boundary of the square are ≥ 1.

If $x < \delta, y < \delta$, then

$$\begin{aligned} F(x, y) &\geq (1-x)^{2/r}(1-y)^{3-2/r} \geq (1-x)^{2/r} \geq (1-\delta)^{2/r} && \text{for } r > 0, \\ F(x, y) &\geq (1-x)^{2/r}(1-y)^{3-2/r} \geq (1-y)^{3-2/r} \geq (1-\delta)^{3-2/r} && \text{for } r < 0, \end{aligned}$$

and therefore $F(x, y) > 3/4$, if δ is small enough.

The same is true for a small neighborhood of the points $(0, 1)$, $(1, 0)$, and $(1, 1)$. On the other hand, we have $F(1/2, 1/2) = 1/4$.

Therefore, the function F has an inner minimum point (x_0, y_0). We have at this point

$$\begin{aligned} \partial F(x_0, y_0)/\partial x &= 2/r[x_0^{2/r-1} y_0^{3-2/r} - (1-x_0)^{2/r-1}(1-y_0)^{3-2/r}] = 0; \\ \partial F(x_0, y_0)/\partial y &= (3 - 2/r)[x_0^{2/r} y_0^{2-2/r} - (1-x_0)^{2/r}(1-y_0)^{2-2/r}] = 0. \end{aligned}$$

Then

$$\begin{aligned} x_0^{2/r-1} y_0^{3-2/r} &= (1-x_0)^{2/r-1}(1-y_0)^{3-2/r}, \\ x_0^{2/r} y_0^{2-2/r} &= (1-x_0)^{2/r}(1-y_0)^{2-2/r}. \end{aligned} \tag{1}$$

Dividing these equalities term by term we obtain $y_0/x_0 = (1-y_0)/(1-x_0)$, i.e. $x_0 = y_0$. Then (1) implies that $x_0^2 = (1-x_0)^2$ and thus $x_0 = 1/2$. □

Lemma 27. *Let $r < 0$ and $m = \sup_{y \in K_r(0,h)} G[y]$, where $K_r(0,h)$ is the set of functions y absolutely continuous in $[0,h]$ and such that $y(0) = 0$, $y'(x) \geq 0$, $\int_0^h y'(x)^r\,dx < \infty$, $\int_0^h y(x)^2\,dx < \infty$. Then*

$$m = \frac{1}{4}\Big(\frac{2-2r}{2-3r}\Big)^{2/r}(3 - \frac{2}{r})h^{2/r-3}B(1/2, 1-1/r)^2$$

and there exists a function $y_0 \in K_r(0,h)$ such that $G[y_0] = m$. Moreover, as $x \to h$ we have

$$y_0(x) = y_0(h) + c_1(h-x)^{r/(r-1)}[1+o(1)],$$
$$y_0'(x) = c_2(h-x)^{1/(r-1)}[1+o(1)].$$

Proof. Let at first $h = 1$ and $\{y_k\}$ be a sequence of functions in K such that $\int_0^1 y_k'(x)^r\,dx = 1$ and $G[y_k] \to m$. As we saw in the beginning of step 5 of the proof of Theorem 25, the value of m is finite and positive.

Let us show that we can assume all functions y_k to be smooth. Let $y \in K_r(0,1)$. Let us define y on the whole line, setting $y_1(x) = 0$ for $x < 0$, $y_1(x) = y(x)$ for $0 < x < 1-\varepsilon$, and $y_1(x) = y(1-\varepsilon) + \varepsilon^{-1/2r}(x - 1 + \varepsilon)$ for $x > 1-\varepsilon$, and set $u_k(x) = y_1(x) + \varepsilon_k x$, where $\varepsilon_k \to 0$ as $k \to \infty$. Obviously, $\int_0^1 |u_k(x) - y(x)|^2\,dx \to 0$ and by the Lebesgue theorem, $\int_0^1 |u_k'(x)|^r\,dx \to \int_0^1 |y'(x)|^r\,dx$. So we can assume that $y_k'(x) > \varepsilon_k > 0$.

Let now $y \in K_r(0,1)$ and $y'(x) > \varepsilon > 0$. Set $z_k(x) = y_k(x) - y_h(0)$, where y_k is the averaging of y with a positive kernel ω, so that

$$y_k(x) = k\int \omega(k(x-t))y(t)\,dt, \int \omega(t)\,dt = 1,\, \omega(t) \geq 0,$$

and $\omega \in C_0^\infty(-1,1)$. Observe that $|y_k(0)| \leq \delta_k \to 0$ as $k \to \infty$ so that z_k converges to $y(x)$ uniformly and

$$\int_0^1 |z_k(x) - y(x)|^2\,dx \to 0, \quad \int_0^1 |z_k'(x) - y'(x)|\,dx \to 0.$$

Therefore, a subsequence z_{n_k}' converges to y' almost everywhere and z_{n_k}' converges to y' almost everywhere. Since $0 \leq y(x) \leq C$ for $0 \leq x \leq 1$ and $y'(x) > 0$, we see that $\int_0^1 |y'(x)|\,dx \leq C$. Since $|z_{n_k}'|^r \leq \varepsilon^r$, we have by the Lebesgue theorem $\int_0^1 z_{n_k}'(x)^r\,dx \to \int_0^1 y'(x)^r\,dx$. This allows us to assume that all y_k are smooth.

We will call a function y *convex*, if its derivative is decreasing, and *concave*, if its derivative is increasing.

If y_k is convex in an interval (x_1, x_2), where $0 \leq x_1 \leq x_2 \leq 1$, then it is possible to substitute it for the linear function

$$z(x) = y_k(x_1) + \gamma(x - x_1), \text{ where } \gamma = [y_k(x_2) - y_k(x_1)]/(x_2 - x_1)$$

so that the value of the functional G will increase.

Indeed, then $y_k(x_1) = z(x_1)$, $y_k(x_2) = z(x_2)$, and

$$\int_{x_1}^{x_2} y_k(x)^2\,dx \geq \int_{x_1}^{x_2} z(x)^2\,dx.$$

On the other hand, by the Hölder inequality

$$\begin{aligned} x_2 - x_1 &= \int_{x_1}^{x_2} y_k'(x)^{r/(r-1)} y_k'(x)^{r/(1-r)}\,dx \\ &\leq \Big(\int_{x_1}^{x_2} y_k'(x)\,dx\Big)^{r/(r-1)} \Big(\int_{x_1}^{x_2} y_k'(x)^r\,dx\Big)^{1/(1-r)} \end{aligned}$$

and therefore,

$$\begin{aligned} \int_{x_1}^{x_2} y_k'(x)^r\,dx &\geq \Big(\int_{x_1}^{x_2} y_k'(x)\,dx\Big)^r (x_2 - x_1)^{1-r} \\ &= [y_k(x_2) - y_k(x_1)]^r (x_2 - x_1)^{1-r}. \end{aligned}$$

Since

$$\int_{x_1}^{x_2} z'(x)^r\,dx = \gamma^r (x_2 - x_1) = [y_k(x_2) - y_k(x_1)]^r (x_2 - x_1)^{1-r},$$

we see that $G[z] \geq G[y_k]$, if z coincides with y_k outside the interval (x_1, x_2).

We can do the same for all other intervals where y_k is convex. As a result we shall obtain a function z_k, which can be described in the following way. Consider the set of points (x, y) such that $0 < x < 1$ and $y > y_k(x)$, and take the convex hull of this set. The lower boundary of the hull serves as the graph of the function z_k. Let us show that $z_k \in C^1(0,1)$. Indeed, if a point $(x, z_k(x))$ belongs to an interval of the straight line, which is a part of the graph of z_k, then it is obvious that z_k is smooth at the point x. The same is true in the case when the point $(x, z_k(x))$ belongs to a part of the graph of the function y_k. If z_k is linear on one side of x and coincides with y_k on the other side, then z_k is regular at x since the graph of y_k is lying on one side of the straight line, obtained by the continuation of the linear function. Finally, if the point $(x, z_k(x))$ is a limit point for a sequence of such points, then it is also a limit point for a sequence of points belonging to the graph of y_k and the derivative $z_k'(x)$ exists.

Therefore, if one replaces every function y_k by a concave function z_k in the indicated way, then the sequence of these new functions z_k will be maximizing, too. This allows us to consider the maximizing sequence as a sequence of monotone concave functions, i.e. to suppose that the functions $y_k(x)$ and their derivatives $y_k'(x)$ are increasing.

For large k we have $\int_0^1 y_k(x)^2\,dx \leq m^{-1} + 1$. Let $\varepsilon > 0$ be small enough. There is a point $\theta_k \in (1-\varepsilon, 1)$ such that $|y_k(\theta_k)|^2 \leq (m^{-1}+1)\varepsilon^{-1}$. Since the functions y_k are monotone, they are uniformly bounded for $0 \leq x \leq 1-\varepsilon$. This yields $\int_{1-2\varepsilon}^{1-\varepsilon} y_k'(x)\,dx \le C$ with a constant C independent of k. Therefore, there exists a $\tilde{\theta}_k$ such that $1-2\varepsilon < \tilde{\theta}_k < 1-\varepsilon$ and $y_k'(\theta_k) \le C/\varepsilon$. Therefore, the sequence $\{y_k'(x)\}$ is uniformly bounded in $[0, 1-2\varepsilon]$. Since $\int_0^1 y_k'(x)^r\,dx = 1$ there exists a $\theta_k' \in (0,\varepsilon)$ such that $|y_k'(\theta_k')^r| \le 1/\varepsilon$ and therefore $|y_k'(x)^r| \le 1/\varepsilon$ for $\varepsilon \le x \le 1-2\varepsilon$.

By the Arzela theorem one can choose a uniformly in $[\varepsilon, 1-2\varepsilon]$ converging subsequence $\{y_{n_k}(x)\}$, and by the Helly theorem one can suppose that the sequence $\{y_{n_k}'(x)\}$ converges almost everywhere in $[\varepsilon, 1-2\varepsilon]$. Using diagonalization, one can find a subsequence converging in $(0,1)$ to a function $y_0 \in K_r(0,1)$ such that $y_{n_k} \to y_0$, $y_{n_k}'^r \to y_0'^r$ everywhere in $(0,1)$. By the Fatou theorem we have

$$\int_0^1 y_0(x)^2\,dx \le m^{-1}, \quad \int_0^1 y_0'(x)^r\,dx \le 1.$$

Therefore

$$\left(\int_0^1 y_0'(x)^r\,dx\right)^{2/r} \ge 1,$$

so that $G[y_0] \ge m$. However, m is the maximal possible value of G, so that $\int_0^1 y_0(x)^2\,dx = m^{-1}$ and $\int_0^1 y_0'(x)^r\,dx = 1$.

The Euler-Lagrange equation has the form

$$(y_0'^{r-1})' + my_0 = 0,\ y_0(0) = 0 \quad \text{so that} \quad (r-1)y_0'^{r-1}y_0'' + my_0y_0' = 0$$

or $y_0'^r + m_1 y_0^2 = C$, where $m_1 = mr/[2(r-1)] > 0$. Integrating this equality over $(0,1)$, we obtain that $C = 1 + m_1 m^{-1} = 1 + r/2(r-1) > 0$. Therefore

$$\int_0^{y_0} \frac{dz}{(C - m_1 z^2)^{1/r}} = x.$$

We have for any $\varphi \in K_r(0,1)$ the equality

$$\int_0^1 (y_0'(x)^{r-1}\varphi'(x) - my_0(x)\varphi(x))\,dx = 0.$$

Let $\varphi(x) = 1$ for $1-\varepsilon < x < 1$. Then integrating by parts gives the equality

$$y_0'(1-\varepsilon)^{r-1}\varphi(1-\varepsilon) - m\int_{1-\varepsilon}^1 y_0(x)\varphi(x)\,dx = 0.$$

For ε tending to zero, we obtain that $y_0'(1) = \infty$. Let $M = \max y_0(x) = y_0(1)$. Then $M^2 m_1 = C$, i.e. $M = (C/m_1)^{1/2} = ((3r-2)/mr)^{1/2}$ and

$$\int_0^M \frac{dz}{(C - m_1 z^2)^{1/r}} = 1.$$

Set $z = (C/m_1)^{1/2}t$ so that

$$\int_0^1 \frac{dt}{(1-t^2)^{1/r}} = C^{1/r-1/2}m_1^{1/2} = \frac{(2-3r)^{1/r-1/2}(-rm)^{1/2}}{(2-2r)^{1/r}},$$

and therefore

$$m^{1/2} = \frac{(2-2r)^{1/r}}{2(2-3r)^{1/r-1/2}(-r)^{1/2}} B(\frac{1}{2}, 1-\frac{1}{r}).$$

Since

$$\int_{y_0}^{M} \frac{dz}{(C-m_1 z^2)^{1/r}} = 1-x,$$

we can set $z = M-u$ and see that

$$\int_0^{y_0-M} \frac{du}{u^{1/r}(2M-u)^{1/r}} = m^{1/r}(1-x).$$

If $y_0 \to M$ we have $c(y_0-M)^{1-1/r} = 1-x+o(1)$, so that $y_0(x) = M + c_1(1-x)^\gamma[1+o(1)]$ as $x \to 1$, where $\gamma = r/(r-1)$.

In order to find the value m for any $h > 0$ it suffices to substitute in the obtained estimate the function $y_0(xh)$. □

Lemma 28. *Let $r < 0$ and*

$$m_1 = \sup_{y \in K_r(0,h,l)} G[y],$$

where $K_r(0,h,l)$ is the set of real functions y defined on $[0,l]$ and such that $y(x) \in K_r(0,h)$, $y(-x) \in K_r(-l,-h)$. Then there is a function $y_0 \in K_r(0,h,l)$ such that $y_0(x) = y_0(l-x)$, $G[y_0] = m_1$, and $m_1 = 4ml^{2/r-3}$, where the value of m was indicated in Lemma 27 for $h = 1$.

Proof. The existence of the extremal function for any fixed $h \in [0,1]$ follows from Lemma 27. Furthermore, we can suppose that $y(x)$ is monotone and concave in $[0,h]$ and in $[h,l]$.

Let at first $l = 1$. By Lemma 27 we have for $y \in K_r(0,h,1)$ the inequalities

$$h^{3-2/r}(\int_0^h |y'(x)|^r\,dx)^{2/r} \le m\int_0^h y(x)^2\,dx,$$

$$(l-h)^{3-2/r}(\int_h^l |y'(x)|^r\,dx)^{2/r} \le m\int_h^l y(x)^2\,dx,$$

where the value of m, corresponding to $h = 1$, was found in Lemma 27. Let $\int_0^l |y'(x)|^r\,dx = 1$ and $\int_0^h |y'(x)|^r\,dx = a$. Then

$$a^{2/r}h^{3-2/r} + (1-a)^{2/r}(l-h)^{3-2/r} \le m\int_0^1 y(x)^2\,dx.$$

By Lemma 26 the function $F(a,h) = a^{2/r}h^{3-2/r} + (1-a)^{2/r}(l-h)^{3-2/r}$ defined in the square $0 < a < 1$, $0 < h < 1$, has a minimal value $1/4$ at the point $a = h = 1/2$. Therefore, $\int_0^1 y(x)^2\,dx \geq 1/4m$, i.e. $G[y] \leq 4m$ for all admissible y. On the other hand, if $a = h = 1/2$ and if the function y coincides on $(0,1/2)$ with the function y_0, found in Lemma 27 for $h = 1/2$, and is even with respect to the point $x = 1/2$, then $G[y] = 4m$.

In order to find the value m for any $l > 0$ it suffices to substitute in the obtained estimate the function $y_0(xl)$. □

Corollary 29. *Let $0 < \alpha < 1, \alpha \neq 0$, and y_0 be the function found in Lemma 28. Set $p(x) = |y_0'(x)|^{2/(\alpha-1)}$. Then $p(x) = p(l-x)$* for $0 < x < l$,

$$p(x) = a(l/2 - x)^{\gamma}[1 + o(1)], \quad p'(x) = -a\gamma(l/2 - x)^{\gamma-1}[1 + o(1)],$$

as $x \to l/2-$, where $\gamma = 2/(\alpha+1) \in]1,2[$.

Lemma 30. *Let $p(x)$ be a smooth positive function on $[0,d[$, such that*

$$\lim_{x\to d} p(x)(d-x)^{-\gamma} = a, \; p'(x) = O((d-x)^{\gamma-1}),$$

where $1 < \gamma \leq 2$. Let $y(x)$ be a solution of the equation

$$(p(x)y')' + my(x) = 0, \; 0 < x < d, \quad \textit{such that} \quad \int_0^d p(x)y'(x)^2\,dx < \infty.$$

Then $\lim_{x\to d} p(x)y(x)y'(x) = 0$. Moreover, as $x \to d^-$ we have

$$y(x) = C(1 + o(1)), \; y'(x) = C(d-x)^{1-\gamma}(m/a + o(1)),$$

if $1 < \gamma < 2$, and

$$y(x) = (d-x)^{\rho}(C + o(1)), \; y'(x) = (d-x)^{\rho-1}(\rho C + o(1)),$$

if $\gamma = 2$ with $\rho > -1/2$.

Proof. Set $I[y] = \int_0^d p(x)y'(x)^2\,dx$.

If $1 < \gamma < 2$, then by solving the Cauchy problem we can find two linearly independent solutions $y(x)$ and $z(x)$ such that as $x \to d^-$

$$y(x) = 1 + c_1(d-x)^{2-\gamma} + \ldots, \qquad c_1 = \frac{m}{a(2-\gamma)},$$

$$z(x) = (d-x)^{1-\gamma} + c_2(d-x)^{3-2\gamma} + \ldots, \quad c_2 = \frac{m}{a(3-2\gamma)(2-\gamma)}.$$

However $z'(x) = (d-x)^{-\gamma}(1-\gamma+o(1))$ and the integral $I[z] = \int_0^d p(x)z'(x)^2\,dx$ is divergent. So the solution with a finite value of I is proportional to y and pyy' vanishes at $x = d$.

If $\gamma = 2$, the corresponding solutions have the form

$$y(x) = (d-x)^{\kappa_1}(1+o(1)),\ z(x) = (d-x)^{\kappa_2}(1+o(1)),$$

where κ_j are the roots of the characteristic equation $a\kappa(\kappa+1)+m = 0$ so that $\kappa_{1,2} = -\frac{1}{2} \pm \sqrt{\frac{1}{4} - \frac{m}{a}}$. If $4m > a$, then both roots are complex and $Re\ \kappa_{1,2} = -\frac{1}{2}$. The integrals $I[y]$ and $I[z]$ diverge. If $4m < a$, then both roots are real, the integral $I[y]$ is finite but $I[z]$ is divergent. We have $pyy'(d) = 0$.

Finally, if $\kappa_1 = \kappa_2$, i.e. $a = 4m$, the solutions have the form

$$y(x) = (d-x)^{-1/2}[1+o(1)],\ z(x) = (d-x)^{-1/2}\ln(d-x)[1+o(1)],$$

so that both integrals $I[y]$ and $I[z]$ are divergent.

Therefore, if $\gamma = 2$ and $I[y]$ is finite, then $4m < a$ and the function pyy' vanishes at $x = d$. □

Lemma 31. *Let $p(x)$ be the function found in Corollary 29. Let*

$$m_1 = \inf_{y\in H_0^1(0,1)} \frac{\int_0^1 p(x)y'(x)^2\,dx}{\int_0^1 y(x)^2\,dx}.$$

Then m_1 coincides with the value indicated in Lemma 28 for $l = 1$. The minimal value is attained on the function y_0 also indicated in Lemma 28.

Proof. Consider a minimizing sequence of functions $y_k(x)$ such that $\int_0^1 y_k(x)^2\,dx = 1$. We can assume that $y_k(x) \geq 0$ in $(0,1)$. The integrals $\int_0^{1/2-\varepsilon} y_k'(x)^2\,dx$ and $\int_{1/2+\varepsilon}^1 y_k'(x)^2\,dx$ are bounded and one can choose a subsequence converging almost everywhere in $(0,1/2-\varepsilon)$ and $(1/2+\varepsilon,1)$, in $L_2(0,1)$ and weakly in $H^1(0,1/2-\varepsilon)$ and $H^1(1/2+\varepsilon,1)$. Using diagonalization, one can find a subsequence converging almost everywhere in $(0,1)$ to $u(x)$ and $u(x) \geq 0$. Then

$$\int_0^1 u(x)^2\,dx = 1,\ \int_0^1 p(x)u'(x)^2\,dx \leq m_1.$$

But m_1 is the minimal value of the latter integral. Therefore, $\int_0^1 p(x)u'(x)^2\,dx = m_1$. The function u satisfies the equation

$$(p(x)u(x)')' + m_1 u(x) = 0,$$

$u(0) = 0,\ u(1) = 0,\ u(x) \geq 0$ in $(0,1)$.

On the other hand,

$$(p(x)y_0')' + my_0 = 0,\ y_0(0) = y_0(1) = 0$$

and $y_0 > 0$ in $(0,1)$. We have by Lemma 30

$$(m-m_1)\int_0^1 u(x)y_0(x)\,dx = p(x)(z'(x)y_0(x) - z(x)y_0'(x))|_{x=0}^{x=1} = 0$$

and therefore $m_1 = m$, $u = Cy_0$. □

Proof of Theorem 25 (*continuation*). Let p be an arbitrary positive function of the class R_α. As we have seen before, there exists a function $u(x) \in K_r(0,b,1)$ such that $p(x) = |u'(x)|^{2/(\alpha-1)}$, where $0 < b < 1$. Therefore, $L[p,u] = G[u]$ and $M_\alpha \leq m_1$. On the other hand, we have the equality $L[p_0, y_0] = G[y_0]$, if $p_0(x) = |y_0(x)|^{2/(\alpha-1)}$. By Lemma 31 we have

$$m_1 = \inf_{y \in H_0^1(0,1)} \frac{\int_0^1 p_0(x) y'(x)^2 \, dx}{\int_0^1 y(x)^2 \, dx} = L[p_0, y_0].$$

6. Let now $0 > \alpha > -1/2$. We will use the same function u as above, in the beginning of step 5. Let $r = 2\alpha/(\alpha - 1)$. Then $0 < r < 2/3$. By the Hölder inequality

$$\int_0^1 u'(x)^r \, dx \leq (\int_0^1 |u'(x)| \, dx)^r (\int_0^1 dx)^{1-r} \leq (\int_0^1 |u'(x)| \, dx)^r.$$

Since $\int_0^1 u'(x)^r \, dx = 1$, it follows that $\int_0^1 |u'(x)| \, dx \geq 1$ and therefore

$$M = \max u(x)^2 = u(b)^2 \geq 1/4.$$

We can assume that $\int_0^r u'(x)^r \, dx \geq 1/2$. Let x_1 be such a point of $(0,b)$ that $u(x_1) = 1/4^{1/r}$. Then as above we have

$$\int_0^{x_1} |u'(x)|^r \, dx \leq (\int_0^{x_1} |u'(x)| \, dx)^r = 1/4.$$

Therefore, $\int_{x_1}^b |u'(x)|^r \, dx \geq 1/4$. On the other hand, by the Hölder inequality

$$\begin{aligned} 1/4 &\leq \int_{x_1}^b |u'(x)|^r \, dx = \int_{x_1}^b |u'(x)|^r u(x)^{2r-2} u(x)^{2-2r} \, dx \\ &\leq (\int_{x_1}^b u'(x) u(x)^{2-2/r} \, dx)^r (\int_{x_1}^r u(x)^2 \, dx)^{1-r} \\ &= [(u(x_1))^{3-2/r} - u(b)^{3-2/r}]^r / (2/r - 3)^r (\int_{x_1}^b u(x)^2 \, dx)^{1-r} \\ &\leq C_1(\alpha) (\int_0^1 u(x)^2 \, dx)^{1-r}. \end{aligned}$$

Therefore, $\int_0^1 u(x)^2 \, dx \geq (4C_1(\alpha))^{1/(r-1)}$ and $\lambda_1 \leq (4C_1(\alpha))^{1/(1-r)}$.

As above to prove the attainment of the optimal value we need some lemmas.

Lemma 32. *Let $0 < r < 2/3$ and let K be the set of absolutely continuous non-negative in $[0,h]$ functions y of the space $W_r^1(0,h)$ such that $y(0) = 0$, $y'(x) \geq 0$. Let*

$$G[y] = \frac{(\int_0^h y'(x)^r \, dx)^{2/r}}{\int_0^h y(x)^2 \, dx}$$

and $m = \sup_{y\in K} G[y]$. *Then*

$$m = \Big(\frac{2-2r}{2-3r}\Big)^{2/r}\big(\frac{2}{r}-3\big)h^{2/r-3}\Big(\int_0^\infty \frac{dt}{(1+t^2)^{1/r}}\Big)^2$$

and there exists a function $y_0(x)$ *absolutely continuous and such that* $y_0(0) = 0$, $y'(x) \geq 0$ *and* $G[y_0] = m$. *As* $x \to h$ *we have*

$$y_0(x) = y_0(1) + c_1(h-x)^{r/(r-2)}[1+o(1)],\ y_0'(x) = c_2(h-x)^{2/(r-2)}[1+o(1)].$$

Proof. Let at first $h = 1$ and $\{y_k\}$ be such a sequence of functions of $K_r(0,1)$ that $\int_0^1 y_k'(x)^r\,dx = 1$ and $G[y_k] \to m$.

Let us show that we can assume all functions y_k to be smooth. Let $y \in K_r(0,1)$. Let us define y on the whole line, setting $y_1(x) = 0$ for $x < 0$, $y_1(x) = y(x)$ for $0 < x < 1-\varepsilon$ and $y_1(x) = y(1-\varepsilon)$ for $x > 1-\varepsilon$. Obviously, $\int_0^1 |y_1(x) - y(x)|^2\,dx \to 0$ as $\varepsilon \to 0$ and $\int_{1-\varepsilon}^1 |y'(x)|^r\,dx \to 0$, as $\varepsilon \to 0$. This allows us to assume that y_k are bounded functions.

Let now $y \in K_r(0,1)$ and $0 \leq y(x) \leq C$. Then $\int_0^1 |y'(x)|\,dx = y(h) \leq C$. Set $z_k(x) = y_k(x) - y_k(0)$, where y_k is the averaging of y with a positive kernel ω such that

$$y_k(x) = k\int \omega(k(x-t))y(t)\,dt,\ \int\omega(t)\,dt = 1,\ \omega(t)\geq 0,$$

and $\omega \in C_0^\infty(-1,1)$. Observe that $|y_k(0)| \leq \delta_k \to 0$ as $k\to\infty$, so that z_k converge to $y(x)$ uniformly, $z_k'(x) \geq 0$, and

$$\int_0^1 |z_k(x) - y(x)|^2\,dx \to 0, \quad \int_0^1 |z_k'(x) - y'(x)|\,dx \to 0.$$

We have

$$\int_0^1 |z_k'(x) - y'(x)|^r\,dx \leq \Big(\int_0^1 |z_k'(x) - y'(x)|\,dx\Big)^r.$$

The elementary inequality $|a^r - b^r| \leq |a-b|^r$, valid for all a, b such that $a\geq 0, b\geq 0$, implies that $\int_0^1 z_k'(x)^r\,dx \to \int_0^1 y'(x)^r\,dx$. This allows us to assume that all y_k are smooth functions.

As above we call a function y *convex*, if its derivative is decreasing, and *concave*, if its derivative is increasing.

Let us show that if y_k is convex in an interval (x_1, x_2), where $0 \leq x_1 \leq x_2 \leq 1$, then it is possible to replace it by the linear function

$$z(x) = y_k(x_1) + \gamma(x - x_1),\ \text{where } \gamma = [y_k(x_2) - y_k(x_1)]/(x_2 - x_1)$$

and the value of $G[z]$ will be bigger than $G[y_k]$.

Indeed, we have $y_k(x_1) = z(x_1),\ y_k(x_2) = z(x_2)$, and

$$\int_{x_1}^{x_2} y_k(x)^2\,dx \geq \int_{x_1}^{x_2} z(x)^2\,dx;$$

$$\int_{x_1}^{x_2} y_k'(x)^r\,dx \leq \Big(\int_{x_1}^{x_2} y_k'(x)\,dx\Big)^r \Big(\int_{x_1}^{x_2} dx\Big)^{1-r}$$

$$= \gamma^r(x_2 - x_1) = \int_{x_1}^{x_2} z'(x)^r\,dx.$$

If z coincides with y_k outside the interval (x_1, x_2), then $G[z] \geq G[y_k]$.

We can do the same for all other intervals where y_k is convex. As a result we shall obtain a function z_k, which can be described in the following way. Consider the set of points (x, y) such that $0 < x < 1$ and $y > y_k(x)$, and take the convex hull of this set. The lower boundary of the hull serves as the graph of the function z_k. Let us show that $z_k \in C^1(0,1)$. Indeed, if a point $(x, z_k(x))$ belongs to an interval of the straight line, which is a part of the graph of z_k, then it is obvious that z_k is smooth at the point x. The same is true in the case when the point $(x, z_k(x))$ belongs to a part of the graph of the function y_k. If z_k is linear on one side of x and coincides with y_k on the other side, then z_k is regular at x since the graph of y_k is lying on one side of the straight line, obtained by the continuation of the linear function. Finally, if the point $(x, z_k(x))$ is a limit point for a sequence of such points, then it is also a limit point for a sequence of points belonging to the graph of y_k and the derivative $z_k'(x)$ exists.

Therefore, if one replaces every function y_k by a concave function z_k in the indicated way, then the sequence of new functions will be maximizing. This allows us to consider as maximizing the sequences of increasing concave functions, i.e. to suppose that the functions $y_k(x)$ and their derivatives $y_k'(x)$ are increasing.

For large k we have $\int_0^1 y_k(x)^2\,dx \leq m^{-1} + 1$. Let $\varepsilon > 0$ be small enough. There is a point $\theta_k \in (1-\varepsilon, 1)$ such that $|y_k(\theta_k)|^2 \leq (m^{-1}+1)\varepsilon^{-1}$. Since the functions y_k are monotone, they are uniformly bounded for $0 \leq x \leq 1-\varepsilon$.

Analogously, we can deduce from the equality $\int_0^1 y_k'(x)^r\,dx = 1$ that the sequence $\{y_k'(x)\}$ is uniformly bounded in $[0, 1-\varepsilon]$. By the Arzela theorem one can choose an uniformly converging subsequence $\{y_{n_k}(x)\}$, and by the Helly theorem one can suppose that the subsequence $\{y_{n_k}'(x)\}$ converges everywhere in $[0, 1-\varepsilon]$. Using the diagonalization, one can find a subsequence converging to a function $y_0 \in K_r(0,h)$ such that the sequence of the first derivatives converges almost everywhere in $[0,1)$ and y_0 satisfies the Lipschitz condition on the interval $[0, 1-\varepsilon]$ for any $\varepsilon > 0$. It is easy to see that the estimate obtained in the beginning of step 6 implies that

$$\Big(\int_{1-\varepsilon}^{1} y_k'(x)^r\,dx\Big)^{1/r} \leq C\varepsilon^{1/r-3/2},$$

where C is independent of k and, since $1/r-3/2>0$, we obtain that $\int_0^1 y_0'(x)^r dx=1$.

Besides, we have $\int_0^1 y_0(x)^2\,dx \le m^{-1}$. Since this integral cannot be less than m^{-1}, we see that $\int_0^1 y_0(x)^2\,dx = m^{-1}$.

If $y_0'(x_0) > 0$, then $y_0'(x) > 0, y_0(x) > 0$, for all $x > x_0$. Let us assume at first that $y_0'(x) > 0$ for $x > 0$. Then the function $G[y_0 + tz]$ for any $z \in H_0^1(0,1)$ has a minimal value at $t = 0$ and hence $\frac{d}{dt}G[y_0 + tz] = 0$ if $t = 0$. This gives the Euler-Lagrange equation of the form

$$(y_0'^{r-1})' + my_0 = 0,$$

so that $(r-1)y_0'^{r-1}y_0'' + my_0y_0' = 0$ or $y_0'^r - m_1y_0^2 = C$, where $m_1 = -mr/[2(r-1)] > 0$. Integrating this equality over $(0,1)$, we obtain that $C = 1 - m_1m^{-1} = 1 + r/2(r-1) > 0$. Therefore

$$\int_0^y \frac{dz}{(C + m_1z^2)^{1/r}} = x.$$

We have for any $\varphi \in K_r(0,1)$ the equality

$$\int_0^1 (y_0'(x)^{r-1}\varphi'(x) - my_0(x)\varphi(x))\,dx = 0.$$

Let $\varphi(x) = 1$ for $1 - \varepsilon < x < 1$. Then integrating by parts gives the equality

$$y_0'(1-\varepsilon)^{r-1}\varphi(1-\varepsilon) - m\int_{1-\varepsilon}^1 y_0(x)\varphi(x)\,dx = 0.$$

For ε tending to zero, we obtain that $y_0'(1) = \infty$. The equality $y_0'^r - m_1y_0^2 = C$ yields that $y_0(1) = \infty$, too.

Therefore

$$\int_0^\infty \frac{dz}{(C + m_1z^2)^{1/r}} = 1.$$

Let $z = (C/m_1)^{1/2}t$ so that

$$\int_0^\infty \frac{dt}{(1+t^2)^{1/r}} = C^{1/r-1/2}m_1^{1/2} = \frac{(2-3r)^{1/r-1/2}r^{1/2}m^{1/2}}{(2-2r)^{1/r}}$$

and therefore

$$m^{1/2} = \frac{(2-2r)^{1/r}}{r^{1/2}(2-3r)^{1/r-1/2}}\int_0^\infty \frac{dt}{(1+t^2)^{1/r}}.$$

If $y_0(x) \equiv 0$ for $0 < x < x_0$ and $y_0'(x) > 0$ for $x > x_0$, then $m = C_1(1-x_0)^{1/r-3/2}$, where C_1 does not depend on x_0 and therefore the optimal value of x_0 is 0. Observe that

$$\int_{y_0}^\infty \frac{dz}{(C + m_1z^2)^{1/r}} = 1 - x.$$

Therefore, for big values of y_0 we have

$$y_0^{1-2/r}(1+o(1)) = C_1(1-x),$$

which implies that $y_0(x) = A(1-x)^\gamma[1+o(1)]$ with $\gamma = r/(r-2) < 0$.

In order to find the value m for an arbitrary $h > 0$ it suffices to substitute in the obtained estimate the function $y_0(xh)$. □

Lemma 33. *Let* $0 < r < 2/3$ *and* $m_1 = \sup_{y\in\mathcal{L}} G[y]$, *where*

$$G[y] = \frac{(\int_0^l |y'(x)|^r\,dx)^{2/r}}{\int_0^l y(x)^2\,dx},$$

and $\mathcal{L}$ *is the class of nonnegative functions* y *of* $W^1_{2,0}(0,d)$, *having a finite value of the integral* $\int_0^l |y'(x)|^r\,dx$, *which are monotone for* $0 < x < h$ *and for* $h < x < l$. *Then there is a function* $y_0 \in \mathcal{L}$ *such that* $y_0(x) = y_0(l-x)$, $G[y_0] = m_1$, *and* $m_1 = 4ml^{2/r-3}$, *where the value* m *was indicated in Lemma 32 for* $h = 1$.

Proof. Let at first $l = 1$. The existence of the extremal function for any fixed $h \in [0,1]$ follows from Lemma 32. Furthermore, we can suppose that $y(x)$ is monotone and concave in $[0,h]$ and in $[h,1]$.

By Lemma 32 we have for $y \in \mathcal{L}$ the inequalities

$$h^{3-2/r}\Big(\int_0^h |y'(x)|^r\,dx\Big)^{2/r} \le m\int_0^h y(x)^2\,dx,$$
$$(1-h)^{3-2/r}\Big(\int_h^1 |y'(x)|^r\,dx\Big)^{2/r} \le m\int_h^1 y(x)^2\,dx,$$

where the value of m, corresponding to $h = 1$, was found in Lemma 32. Let $\int_0^1 |y'(x)|^r\,dx = 1$ and $\int_0^h |y'(x)|^r\,dx = a$. Then

$$a^{2/r}h^{3-2/r} + (1-a)^{2/r}(1-h)^{3-2/r} \le m.$$

By Lemma 26 the function $F(a,h) = a^{2/r}h^{3-2/r} + (1-a)^{2/r}(1-h)^{3-2/r}$ defined in the square $0 \le a \le 1$, $0 \le h \le 1$, has a minimal value $1/4$ at the point $a = h = 1/2$. Therefore,

$$\int_0^1 y(x)^2\,dx \ge 1/4m,$$

i.e. $G[y] \le 4m$ for all admissible y. On the other hand, if $a = h = 1/2$ and the function y coincides on $(0,1/2)$ with the function y_0, found in Lemma 32 for $h = 1/2$, and is even with respect to the point $x = 1/2$, then $G[y] = 4m$.

In order to find the value m for an arbitrary $l > 0$ it suffices to substitute in the obtained estimate the function $y_0(xl)$. □

Corollary 34. *Let $-1/2 < \alpha < 0$, $\alpha \neq 0$. Set $p(x) = |y_0'(x)|^{2/(\alpha-1)}$, where y_0 is the function found in Lemma 33. Then $p(x) = p(l-x)$ for $0 < x < l$,*

$$p(x) = a(x-l/2)^2[1+o(1)], \quad p'(x) = 2a(l/2-x)[1+o(1)] \text{ as } x \to l/2.$$

Lemma 35. *Let $-1/2 < \alpha < 0$, and $p(x), y_0(x)$ be the functions indicated in Lemma 33 and Corollary 34. Let*

$$m_1 = \inf_{y \in H_0^1(0,1)} \frac{\int_0^1 p(x)y'(x)^2\,dx}{\int_0^1 y(x)^2\,dx}.$$

Then

$$m_1 = -4\frac{2\alpha+1}{\alpha}\left(\frac{\alpha+1}{2\alpha+1}\right)^{1-1/\alpha}\left(\int_0^\infty \frac{dt}{(1+t^2)^{1/2-1/2\alpha}}\right)^2.$$

The minimal value is attained when $y = y_0$.

Proof. Consider a minimizing sequence $\{y_k(x)\}$ such that $\int_0^1 y_k(x)^2\,dx = 1$. We can assume that $y_k(x) \geq 0$. Since the integrals $\int_0^{1/2-\varepsilon} y_k'(x)^2\,dx$ and $\int_{1/2+\varepsilon}^1 y_k'(x)^2\,dx$ are bounded, one can choose a subsequence converging almost everywhere in $(0, 1/2-\varepsilon)$ and $(1/2+\varepsilon, 1)$, in $L_2(0,1)$ and weakly in $H^1(0,1/2-\varepsilon)$ and $H^1(1/2+\varepsilon, 1)$. Using diagonalization, one can find a subsequence converging almost everywhere in $(0,1)$ to $u(x)$ and $u(x) \geq 0$ in $(0,1)$. Then

$$\int_0^1 u(x)^2\,dx = 1, \quad \int_0^1 p(x)u'(x)^2\,dx \leq m_1.$$

But m_1 is the minimal value of the latter integral. Therefore $\int_0^1 p(x)u'(x)^2\,dx = m_1$. The function u satisfies the equation

$$(p(x)u(x)')' + m_1 u(x) = 0, \; u(0) = 0, \; u(1) = 0.$$

On the other hand,

$$(p(x)y_0')' + my_0 = 0, \; y_0(0) = y_0(1) = 0$$

and $y_0(x) > 0$ in $(0,1)$. We have by Lemma 30

$$(m - m_1)\int_0^1 u(x)y_0(x)\,dx = p(x)(u'(x)y_0(x) - u(x)y_0'(x))|_{x=0}^{x=1} = 0$$

and $m_1 = m$, $u = Cy_0$. □

7. Let $\alpha < -1/2$. Set

$$p(x) = \begin{cases} k^{1/\alpha}, & \text{if } 0 < x < \varepsilon, \\ k, & \text{if } \varepsilon < x < 1, \end{cases}$$

where k is such a number that $k\varepsilon + k^\alpha(1-\varepsilon) = 1$, so that $k \sim \varepsilon^{-1}$ and $\int_0^1 p(x)^\alpha\,dx = 1$. For any $\varepsilon > 0$ we have $\int_0^\varepsilon y(x)^2\,dx \le \varepsilon^2 \int_0^1 y'(x)^2\,dx$. On the other hand,

$$\int_\varepsilon^1 y(x)^2\,dx \le (1-\varepsilon)^2 \int_\varepsilon^1 y'(x)^2\,dx.$$

Therefore

$$\int_0^1 y(x)^2\,dx \le \varepsilon^2 \int_0^1 y'(x)^2\,dx + (1-\varepsilon)^2 \int_\varepsilon^1 y'(x)^2\,dx \le \delta \int_0^1 p(x) y'(x)^2\,dx,$$

where

$$\delta = \max(\varepsilon^2 k^{-1/\alpha},\ (1-\varepsilon)^2 k^{-1}) \le \max(\varepsilon^{2+1/\alpha},\ (1-\varepsilon)^2\varepsilon),$$

so that $\delta \to 0$ as $\varepsilon \to 0$. Hence $\lambda_1 \ge \frac{1}{\delta} \to \infty$ as $\varepsilon \to 0$ and $M_\alpha = \infty$.

8. Finally, consider the case when $\alpha = -1/2$. Set $p(x) = \max(x^2/\varepsilon^2,\ \delta^2/\varepsilon^2)$ for $0 < x < 1$, where $\delta = \exp(1 - 1/\varepsilon)$. Observe that

$$\int_0^1 p(x)^{-1/2}\,dx = \delta \cdot \varepsilon/\delta + \int_0^1 \varepsilon/x\,dx = \varepsilon - \varepsilon \ln \delta = 1.$$

On the other hand, from the well-known estimate

$$\int_0^1 y(x)^2\,dx \le 4 \int_0^1 x^2 y'(x)^2\,dx,$$

valid for all functions $y \in C^1$, vanishing in 0, it follows that

$$\int_0^1 y(x)^2\,dx \le 4\delta^2 \int_0^\delta y'(x)^2\,dx + 4 \int_\delta^1 y(x)^2\,dx \le 4\varepsilon^2 \int_0^1 p(x) y'(x)^2\,dx.$$

This means that for the chosen function p

$$\lambda_1 = \inf_y \frac{\int_0^1 p(x) y'(x)^2\,dx}{\int_0^1 y(x)^2\,dx} \ge \frac{1}{4\varepsilon^2},$$

so that $M_{-1/2} = \infty$. The proof of Theorem 25 is complete. □

Corollary 36.

If $\alpha > -1/2,\ \alpha \ne 0,$ *then* $\lambda_1 \le C(\alpha)(\int_0^1 p(x)^\alpha\,dx)^{1/\alpha};$

if $\alpha \le -1,$ *then* $\lambda_1 \ge C(\alpha)(\int_0^1 p(x)^\alpha\,dx)^{1/\alpha};$

where $C(\alpha)$ *is indicated in Theorem 25.*

5.4 On a more general estimate of the first eigenvalue of the Sturm-Liouville operator

In this section the Sturm-Liouville problem

$$(P(x)y')' + \lambda Q(x)y = 0, \qquad 0 < x < 1,\ y(0) = y(1) = 0.$$

is considered. Our aim is to estimate the minimal eigenvalue λ_1 of this problem under the condition that the nonnegative measurable functions $P(x)$ and $Q(x)$ are such that

$$\int_0^1 P(x)^\alpha\, dx = 1, \int_0^1 Q(x)^\beta\, dx = 1,$$

where α and β are nonzero real numbers, i.e. $P \in R_\alpha$, $Q \in R_\beta$. The variational principle implies that

$$\lambda_1 = \inf_{y \in H_0^1(0,1)} \frac{\int_0^1 P(x)y'(x)^2\, dx}{\int_0^1 Q(x)y(x)^2\, dx},$$

where the greatest lower bound is taken in the class of all nonzero functions from $C_0^1[0,1]$.

Let us set

$$M_{\alpha,\beta} = \sup_{P \in R_\alpha, Q \in R_\beta} \lambda_1,\ m_{\alpha,\beta} = \inf_{P \in R_\alpha, Q \in R_\beta} \lambda_1.$$

The main result of this section is the following.

Theorem 37. *If $\alpha > -1/2$, $\beta - \alpha + 2\alpha\beta < 0$, then $M_{\alpha,\beta} \le C(\alpha,\beta)$ and $m_{\alpha,\beta} = 0$.*

If $\alpha \le -1$, $\beta \ge 1$, the value $m_{\alpha,\beta} \ge C(\alpha,\beta) > 0$ and $M_{\alpha,\beta} = \infty$.

If $1/\alpha - 1/\beta + 2 \le 0$ and either $\alpha > -1$ or $\beta < 1$, then $m_{\alpha,\beta} = 0$ and $M_{\alpha,\beta} = \infty$.

Proof.

A. Estimates of $M_{\alpha,\beta}$.

1. Let at first $\alpha > 0$, $\beta > 0$, and $\beta - \alpha + 2\alpha\beta > 0$. We show that $M_{\alpha,\beta} = \infty$. For this we set $P(x) = \varepsilon^{-1/\alpha}$ for $0 < x < \varepsilon$ and $P(x) = 0$ for $\varepsilon < x < 1; Q(x) = \varepsilon^{-1/\beta}$ for $0 \le x \le \varepsilon$ and $Q(x) = 0$ for $\varepsilon \le x \le 1$, where ε is a small positive number. Then for $y \in C_0^1(0,1)$ we have

$$\int_0^1 Q(x)y(x)^2\, dx = \varepsilon^{-1/\beta} \int_0^\varepsilon y(x)^2\, dx; \quad \int_0^1 P(x)y'(x)^2\, dx = \varepsilon^{-1/\alpha} \int_0^\varepsilon y'(x)^2\, dx.$$

Since

$$\int_0^\varepsilon y(x)^2\, dx \le \varepsilon^2 \int_0^\varepsilon y'(x)^2\, dx; y(x) \in C_0^1(0,1),$$

we have that

$$\int_0^1 Q(x)y(x)^2\,dx \le \varepsilon^{2-1/\beta+1/\alpha}\int_0^1 P(x)y'(x)^2\,dx;$$

i.e. $\lambda_1 \ge \varepsilon^{-2+1/\beta-1/\alpha}$ and $\lambda_1 \to \infty$ as $\varepsilon \to 0$. Therefore, in this case $M_{\alpha,\beta} = \infty$.

2. If $\alpha < 0$, $\beta > 0$, we set

$$P(x) = \begin{cases} \varepsilon^{-1/\alpha}(1-\varepsilon)^{1/\alpha} & \text{for } 0 < x < \varepsilon; \\ \varepsilon^{1/\alpha}(1-\varepsilon)^{-1/\alpha} & \text{for } \varepsilon < x < 1; \end{cases}$$

$$Q(x) = \begin{cases} 0 & \text{for } 0 < x < 2\varepsilon; \\ (1-2\varepsilon)^{-1/\beta} & \text{for } 2\varepsilon < x < 1, \end{cases}$$

where $0 < \varepsilon < 1/8$. It is clear that

$$\int_0^1 P(x)^\alpha\,dx = 1,\ \int_0^1 Q(x)^\beta\,dx = 1,$$

and

$$(1-2\varepsilon)^{1/\beta}\int_0^1 Q(x)y(x)^2\,dx = \int_{2\varepsilon}^1 y(x)^2\,dx \le \varepsilon^{-1/\alpha}(1-\varepsilon)^{1/\alpha}\int_{2\varepsilon}^1 P(x)y'(x)^2\,dx.$$

Hence it follows that $\lambda_1 \ge \varepsilon^{1/\alpha}/2$ and so $M_{\alpha,\beta} = \infty$.

3. Let now $\alpha < 0$, $\beta < 0$ and $\beta - \alpha + 2\alpha\beta > 0$.

We set

$$P(x) = \begin{cases} \varepsilon^{-1/\alpha}(1-\varepsilon)^{1/\alpha} & \text{for } 0 < x < \varepsilon; \\ \varepsilon^{1/\alpha}(1-\varepsilon)^{-1/\alpha} & \text{for } \varepsilon < x < 1; \end{cases}$$

$$Q(x) = \begin{cases} \varepsilon^{-1/\beta}(1-\rho)^{1/\beta} & \text{for } 0 < x < \varepsilon; \\ \rho^{1/\beta}(1-\varepsilon)^{-1/\beta} & \text{for } \varepsilon < x < 1; \end{cases}$$

where $0 < \varepsilon < 1/8$, $0 < \rho < 1/8$. It is clear that

$$\int_0^1 P(x)^\alpha\,dx = 1,\ \int_0^1 Q(x)^\beta\,dx = 1,$$

and

$$\begin{aligned}\int_0^1 Q(x)y(x)^2\,dx &= \varepsilon^{-1/\beta}(1-\rho)^{1/\beta}\int_0^\varepsilon y(x)^2 + \rho^{1/\beta}(1-\varepsilon)^{-1/\beta}\int_\varepsilon^1 y(x)^2\,dx \\ &\le \varepsilon^{2-1/\beta+1/\alpha}(1-\rho)^{1/\beta}(1-\varepsilon)^{-1/\alpha}\int_0^\varepsilon P(x)y'(x)^2\,dx \\ &\quad + \rho^{1/\beta}\varepsilon^{-1/\alpha}(1-\varepsilon)^{1/\alpha-1/\beta+2}\int_\varepsilon^1 P(x)y'(x)^2\,dx \\ &\le C_1\int_0^1 P(x)y'(x)^2\,dx,\end{aligned}$$

where

$$C_1 = \max(\varepsilon^{2-1/\beta+1/\alpha}(1-\rho)^{1/\beta}(1-\varepsilon)^{-1/\alpha}, \rho^{1/\beta}\varepsilon^{-1/\alpha}(1-\varepsilon)^{1/\alpha-1/\beta+2}).$$

If we set $\rho = \varepsilon^{\beta/2\alpha}$, then for small ε we have $C_1 \le 2\varepsilon^{\gamma}$, where

$$\gamma = \min(2 - 1/\beta + 1/\alpha, \beta/2\alpha) > 0.$$

Since $\lambda_1 \ge C_1^{-1}$, we see that in this case $M_{\alpha,\beta} = \infty$.

4. Now we show that $M_{\alpha,\beta} = \infty$ if

$$\beta - \alpha + 2\alpha\beta = 0,\ \alpha > -1/2.$$

Observe that for $\alpha \ne 0$, $\beta \ne 0$, $\beta - \alpha + 2\alpha\beta = 0$, and $\alpha > -1/2$, α and β have the same signs. Set for $\alpha > 0$

$$P(x) = C_1 x^{\varepsilon-1/\alpha}, \quad Q(x) = C_2 x^{\varepsilon-1/\beta};\ \varepsilon > 0,$$

and for $\alpha < 0$

$$P(x) = C_1 x^{-\varepsilon-1/2\alpha}, \quad Q(x) = C_2 x^{-\varepsilon-1/\beta}.$$

The constants C_1, C_2 are chosen so that

$$\int_0^1 P(x)^{\alpha}\,dx = 1,\ \int_0^1 Q(x)^{\beta}\,dx = 1,$$

i.e. $C_1 = |\alpha\varepsilon|^{1/\alpha}$, $C_2 = |\beta\varepsilon|^{1/\beta}$. From the Hardy inequality

$$\int_0^1 x^{\varepsilon-1/\beta} y(x)^2\,dx \le C_0 \int_0^1 x^{\varepsilon-1/\beta+2} y'(x)^2\,dx;\ y(x) \in C^1(0,1),$$

where $C_0 = 4/[1 + (\varepsilon - 1/\beta)^2]$, it follows that

$$\int_0^1 Q(x)y(x)^2\,dx \le C_0 C_2/C_1 \int_0^1 P(x)y'(x)^2\,dx.$$

It remains to note that

$$C_2/C_1 = |\alpha\varepsilon|^{-1/\alpha}|\beta\varepsilon|^{1/\beta} = |\alpha|^{-1/\alpha}|\beta|^{1/\beta}\varepsilon^2 \to 0$$

as $\varepsilon \to 0$. This means that

$$\lambda_1 \ge |\alpha|^{1/\alpha}|\beta|^{-1/\beta}\varepsilon^{-2},$$

i.e. $M_{\alpha,\beta} = \infty$.

5. Now we show that $M_{\alpha,\beta} < \infty$ if

$$-1/2 < \alpha < 0,\ \beta - \alpha + 2\alpha\beta < 0.$$

Set $y_0'(x)^2 = P(x)^{\alpha-1}$ and define the function $y_0(x)$ so that it increases on the segment $[0, x_0]$ from 0 to some $m > 0$ and decreases on the segment $[x_0, 1]$ from m to 0. By the Hölder inequality

$$1 = \int_0^1 |y_0'(x)|^{2\alpha/(\alpha-1)}\,dx \le \Big(\int_0^1 |y_0'(x)|\,dx\Big)^{2\alpha/(\alpha-1)}.$$

Therefore

$$\int_0^1 |y_0'(x)|\,dx \ge 1,$$

and so $m \ge 1/2$. For definiteness, let

$$\int_0^{x_0} |y_0'(x)|^{2\alpha/(\alpha-1)}\,dx \ge 1/2.$$

Let $x_1 \in (0,\, x_0)$ be a point such that $y_0(x_1)^{2\alpha} = 4^{1-\alpha}$. Then

$$\int_0^{x_1} y_0'(x)^{2\alpha/(\alpha-1)}\,dx \le \Big(\int_0^{x_1} y_0'(x)\,dx\Big)^{2\alpha/(\alpha-1)} = 1/4.$$

Therefore

$$\int_{x_1}^{x_0} y_0'(x)^{2\alpha/(\alpha-1)}\,dx \ge 1/4.$$

On the other hand, by the Hölder inequality

$$\int_{x_1}^{x_0} y_0'(x)^{2\alpha/(\alpha-1)}\,dx$$
$$\le \Big(\int_{x_1}^{x_0} y_0(x)^{2\beta/(\beta-1)}\,dx\Big)^{(1+\alpha)/(1-\alpha)}\Big(\int_{x_1}^{x_0} y_0' y^{\beta(1+\alpha)/\alpha(\beta-1)}\,dx\Big)^{2\alpha/(\alpha-1)}$$

and so

$$1/4 \le \int_{x_1}^{x_0} y_0'(x)^{2\alpha/(\alpha-1)}\,dx$$
$$\le \Big(\int_{x_1}^{x_0} y_0^{2\beta/(\beta-1)}\,dx\Big)^{(1+\alpha)/(1-\alpha)} \cdot [(m^{s+1} - 4^{-s-1})/s]^{2\alpha/(\alpha-1)},$$

where

$$1 + s = \frac{\beta(1+\alpha)}{\alpha(\beta-1)} + 1 = \frac{2\alpha\beta - \alpha + \beta}{\alpha(\beta-1)} < 0.$$

If $\beta < 0$, then

$$\int_0^1 |y(x)|^{2\beta/(\beta-1)}\,dx \le \Big(\int_0^1 Q(x)y^2(x)\,dx\Big)^{\beta/(\beta-1)}\Big(\int_0^1 Q(x)^\beta\,dx\Big)^{1/(1-\beta)}$$

and therefore

$$\int_0^1 Q(x)y(x)^2\,dx \ge \Big(\int_0^1 |y(x)|^{2\beta/(\beta-1)}\,dx\Big)^{(\beta-1)/\beta}.$$

Thus the obtained estimate $\int_0^{x_1} y_0^{2\beta/(\beta-1)}\,dx \ge C_1 > 0$ implies that $\lambda_1 \le C_1^{(1-\beta)/\beta}$.

6. Let now $0<\alpha\le 1$, $\beta-\alpha+2\alpha\beta<0$. We obtain the estimate above for λ_1.

The function $P(x)$ is nonnegative and $\int P(x)^\alpha\,dx = 1$. It is sufficient to get the uniform estimate above for the functions $P(x)$, taking positive values only. Set $y_0'(x)^2 = P(x)^{\alpha-1}$ and define the function $y_0(x)$ so that it increases on the segment $[0,x_0]$ from 0 to some $m>0$ and decreases on the segment $[x_0,1]$ from m to 0. We have for $0<\alpha<1$

$$\int_0^1 P(x)y_0'(x)^2\,dx = \int_0^1 |y_0'(x)|^{2\alpha/(\alpha-1)}\,dx = \int_0^1 P(x)^\alpha\,dx = 1.$$

If $\alpha=1$, then $y_0(x)=|x-1/2|-1/2$ and $\int_0^1 P(x)y_0'(x)^2\,dx=1$, too.

By the Hölder inequality we have for $0<\beta<1$

$$\int_0^1 Q(x)^\beta\,dx \le \Big(\int_0^1 Q(x)y^2(x)\,dx\Big)^\beta\Big(\int_0^1 y(x)^{2\beta/(\beta-1)}\,dx\Big)^{1-\beta},$$

so that

$$\int_0^1 Q(x)y(x)^2\,dx \ge \Big(\int_0^1 y(x)^{2\beta/(\beta-1)}\,dx\Big)^{(\beta-1)/\beta}.$$

Thus the estimate from above follows from the inequality

$$\int_0^1 y(x)^{2\beta/(\beta-1)}\,dx \le C_1,$$

which will be proved.

We note that for $0<x<x_0$ we have

$$\begin{aligned} x = \int_0^x dt &= \int_0^x y_0'(x)^s y_0'(x)^{-s}\,dt \\ &\le \Big(\int_0^1 y_0'(x)^{sp_1}\,dt\Big)^{1/p_1}\cdot\Big(\int_0^1 y_0'(x)^{-sp_2}\,dt\Big)^{1/p_2}\cdot x^{1/p_3}, \end{aligned}$$

where

$$1>s>0,\quad p_1=\frac{1}{s},\quad p_2=\frac{2\alpha}{s(1-\alpha)},\quad p_3=\frac{1}{1-s/2-s/2\alpha}.$$

Since $\int_0^1 y_0'(x)^{-sp_2}\,dt = 1$, it follows that $x^{s/2+s/2\alpha} \le y_0(x)^{1/p_1}$, i.e. $y_0(x)\ge x^{1/2+1/2\alpha}$ if $0<x<x_0$. Therefore

$$\int_0^{x_0} y_0(x)^{2\beta/(\beta-1)}\,dx \le \int_0^{x_0} x^{(\alpha+1)\beta/\alpha(\beta-1)}\,dx = C_1,$$

since $\frac{(\alpha+1)\beta}{\alpha(\beta-1)} > -1$ in virtue of our conditions. Analogously, one can show that $y_0(x) \geq (1-x)^{1/2+1/2\alpha}$ for $x_0 < x < 1$, and thus $y_0(x)^{2\beta/(\beta-1)}\,dx \leq 2C_1$.

If $\beta < 0$, then

$$\int_0^1 y(x)^{2\beta/(\beta-1)}\,dx \leq \Big(\int_0^1 Q(x)y^2(x)\,dx\Big)^{\beta/(\beta-1)}\Big(\int_0^1 Q(x)^\beta\,dx\Big)^{1/(1-\beta)},$$

and therefore

$$\int_0^1 Q(x)y(x)^2\,dx \geq \Big(\int_0^1 y(x)^{2\beta/(\beta-1)}\,dx\Big)^{(\beta-1)/\beta}.$$

Thus in this case it is sufficient to prove that

$$\int_0^1 y(x)^{2\beta/(\beta-1)}\,dx \geq C > 0.$$

As above we have the estimates

$$y_0(x) \geq x^{1/2+1/2\alpha} \text{ for } 0 < x < x_0,\ y_0(x) \geq (1-x)^{1/2+1/2\alpha} \text{ for } x_0 < x < 1$$

and thus

$$\int_0^1 y_0(x)^{2\beta/(\beta-1)}\,dx \geq \int_0^{x_0} x^{(\alpha+1)\beta/\alpha(\beta-1)}\,dx + \int_{x_0}^1 (1-x)^{(\alpha+1)\beta/\alpha(\beta-1)}\,dx \geq C > 0.$$

7. Let $\alpha > 1$, $\beta - \alpha + 2\alpha\beta < 0$. We show that $\lambda_1 \leq C(\alpha, \beta)$. We have

$$\int_0^1 P(x)y'(x)^2\,dx \leq \Big(\int_0^1 |y'(x)|^{2\alpha/(\alpha-1)}\,dx\Big)^{(\alpha-1)/\alpha}\Big(\int_0^1 P(x)^\alpha\,dx\Big)^{1/\alpha}$$

and $$\int_0^1 Q(x)y(x)^2\,dx \geq \Big(\int_0^1 |y(x)|^{2\beta/(\beta-1)}\,dx\Big)^{(\beta-1)/\beta}.$$

Therefore

$$\lambda_1 \leq \inf_y \frac{(\int_0^1 |y'(x)|^{2\alpha/(\alpha-1)}\,dx)^{\alpha-1)/\alpha}}{(\int_0^1 |y(x)|^{2\beta/(\beta-1)}\,dx)^{(\beta-1)/\beta}} \leq C.$$

Let $y_0(x) = x^\rho$ for $0 \leq x \leq 1/2$ and $y_0(x) = (1-x)^\rho$ for $1/2 \leq x \leq 1$. The number ρ must be such that $2\beta/(\beta-1)\rho > -1$ and $2\alpha(\rho-1)/(\alpha-1) > -1$.

If $\alpha > 1, 0 < \beta < 1$, then such ρ exists if $(1+\alpha)/\alpha < (1-\beta)/\beta$, i.e. for $\beta - \alpha + 2\alpha\beta < 0$. And if $\alpha > 1$, but $\beta < 0$, then as ρ one can take any number, greater than $(1+\alpha)/2\alpha$, since the first condition is satisfied for any $\rho > 0$.

B. Estimates of $m_{\alpha,\beta}$.

1. We prove that $m_{\alpha,\beta} = 0$ for $\beta < 1$. For that we set $P(x) = 1$. If $\beta < 0$, we set $Q(x) = \varepsilon^{-1/\beta}(1-\varepsilon)$ for $x - 1/2 < \varepsilon/2$ and $Q(x) = N$ for $0 < x < 1/2 - \varepsilon/2$ and for $1/2 + \varepsilon/2 < x < 1$, assuming that $0 < \varepsilon < 1/4$ and N is a constant such that

$$N^{\beta}(1-\varepsilon) + \varepsilon^{-1}(1-\varepsilon)^{\beta}\varepsilon = 1,$$

so that $N \to \infty$ as $\varepsilon \to 0$. We note that $b_\varepsilon = \int_0^1 Q(x)^{1/4}\,dx \to \infty$ as $\varepsilon \to 0$. Let $q_\varepsilon(x) = Q_\varepsilon(x) b_\varepsilon^{-4}$. Then from the equation

$$y'' + \lambda_1 Q_\varepsilon(x) y = 0$$

it follows that

$$y'' + m q_\varepsilon(x) y = 0, \quad \text{where } m = \lambda_1 b_\varepsilon^4,$$

and $\int_0^1 q_\varepsilon^{1/4}\,dx = 1$. In virtue of the first part of our theorem, if $\alpha = 2$, $\beta = 1/4$, the first eigenvalue is bounded above, i.e. $m \le C$ and C is independent of ε. But then $\lambda_1 \le C b_\varepsilon^{-4}$, and therefore, $m_{\alpha,\beta} = 0$.

If $0 < \beta < 1$, then we set $P(x) = 1$ and $Q(x) = \varepsilon^{-1/\beta}(1-\varepsilon)$ for $x - 1/2 < \varepsilon/2$ and $Q(x) = 0$ for $0 < x < 1/2 - \varepsilon/2$ and for $x > 1/2 + \varepsilon/2$, assuming that $0 < \varepsilon < 1/4$. Let $y_0 \in C_0^\infty(0,1)$ and $y_0(x) = 1$ for $1/2 - x < \varepsilon$. Then

$$\lambda_1 = \inf_y \frac{\int_0^1 P(x) y'(x)^2\,dx}{\int_0^1 Q(x) y(x)^2\,dx} \le \frac{\int_0^1 y_0'(x)^2\,dx}{\int_0^1 Q(x) y_0(x)^2\,dx} = C\varepsilon^{1/\beta - 1}.$$

Therefore, $\lambda_1 \to 0$ as $\varepsilon \to 0$.

2. Show that $m_{\alpha,\beta} = 0$ for $\alpha > 0$. To this end we set $Q(x) \equiv 1$. Since $\alpha > 0$, the function $P(x)$ can vanish on the segment $[0, 1/2]$, when $y(x) = 0$ on the segment $[1/2, 1]$, so that $\lambda_1 = 0$.

3. Let $0 > \alpha > -1$. Let $Q(x) \equiv 1$. Let us set

$$P(x) = \begin{cases} \delta & \text{for } 0 < x < \varepsilon; \\ \varepsilon^{-1} & \text{for } \varepsilon < x < 1 - \varepsilon; \\ \delta & \text{for } 1 - \varepsilon < x < 1, \end{cases}$$

and

$$y(x) = \begin{cases} x/\varepsilon & \text{for } 0 < x < \varepsilon; \\ 1 & \text{for } \varepsilon < x < 1 - \varepsilon; \\ (1-x)/\varepsilon & \text{for } 1 - \varepsilon < x < 1, \end{cases}$$

where δ is a number such that

$$\int_0^1 P(x)^\alpha\,dx = 2\varepsilon\delta^\alpha + (1 - 2\varepsilon)\varepsilon^{-\alpha} = 1, \quad \text{i.e.} \quad \delta \approx \varepsilon^{-1/\alpha}.$$

It is obvious that

$$\int_0^1 y^2(x)\,dx = 1 - 2\varepsilon + 2\varepsilon/3 = 1 - 4\varepsilon/3.$$

On the other hand,

$$\int_0^1 P(x)y_0'(x)^2\,dx = 2\delta/\varepsilon \approx 2^{1-1/\alpha}\varepsilon^{-1-1/\alpha}.$$

Therefore, $\lambda_1 \le C\varepsilon^{-1-1/\alpha}$ and since $-1-1/\alpha > 0$, the value λ_1 can be arbitrarily small.

4. Let now $\alpha \le -1$, $\beta \ge 1$. Using the Hölder inequality we get that

$$\int_0^1 Q(x)y(x)^2\,dx \le (\int_0^1 |y(x)|^{2\beta/(\beta-1)}\,dx)^{(\beta-1)/\beta},$$

$$\int_0^1 |y'(x)|^{2\alpha/(\alpha-1)}\,dx \le \int_0^1 (P(x)y'(x)^2\,dx)^{\alpha/(\alpha-1)}.$$

Thus

$$\lambda_1 \ge \inf_y \frac{(\int_0^1 |y'(x)|^{2\alpha/(\alpha-1)}\,dx)^{\alpha-1)/\alpha}}{(\int_0^1 |y(x)|^{2\beta/(\beta-1)}\,dx)^{(\beta-1)/\beta}} \ge 1.$$

The last inequality follows from the estimate

$$(\int_0^1 |y(x)|^q\,dx)^{1/q} \le (\int_0^1 |y'(x)|^p\,dx)^{1/p},$$

where

$$2 < q = 2\beta/(\beta-1) < \infty,\ 1 \le p = 2\alpha/(\alpha-1) < 2.$$

This turn is implied by the inequality $\max|y(x)| \le \int_0^1 |y'(x)|\,dx$ following from the formula $y(x) = \int_0^x y'(t)\,dt$. If $\alpha \le -1$, $\beta = 1$, we can use instead of the Hölder inequality the estimate

$$\int_0^1 Q(x)y(x)^2\,dx \le \max y(x)^2$$

and since

$$\max y(x)^2 \le (\int_0^1 |y'(x)|^p\,dx)^{2/p} \text{ for } 1 \le p < 2,$$

we can see that $\lambda_1 \ge 1$. The proof is complete. □

The proved theorem can be stated in the following way:

Theorem 38. *If* $\alpha > -1/2,\ \beta - \alpha + 2\alpha\beta < 0$, *then*

$$\lambda_1 \le C(\alpha, \beta)\frac{(\int_0^1 P(x)^\alpha\, dx)^{1/\alpha}}{(\int_0^1 Q(x)^\beta\, dx)^{1/\beta}};$$

if $\alpha \le -1, \beta \ge 1$, *then*

$$\lambda_1 \ge C(\alpha, \beta)\frac{(\int_0^1 P(x)^\alpha\, dx)^{1/\alpha}}{(\int_0^1 Q(x)^\beta\, dx)^{1/\beta}},$$

where $C(\alpha,\ \beta)$ *is a positive constant, depending on* α *and* β *only.*[1]

5.5 On estimates of all eigenvalues

Once more consider the Sturm-Liouville problem:

$$y'' + \lambda Q(x)y = 0,\ y(0) = 0,\ y(1) = 0$$

under the condition that

$$\int_0^1 Q(x)^\beta\, dx = 1$$

and estimate the kth eigenvalue λ_k. Our main result is the following.

Theorem 39. *If* $\beta \ge 1$, *then* $\lambda_k \ge C_0(\beta)k^2$. *If* $\beta < \frac{1}{2}$, $\beta \ne 0$, *then* $\lambda_k \le C_0(\beta)k^2$. *The constant* $C_0(\beta)$ *here is independent of* k.

To prove this we need the following elementary lemma, whose proof we leave to the reader.

Lemma 40. *Let* $x_1, \dots, x_k$ *be positive numbers and* $x_1 + \cdots + x_k = 1$. *Then if* $0 \le \gamma \le 1$ *the inequality*

$$x_1^\gamma + \cdots + x_k^\gamma \le k^{1-\gamma}$$

holds. If $\gamma \ge 1$ *or* $\gamma \le 0$, *then*

$$x_1^\gamma + \cdots + x_k^\gamma \ge k^{1-\gamma}.$$

So the extremum of the function $x_1^\gamma + \cdots + x_k^\gamma$ *is attained at the point* $x_1 = \ldots = x_k = 1/k$.

---

[1] S. Karaa has proved that for $\alpha < 0, \beta > 0$ and for $\alpha > 1,\ \beta \le -1$

$$C(\alpha, \beta) = (1 - \frac{1}{\alpha})^2 B(\frac{1}{2} - \frac{1}{2\alpha}, \frac{1}{2} + \frac{1}{2\beta})^2 (\frac{\alpha\beta + \alpha}{\alpha - \beta + 2\alpha\beta})^{1-1/\beta} (\frac{\alpha - \beta + 2\alpha\beta}{\alpha\beta - \beta})^{1-1/\alpha}.$$

Proof of Theorem 39. Let y_k be an eigenfunction of the Sturm-Liouville problem having the number k. This function has $k-1$ zeroes in the interval $(0,1)$: $\nu_1, \dots, \nu_{k-1}$. Let $\nu_0 = 0 < \nu_1 < \dots < \nu_k = 1$ and I be one of the intervals (ν_j, ν_{j+1}), where $j = 0, 1, \dots, k-1$. Consider the function $y_k(x)$ on the interval I. If $\max_I y_k'(x) = y_k'(\xi_1) = 1$ and $y_k'(\xi) = 0,\ \xi \in I$, then

$$1 = |\int_\xi^{\xi_1} y_k''(x)\,dx| = \lambda_k \int_\xi^{\xi_1} Q(x)|y_k(x)|\,dx.$$

Observe that $|y_k(x)| \le \nu_{j+1} - \nu_j$, $|y_k'(x) \le 1$, and we can assume that $y_k(x) \ge 0$ in I. Therefore

$$\int_I Q(x)\,dx \ge (\nu_{j+1} - \nu_j)^{-1}\lambda_k^{-1}.$$

By Lemma 40

$$\int_0^1 Q(x)\,dx = \sum_{j=0}^{k-1}\int_{I_j} Q\,dx \ge \lambda_k^{-1}\sum_{j=0}^{k-1}(\nu_{j+1}-\nu_j)^{-1} \ge \lambda_k^{-1}k^2$$

and if $\beta \ge 1$

$$\lambda_k \ge k^2\Big(\int_0^1 Q\,dx\Big)^{-1} \ge k^2\Big(\int_0^1 Q^\beta\,dx\Big)^{-1/\beta} = k^2.$$

Analogously, if $0 < \beta < \frac{1}{2}$, then by the Hölder inequality

$$\int_I Q^\beta\,dx = \int_I Q^\beta y_k^{2\beta} y_k^{-2\beta}\,dx \le \Big(\int_I Q y_k^2\,dx\Big)^\beta\Big(\int_I y_k^{2\beta/(\beta-1)}\,dx\Big)^{1-\beta}.$$

It is clear that

$$\int_I Q y_k^2\,dx = \lambda_k^{-1}\int_I Q y_k'(x)^2\,dx \le \lambda_k^{-1}(\nu_{j+1}-\nu_j).$$

Therefore

$$\int_I Q^\beta\,dx \le \lambda_k^{-\beta}(\nu_{j+1}-\nu_j)^{1-2\beta}.$$

Summing over j we obtain

$$1 \le \lambda_k^{-\beta}\sum_{j=0}^{k-1}(\nu_{j+1}-\nu_j)^{1-2\beta}.$$

Since $1-2\beta < 1$, we have by Lemma 40, that

$$\sum_{j=0}^{k-1}(\nu_{j+1}-\nu_j)^{1-2\beta} \le k\cdot\Big(\frac{1}{k}\Big)^{1-2\beta} = k^{2\beta}.$$

Thus $1 \le \lambda_k^{-\beta}k^{2\beta}$, i.e. $\lambda_k \le k^2$.

If $\beta < 0$, set

$$l = \nu_{j+1} - \nu_j,\ \rho = \int_{\nu_j}^{\nu_{j+1}} Q(x)^\beta\,dx.$$

Substituting x for $\nu_j + tl$, $Q(x)$ for $q(t)(\rho/l)^{1/\beta}$, and λ_k for $\rho^{-1/\beta} l^{1/\beta-2}\mu$, we obtain that

$$y''_{tt} + \mu y = 0,\ 0 \le t \le 1;\ y(0) = 0,\ y(1) = 0;\ \int_0^1 q(t)^\beta\,dt = 1.$$

Theorem 23 implies that $\mu \le C_0(\beta)$, so that $\rho \ge C_0(\beta)^\beta \lambda_k^{-\beta}(\nu_{j+1} - \nu_j)^{1-2\beta}$. Summing over $j = 0, \dots, k-1$ we obtain that

$$1 \ge C_0(\beta)^\beta \lambda_k^{-\beta} \sum_{j=0}^{k-1} (\nu_{j+1} - \nu_j)^{1-2\beta},$$

and since $1 - 2\beta > 1$, we get by Lemma 40 that $1 \ge C_0(\beta)^\beta \lambda_k^{-\beta} k^{2\beta}$, which gives the inequality $\lambda_k \le C_0(\beta) k^2$. □

Now consider another Sturm-Liouville problem:

$$(P(x)y')' + \lambda y = 0,\ y(0) = 0,\ y(1) = 0$$

under the condition

$$\int_0^1 P(x)^\alpha\,dx = 1.$$

Theorem 41. *If $\alpha > -\frac{1}{2}$, $\alpha \ne 0$, then $\lambda_k \le C_0(\alpha)k^2$. If $\alpha \le -1$, then $\lambda_k \ge C_0(\alpha)k^2$. Here $C_0(\alpha)$ is a positive constant independent of k.*

Proof. Let at first $\alpha > -\frac{1}{2}$, $\alpha \ne 0$. As above, consider the kth eigenfunction $y_k(x)$, corresponding to the eigenvalue λ_k. Let $\nu_0, \dots, \nu_k$ be the zeroes of $y_k(x)$ and $\nu_0 = 0 < \nu_1 < \dots < \nu_k = 1$. Let I be one of the intervals (ν_j, ν_{j+1}) with $j = 0, 1, \dots, k-1$, and set

$$l = \nu_{j+1} - \nu_j,\ \rho = \int_{\nu_j}^{\nu_{j+1}} P(x)^\alpha\,dx.$$

Substituting x for $\nu_j + tl$, $P(x)$ for $p(t)(\rho/l)^{1/\alpha}$, and λ_k for $\rho^{1/\alpha} l^{-2-1/\alpha}\mu$, we obtain that

$$(p(t)y'_t)'_t + \mu y = 0,\ 0 \le t \le 1;\ y(0) = 0,\ y(1) = 0;\ \int_0^1 p(t)^\alpha\,dt = 1.$$

Theorem 25 implies that $\mu \le C_0(\alpha)$, so that $\lambda_k \le C_0(\alpha) l^{-2-1/\alpha} \rho^{1/\alpha}$. If $\alpha > 0$, then it follows that $\lambda_k^\alpha \le C_0(\alpha)^\alpha l^{-2\alpha-1}\rho$ or

$$\lambda_k^\alpha (\nu_{j+1} - \nu_j)^{2\alpha+1} \le C_0(\alpha)^\alpha \int_{\nu_j}^{\nu_{j+1}} P(x)^\alpha\,dx.$$

Summing over j from 0 to $k-1$ we obtain that

$$\lambda_k^\alpha \sum_{j=0}^{k-1} (\nu_{j+1} - \nu_j)^{2\alpha+1} \leq C_0(\alpha)^\alpha.$$

Since $1 + 2\alpha > 1$, we have by Lemma 40 that

$$\sum_{j=0}^{k-1} (\nu_{j+1} - \nu_j)^{2\alpha+1} \geq k(1/k^{2\alpha+1}) = k^{-2\alpha}$$

so that $\lambda_k \leq C_0(\alpha)k^2$.

If $-\frac{1}{2} < \alpha < 0$, then as above $\lambda_k^\alpha \geq C_0(\alpha)^\alpha l^{-2\alpha-1}\rho^2$ or

$$\lambda_k^\alpha (\nu_{j+1} - \nu_j)^{2\alpha+1} \geq C_0(\alpha)^\alpha \int_{\nu_j}^{\nu_{j+1}} P(x)^\alpha \, dx.$$

Summing over j between 0 and $k-1$ we see that

$$\lambda_k^\alpha \sum_{j=0}^{k-1} (\nu_{j+1} - \nu_j)^{1+2\alpha} \geq C_0(\alpha)^\alpha.$$

since $1 + 2\alpha > 0$, Lemma 40 implies that

$$\sum_{j=0}^{k-1} (\nu_{j+1} - \nu_j)^{1+2\alpha} \leq k(1/k^{1+2\alpha}) = k^{-2\alpha}.$$

Therefore, $\lambda_k^\alpha k^{-2\alpha} \geq C_0(\alpha)^\alpha$ and since $\alpha < 0$, we see that $\lambda_k \leq C_0(\alpha)k^2$. For $\alpha \leq -1$ the same arguments lead to the estimate $\lambda_k \geq C_0(\alpha) l^{-2-1/\alpha}\rho^{1/\alpha}$, i.e.

$$(\nu_{j+1} - \nu_j)^{1+2\alpha} \lambda_k^\alpha \leq C_0(\alpha)^\alpha \int_{\nu_j}^{\nu_{j+1}} P(x)^\alpha \, dx.$$

Summing over j from 0 to $k-1$ we get the estimate

$$\lambda_k^\alpha \sum_{j=0}^{k-1} (\nu_{j+1} - \nu_j)^{1+2\alpha} \leq C_0(\alpha)^\alpha.$$

Since $2\alpha + 1 \leq -1$, Lemma 40 implies that

$$\sum_{j=0}^{k-1} (\nu_{j+1} - \nu_j)^{1+2\alpha} \leq k(1/k^{1+2\alpha}) = k^{-2\alpha}$$

and hence $\lambda_k^\alpha k^{-2\alpha} \leq C_0(\alpha)^\alpha$, so that $\lambda_k \geq C_0(\alpha)k^2$. □

The stated theorems can be reformulated in the following forms.

Theorem 42. *Let λ_k be the kth eigenvalue of the Sturm-Liouville problem considered in Theorem 39.*
If $\beta \geq 1$ then

$$\lambda_k \geq C_0(\beta)k^2\Big(\int_0^1 Q(x)^\beta\,dx\Big)^{-1/\beta}.$$

If $\beta < 1/2,\ \beta \neq 0$, then

$$\lambda_k \leq C_0(\beta)k^2\Big(\int_0^1 Q(x)^\beta\,dx\Big)^{-1/\beta}.$$

Theorem 43. *Let λ_k be the kth eigenvalue of the Sturm-Liouville problem considered in Theorem 41.*
If $\alpha > -1/2,\ \alpha \neq 0$, then

$$\lambda_k \leq C_0(\alpha)k^2\Big(\int_0^1 P(x)^\alpha\,dx\Big)^{1/\alpha}.$$

If $\alpha \leq -1$, then

$$\lambda_k \geq C_0(\alpha)k^2\Big(\int_0^1 P(x)^\alpha\,dx\Big)^{1/\alpha}.$$

Chapter 6

Differential Operators of Any Order

6.1 Oscillation of solutions of an equation of any order

We discuss two types of problems in this section.

1. Consider a boundary value problem for a differential equation

$$y^{(n)} + \sum_{i=0}^{n-1} p_i(x) y^{(i)}(x) = 0, \ 0 \le x \le h, \tag{1}$$

where the functions $p_i(x)$ are continuous on $[0, h]$. Let some points $0 = x_1 < x_2 < \ldots < x_n = h$ and some reals $A_1, A_2, \ldots, A_n$ be fixed. We are looking for a solution of the differential equation satisfying the conditions

$$y(x_i) = A_i, \ i = 1, \ldots, n. \tag{2}$$

It is easy to see that this problem is resolvable if and only if the equation has no non-trivial solutions such that $y(x_i) = 0, \ i = 1, \ldots, n$.

Indeed, if $y_1(x), \ldots, y_n(x)$ is a fundamental system of solutions, then any solution is represented as

$$y = \sum_{k=1}^{n} C_k y_k(x)$$

with some constants C_k. The boundary conditions are satisfied if

$$\sum_{k=1}^{n} C_k y_k(x_i) = A_i.$$

This is an algebraic system of equations for C_k, resolvable for any $(A_1, \dots, A_n)$ if and only if $\det(y_k(x_i))_{i,k=1,\dots,n} \neq 0$. The latter condition is equivalent to the following one: in the case when $A_1 = \dots = A_n = 0$, the only solution is $y \equiv 0$.

This boundary value problem was stated by Valle-Poussin.

We will assume that

$$|p_i(x)| \leq L_i = \text{const}$$

and consider the problem of solvability.

Theorem 1 (Valle-Poussin). *If $h \leq h_0$, where h_0 is a positive root of the equation*

$$1 = \sum_{i=0}^{n-1} L_i \frac{h^{n-i}}{(n-i)!},$$

then the boundary value problem is resolvable for any $x_2, \dots, x_{n-1}$ and $A_1, \dots, A_n$.

It is easy to see that only one such root exists. The proof of this theorem is based on the following lemma.

Lemma 2. *Let $\varphi \not\equiv 0$ on $[a, b]$, $b - a = h$, and φ have at least n zeroes, among which at least two are different. If $\varphi \in C^n(a, b)$ and $\max_{a<x<b} |\varphi^{(n)}(x)| \leq \mu$, then*

$$\int_a^b |\varphi(x)|\, dx < \mu \frac{h^{n+1}}{(n+1)!}.$$

Proof. Let $\varphi(a_k) = 0,\ k = 1, 2, \dots, n$. Consider the function

$$\Phi(x) = \varphi(x) - \frac{(x-a_1)(x-a_2)\dots(x-a_n)}{(t-a_1)(t-a_2)\dots(t-a_n)}\varphi(t),$$

where $t \in (a, b)$, $t \neq a_i$, for any $i = 1, 2, \dots, n$. The function $\Phi(x)$ has at least $n+1$ zeroes: $x = a_i$, $i = 1, \dots, n$ and $x = t$. By the Rolle theorem there is a point $\xi \in (a, b)$ such that $\Phi^{(n)}(\xi) = 0$. Then

$$\varphi^{(n)}(\xi) = n! \frac{\varphi(t)}{(t-a_1)(t-a_2)\dots(t-a_n)}.$$

Therefore, $|\varphi(t)| \leq \frac{\mu}{n!} |(t-a_1)\dots(t-a_n)|$. Thus

$$\int_a^b |\varphi(x)|\, dx \leq \frac{\mu}{n!} \int_a^b |(x-a_1)\dots(x-a_n)|\, dx.$$

Set $\psi(x) = |(x-a_2)\dots(x-a_n)|$. We have

$$I = \int_a^b |(x-a_1)(x-a_2)\dots(x-a_n)| = \int_a^{a_1} (a_1 - x)\psi\, dx + \int_{a_1}^b (x-a_1)\psi\, dx.$$

The derivative of the function I with respect to a_1 is equal to $\int_a^{a_1}\psi\,dx - \int_{a_1}^b \psi\,dx$ and therefore is increasing with a_1. It is negative at $a_1 = a$ and positive when $a_1 = b$. Therefore the maximal value of $I(a_1)$ is attained at $a_1 = a$ or $a_1 = b$. The same consideration for $a_2, \dots, a_n$ gives that

$$I \le \max_{0\le p\le n} I_p, \text{ where } I_p = \int_a^b (x-a)^p(b-x)^{n-p}\,dx.$$

Changing the variable of integration: $x = a + ht$ we obtain that

$$I_p = h^{n+1}\int_0^1 t^p(1-t)^{n-p}\,dx = h^{n+1}\frac{p!(n-p)!}{(n+1)!} = \frac{h^{n+1}}{n+1}\binom{n}{p}^{-1} \le \frac{h^{n+1}}{n+1}.$$

Since not all a_i coincide, we have

$$\int_a^b |(x-a_1)\dots(x-a_n)|\,dx < \frac{h^{n+1}}{n+1}.$$ □

Proof of Theorem 1. Consider the homogeneous problem (1), (2) with $A_i = 0$, $i = 1,\dots,n$. Let $y(x) \not\equiv 0$ be a nontrivial solution of the boundary problem, i.e. $y(x_i) = 0$, $i = 1,\dots,n$. In virtue of the Rolle theorem there exists a ξ such that $a < \xi < b$, $y^{(n-1)}(\xi) = 0$. Integrating the equation from ξ to x we obtain

$$-y^{(n-1)}(x) = \sum_{i=0}^{n-1}\int_\xi^x p_i(x)y^{(i)}\,dx.$$

Let $\max|y^{(n-1)}(x)| = 1$. Then $1 \le \sum_{i=0}^{n-1} L_i\int_0^h |y^{(i)}(x)|\,dx$. Since $y^{(i)}$ has at least $n-i$ zeroes, by Lemma 2 we obtain that $1 < \sum_{i=0}^{n-1} L_i\frac{h^{n-i}}{(n-i)!}$. If $h \le h_0$, this inequality is impossible. □

2. Let $Q(x)$ be a continuous function on $[0,a]$, $Q(x)\not\equiv O$. Consider the boundary value problem:

$$y^{(n)} + \lambda Q(x)y = 0, \qquad 0 \le x \le a, \tag{3}$$

$$\begin{aligned} y(x_1) = y'(x_1) = \dots = y^{(k_1-1)}(x_1) = 0,\\ y(x_2) = y'(x_2) = \dots = y^{(k_2-1)}(x_2) = 0,\\ \dots\\ y(x_r) = y'(x_r) = \dots = y^{(k_r-1)}(x_r) = 0, \end{aligned} \tag{4}$$

where

$$0 = x_1 < x_2 < \dots < x_r = a,\ k_1 + k_2 + \dots + k_r = n,\ k_i \ge 1,$$

and $k_2,\dots,k_{r-1}$ are even numbers. Suppose that

$$Q(x) \ge 0 \text{ in } [0,a] \text{ if } k_r \text{ is odd;}$$
$$Q(x) \le 0 \text{ in } [0,a] \text{ if } k_r \text{ is even.}$$

Theorem 3. *There is a $\lambda > 0$ such that the boundary value problem (3), (4) has a nonnegative solution $y(x)$ and $y(x) \neq 0$ for $x \neq x_i$. The solution space for this value of λ is one dimensional.*

The proof is based on the following lemmas.

Lemma 4. *Let $q(x) > 0$ be a continuous function for $0 \leq x \leq a$, and n be an even number. Then there exists an $h > 0$, depending on a, n such that if $q(x) > h$, then every nontrivial solution $y(x)$ of the equation*

$$y^{(n)} + q(x)y = 0,\ 0 \leq x \leq a, \tag{5}$$

changes its sign on $[0, a]$.

Proof. Set $\psi(x) = x^n(a-x)^n$. Multiply both parts of equation (5) by $\psi(x)$ and integrate the result over $[0, a]$. After integrating by parts we obtain

$$\int_0^a y(x)[\psi^{(n)}(x) + q(x)\psi(x)]\,dx = 0.$$

Assume that $y(x) \geq 0$, on $[0, a]$ and $y \not\equiv 0$. It is easy to see that $\psi^{(n)}(0) > 0$, $\psi^{(n)}(a) > 0$. Therefore, $\psi^{(n)}(x) > 0$ on $[0, \delta]$ and on $[a-\delta, a]$ for some $\delta > 0$. Hence it follows that

$$\int_\delta^{a-\delta} y(x)[\psi^{(n)}(x) + h\psi(x)]\,dx \leq 0.$$

On $[\delta, a-\delta]$ the values of $\psi(x)$ are larger than $\delta^n(a-\delta)^n$. If

$$h > \max_{[0,a]}[-\psi^{(n)}(x)]\delta^{-n}(a-\delta)^{-n},$$

then $\psi^{(n)} + q\psi > 0$ and we have a contradiction. □

Lemma 5. *Let $q(x) > 0$ be a continuous function for $0 \leq x \leq a$, and n be an odd number. Then there exists an $h > 0$, depending on a, n such that if $q(x) > h$ in $[0, a]$, then every nontrivial solution $y(x)$ of equation (5), such that $y(0) = 0$, changes its sign on $[0, a]$.*

Proof. Set $\psi(x) = x^{n-2}$ on $[0, a/4]$, $\psi(x) = (a-x)^n$ on $[3a/4, a]$, $\psi(x) > 0$ on $]0, a[$, and $\psi(x) \in C^n([0, a])$. Multiplying equation (5) by $\psi(x)$ and integrating over $[0, a]$, we obtain after integrating by parts that

$$\sum_{i=0}^{n-1}(-1)^i\psi^{(i)}(x)y^{(n-i-1)}(x)|_0^a - \int_0^a \psi^{(n)}y\,dx + \int_0^a q\psi y\,dx = 0.$$

Suppose that $y(x) \geq 0$ on $[0, a]$. Then

$$\begin{aligned}(n-2)!y'(0) + \int_{a/4}^{3a/4} [-\psi^{(n)}(x) + q(x)\psi(x)]y(x)\,dx \\ + \int_0^{a/4} q(x)\psi(x)y(x)\,dx + \int_{3a/4}^{a} q(x)\psi(x)y(x)\,dx \\ + \int_{3a/4}^{a} n!y(x)\,dx = 0.\end{aligned}$$

All terms on the left-hand side are positive if

$$q(x) \geq h = \max_{[0,a]} [-\psi^{(n)}(x)]/\min_{[a/4,3a/4]} \psi(x),$$

since $y'(0) \geq 0$. Therefore, $y(x) \equiv 0$, contrary to the conditions. □

Lemma 6. *Let $q(x) < 0$ be a continuous function for $0 \leq x \leq a$, and n be an even number. Then there exists an $h > 0$, depending on a, n such that if $q(x) < -h$, then every non-trivial solution $y(x)$ of equation (5) satisfying the condition $y(0) = 0$, changes its sign on $[0, a]$.*

Proof. Let $y(x) \geq 0$ on $[0, a]$. Set $\psi(x) = x^{n-2}$ on $[0, a/4]$, $\psi(x) = (a-x)^{n-2}$ on $[3a/4, a]$, $\psi(x) > 0$ on $[0, a]$, and $\psi(x) \in C^n([0, a])$. Multiplying equation (5) by $\psi(x)$ and integrating over $[0, a]$, we obtain after integrating by parts that

$$(n-2)!(y'(a) - y'(0)) + \int_{a/4}^{3a/4} (\psi^{(n)} + q\psi)y\,dx + \int_0^{a/4} q\psi y\,dx + \int_{3a/4}^{a} q\psi y\,dx = 0.$$

Hence

$$\int_{a/4}^{3a/4} (\psi^{(n)}(x) + q(x)\psi(x))y(x)\,dx \geq 0.$$

We come to a contradiction if

$$q(x) < -h = -\frac{\max_{[a/4,3a/4]} |\psi^{(n)}(x)|}{\min_{[a/4,3a/4]} \psi(x)}.$$ □

Lemma 7. *Let $q(x) < 0$ be a continuous function for $0 \leq x \leq a$, and n be an odd number. Then there exists an $h > 0$, depending on a, n such that if $q(x) < -h$, then every nontrivial solution $y(x)$ of equation (5), such that $y(a) = 0$, changes its sign on $[0, a]$.*

This lemma follows from Lemma 5 after the change $t = -x + a$.

Proof of Theorem 3. Let $y(x,\lambda)$ be a nontrivial solution of equation (3) such that

$$\begin{aligned} y(x_1,\lambda) = y'(x_1,\lambda) = \cdots = y^{(k_1-1)}(x_1,\lambda) = 0, \\ y(x_2,\lambda) = y'(x_2,\lambda) = \cdots = y^{(k_2-1)}(x_2,\lambda) = 0, \\ \cdots \\ y(x_{r-1},\lambda) = y'(x_{r-1},\lambda) = \cdots = y^{(k_{r-1}-1)}(x_{r-1},\lambda) = 0, \\ y(x_r,\lambda) = y'(x_r,\lambda) = \cdots = y^{(k_r-2)}(x_r,\lambda) = 0. \end{aligned} \tag{6}$$

Such a solution exists for any λ. Indeed, let $y_1(x,\lambda),\ldots,y_n(x,\lambda)$ be a fundamental system of solutions of (3) such that $y_i^{(j)}(0,\lambda) = b_{ij},\ i = 1,\ldots,n,\ j = 0,1,\ldots,n-1,\ det(b_{ij}) \neq 0$, and b_{ij} are chosen independently of λ. Let us look for a solution of (3), (6) in the form $y(x,\lambda) = \sum_{i=1}^{n} c_i y_i(x,\lambda)$. The boundary conditions generate the system $A(\lambda)\vec{c} = 0,\ \vec{c} = (c_1,\ldots,c_n)$, where A is an $(n-1)\times n$–matrix. Such a system always has a nontrivial solution $c = (c_1^{(\lambda)},\ldots,c_n^{(\lambda)})$.

If $\lambda = 0$, then the solution is a polynomial of degree at most $n-1$. We may assume that $y(x,0) \geq 0$ on $[0,a]$, and $y(x,0)$ vanishes only at the points x_i, having there zeroes of order k_i for $i < r$, and of order $k_r - 1$ at the point x_r.

If for some λ the boundary problem (3), (4) has no nontrivial solutions, then the dimension of the space of solutions to the homogeneous problem (3), (6) is 1. Indeed, if $y_1(x,\lambda),\ y_2(x,\lambda)$ are two solutions of (3), (6), then

$$y^*(x,\lambda) = y_2^{(k_r-1)}(x_r,\lambda)y_1(x,\lambda) - y_1^{(k_r-1)}(x_r,\lambda)y_2(x,\lambda) \equiv 0,$$

since y^* satisfies the homogeneous equation (3) and the homogeneous boundary conditions (4).

Therefore, if λ is not an eigenvalue of the problem (3), (4), then the rank of the matrix $A(\lambda)$ is $n-1$, and $y(x,\lambda)$ is a continuous function of λ at regular points.

If $0 \leq x_j < a$, then $y^{(k_j)}(x_j,0) > 0$ and $y^{(k_j)}(x_j,\lambda) > 0$ for $0 < \lambda < \lambda_0$. Since $y^{(i)}(x_j,\lambda) = 0,\ i \leq k_j - 1$, we have $y(x,\lambda) > 0$ for $0 < \lambda < \lambda_0,\ x \neq x_j$, if x is in a small neighborhood of x_j. On the other hand, $y^{(k_j-1)}(a,0) < 0,\ y^{(i)}(a,0) = 0,\ i \leq k_j - 2$, and therefore, $y(x,\lambda) > 0$ in a neighborhood of a for small λ. The positiveness of $y(x,\lambda)$ for small λ outside neighborhoods of x_j follows from the continuity of $y(x,\lambda)$ with respect to λ.

Set $\Lambda = \sup\lambda$, where sup is taken over λ such that $y(x,\lambda) \geq 0,\ 0 < x < a,\ y(x,\lambda) > 0$ for $x \neq x_j$. The boundedness of the set of such λ follows from Lemmas 4–7 applied to equation (3) on an interval $[\alpha_1,\alpha_2] \subset [0,1]$ such that $Q(x) \neq 0$ in $[\alpha_1,\alpha_2]$.

The function $y(x,\Lambda)$ has at the points x_j zeroes of order at least $k_j,\ y(x,\Lambda) \geq 0$ on $[0,a]$. If $y(x^*,\Lambda) = 0,\ x^* \neq x_j,\ 0 < x^* < a$, then the order of zero at x^* of the function $y(x,\Lambda)$ is even and the total number of its zeroes is more than n, so that by the Rolle theorem the function $y^{(n-1)}(x,\Lambda)$ has at least two different zeroes and the function $y^{(n)}(x, \Lambda)$ has to change its sign. Since this is impossible, $y(x, \Lambda) > 0$ for $0 < x < a$, $x \neq x_j$.

Show that at each point x_j, $1 < j < r$, the order of zero of $y(x, \Lambda)$ is equal to k_j. If it were higher, then because of the inequality $y(x, \Lambda) \geq 0$ it would be at least of order $k_j + 2$ and the function $y(x, \Lambda)$ would then have more than n zeroes, which is impossible if the function Q has a constant sign.

At the point x_1 we have $y^{(r_1)}(x_1, 0) > 0$. This inequality holds for small λ, i.e. $y^{(r_1)}(x_1, \lambda) > 0$ for $0 < \lambda \leq \lambda_0$.

Let us show that $y(x, \Lambda)$ has at x_r zero of order k_r. It cannot be $k_r + 2$, since $y^{(n)}$ has a constant sign on $[0, a]$ and $y(x, \Lambda)$ cannot have more than n zeroes. So it remains to consider the case when the order of zero of the function $y(x, \Lambda)$ at x_r is equal to $(k_r + 1)$. Let, for example, n be even and k_r be even. Then $Q \geq 0$, $y^{(n)} \leq 0$. By the Rolle theorem y' has on $[0, a]$ at least $n - 1$ zeroes and the order of the rightmost root is odd. Then y'' has at least $n - 2$ zeroes and the order of the rightmost root is even. In general, $y^{(k)}$ has at least $n - k$ zeroes and the order of the rightmost root is even if k is even and odd if k is odd. In particular, $y^{(n-1)}$ has a zero ξ such that $y^{(n)}(\xi) < 0$. However this contradicts equation (3).

Since $y(x, \Lambda)$ has n zeroes and $y(x, \Lambda) \geq 0$, the number of its zeroes is equal to n only if the order of its zero at x_r is more than that of the functions $y(x, \lambda)$ with $\lambda < \Lambda$, i.e. is k_r. Thus Λ is an eigenvalue and $y(x, \Lambda)$ is the corresponding eigenfunction. If $y_1(x, \Lambda)$ is another eigenfunction, then we consider the function $y(x, \Lambda) - \varepsilon y_1(x, \Lambda) = \overline{y}(x, \Lambda)$. For small $\varepsilon > 0$ the function $\overline{y}(x, \Lambda)$ has n zeroes and $\overline{y}(x, \Lambda) \geq 0$. If ε_1 is the least upper bound of such ε, then $y(x, \Lambda) - \varepsilon_1 y_1(x, \Lambda)$ has more than n zeroes, $y(x, \Lambda) - \varepsilon_1 y_1(x, \Lambda) \geq 0$. However, as we have seen above, this contradicts the Rolle theorem. □

6.2 On estimates of the first eigenvalue for operators of higher order

Let us consider the Sturm-Liouville problem

$$(-1)^{n+1} y^{(2n)}(x) + \lambda q(x) y(x) = 0 \tag{7}$$

on the segment $0 \leq x \leq 1$, with the boundary conditions

$$\begin{cases} y(0) = y'(0) = \cdots = y^{(n-1)}(0) = 0, \\ y(1) = y'(1) = \cdots = y^{(n-1)}(1) = 0. \end{cases} \tag{8}$$

We are interested in studying the dependence of the first eigenvalue λ_1 of this problem on the potential q. Denote R_β the set of measurable on $(0, 1)$ functions q with positive values such that $\int_0^1 q(x)^\beta \, dx = 1$, where β is a real number, $\beta \neq 0$. The problem (7),(8) has a discrete spectrum. The variational principle implies that the first eigenvalue λ_1 can be found as

$$\lambda_1 = \inf_{y \in C_0^\infty(0,1)} \frac{\int_0^1 y^{(n)}(x)^2\, dx}{\int_0^1 q(x) y(x)^2\, dx}.$$

It is easy to see that all eigenfunctions of the problem (7),(8) are real and positive. We will estimate the values

$$m_\beta = \inf_{q \in R_\beta} \lambda_1, \quad M_\beta = \sup_{q \in R_\beta} \lambda_1.$$

Set

$$L[q, y] = \frac{\int_0^1 y^{(n)}(x)^2\, dx}{\int_0^1 q(x) y(x)^2\, dx}.$$

Theorem 8. *If $\beta \geq 1$, then $m_\beta \geq 1$.*

Proof. If $y(x)$ is an eigenfunction, corresponding to λ_1, then by the Rolle theorem each function $y'(x), \ldots, y^{(n-1)}(x)$ has at least one zero on $(0,1)$. Therefore, $y^{(n-1)}(x) = \int_\xi^x y^{(n)}(t)\, dt$, where ξ is a zero of the derivative $y^{(n-1)}$ and $|y^{(n-1)}(x)| \leq \int_0^1 |y^{(n)}(t)|\, dt$. Similarly, $|y^{(n-i)}(x)| \leq \int_0^1 |y^{(n-i+1)}(t)|\, dt$. Induction implies the inequality $|y(x)|^2 \leq (\int_0^1 |y^{(n)}(t)|\, dt)^2 \leq \int_0^1 |y^{(n)}(t)|^2\, dt$. Hence

$$\frac{\int_0^1 |y^{(n)}(t)|^2\, dt}{\int_0^1 q(t) y(t)^2\, dt} \geq \frac{1}{\int_0^1 q(t)\, dt} \geq \frac{1}{(\int_0^1 q(t)^\beta\, dt)^{1/\beta}} \geq 1. \qquad \square$$

Theorem 9. *If $\beta \geq 1/2n$, then $M_\beta = \infty$.*

Proof. Let $q_\varepsilon(x) = C_\varepsilon (x+\varepsilon)^{-2n}$, where C_ε is such that $\int_0^1 q_\varepsilon(x)^\beta\, dx = 1$. It is easy to see that $C_\varepsilon \to 0$ as $\varepsilon \to 0$. Let $y \in W_2^n(0,1)$ and y satisfy the conditions (8). Set $y(x) = 0$ outside of $(0,1)$. The Hardy inequality

$$\int_0^1 y(x)^2 (x+\varepsilon)^{-2n}\, dx \leq C_1 \int_0^1 |y^{(n)}(t)|^2\, dt$$

implies the inequality

$$\int_0^1 q_\varepsilon(t) y(t)^2\, dt \leq C_\varepsilon C_1 \int_0^1 |y^{(n)}(t)|^2\, dt.$$

Therefore

$$L[q_\varepsilon, y] = \frac{\int_0^1 y^{(n)}(t)^2\, dt}{\int_0^1 q_\varepsilon(t) y(t)^2\, dt} \geq \frac{1}{C_\varepsilon C_1} \to \infty \text{ as } \varepsilon \to 0.$$

Thus $M_\beta = \infty$. $\square$

Theorem 10. *If $\beta < 1/2n$, then $M_\beta = C(\beta) < \infty$.*

Proof. Let at first $0 < \beta < 1/2n$. Using the Hölder inequality we obtain

$$1 = \int_0^1 q(x)^\beta \, dx \le \Big(\int_0^1 q(x) y(x)^2 \, dx\Big)^\beta \Big(\int_0^1 |y(x)|^p \, dx\Big)^{1-\beta},$$

where $p = 2\beta/(\beta - 1)$ so that $0 > p > -2(n-1)$. Therefore

$$L[q, y] \le \frac{\int_0^1 y^{(n)}(x)^2 \, dx}{(\int_0^1 |y(x)|^p \, dx)^{2/p}}.$$

Set $y_0(x) = x^{n+\delta-1}(1-x)^{n+\delta-1}$. Then $\int_0^1 y^{(n)}(x)^2 \, dx = c_1$ if $\delta > 1/2$ and $\int_0^1 |y(x)|^p \, dx = c_2 < \infty$, if $p(n + \delta - 1) + 1 > 0$, i.e. if $\delta < (1 + \beta - 2\beta n)/2\beta$. Since $(1 + \beta - 2\beta n)/2\beta > 1/2$, there exists δ satisfying all the conditions. Thus $M_\beta \le L[q, y_0] < c_3$.

Now let $\beta < 0$. Then $0 < p < 2$. Using the Hölder inequality we obtain

$$\begin{aligned}\int_0^1 |y(x)|^p \, dx &= \int_0^1 q(x)^{p/2} |y(x)|^p q(x)^{-p/2} \, dx \\ &\le \Big(\int_0^1 q(x) y(x)^2 \, dx\Big)^{p/2} \Big(\int_0^1 q(x)^\beta \, dx\Big)^{1/(1-\beta)} = \Big(\int_0^1 q(x) y(x)^2 \, dx\Big)^{p/2}.\end{aligned}$$

Therefore,

$$\int_0^1 q(x) y(x)^2 \, dx \ge \Big(\int_0^1 |y(x)|^p \, dx\Big)^{2/p}.$$

Hence

$$L[q, y] \le \frac{\int_0^1 y^{(n)}(x)^2 \, dx}{(\int_0^1 |y(x)|^p \, dx)^{2/p}}.$$

Setting $y_0(x) = x^n (1-x)^n$, we see that $M_\beta \le c$. □

Theorem 11. *If $\beta < 1$, then $m_\beta = 0$.*

Proof. Let at first $\beta < 0$. Set

$$q(x) = \begin{cases} (1-\varepsilon)^{1/\beta} \varepsilon^{-1/\beta}, & \text{if } 0 < x < \varepsilon, \\ (1-\varepsilon)^{-1/\beta} \varepsilon^{1/\beta}, & \text{if } \varepsilon < x < 1, \end{cases}$$

where $\varepsilon > 0$ is a small number. Let $y_0(x) = x^n (1-x)^n$. Then

$$\lambda_1 \le \frac{\int_0^1 y_0^{(n)}(x)^2 \, dx}{\int_0^1 q(x) y_0(x)^2 \, dx} \le C\varepsilon^{-1/\beta}.$$

Therefore λ_1 can be arbitrarily small.

Let $0 < \beta < 1$. Set

$$q(x) = \begin{cases} (2\varepsilon)^{-1/\beta}, & \text{if } |x - 1/2| > \varepsilon, \\ 0, & \text{if } |x - 1/2| < \varepsilon, \end{cases}$$

where $\varepsilon > 0$ is a small number. Let y_0 be a smooth function, vanishing in the points $x = 0$ and $x = 1$, which is equal to 1 in $(1/3, 2/3)$. Then

$$\lambda_1 \leq \frac{C}{\int_{1/2-\varepsilon}^{1/2+\varepsilon} (2\varepsilon)^{-1/\beta}\, dx} = C_1 \varepsilon^{1/\beta - 1} \to 0$$

as $\varepsilon \to 0$. Therefore, λ_1 can be arbitrarily small. □

In the study of the Sturm-Liouville problem for an equation of second order we have obtained the sharp values of the first eigenvalue for the operator $y''(x) + \lambda q(x) y = 0$ under the condition $\int_0^1 q(x)^\beta = 1$. We have also found the potentials q, when these sharp estimates are true. In the case of an operator of a higher order one can write down the differential equations for the function q. However, this is an equation of order n and we cannot find the explicit solution.

Consider in more detail the following problem:

$$-y^{(4)} + \lambda q(x) y = 0,\, y(0) = y'(0) = 0,\, y(1) = y'(1) = 0,\, q(x) \geq 0,\, \int_0^1 q(x)^\beta = 1,\, \beta > 1,$$

where q is a bounded measurable function. As we have shown above the least eigenvalue of this problem is bigger than a positive constant, independent of q. Namely

$$\lambda_1 = \inf_{y \in C_0^\infty(0,1)} \frac{\int_0^1 y'(x)^2\, dx}{\int_0^1 q(x) y(x)^2\, dx} \geq \inf_{y \in C_0^\infty(0,1)} \frac{\int_0^1 y''(x)^2\, dx}{(\int_0^1 |y(x)|^p\, dx)^{2/p}}.$$

Let

$$G[y] = \frac{\int_0^1 y''(x)^2\, dx}{(\int_0^1 |y(x)|^p\, dx)^{2/p}}$$

and $m = \inf_{y \in C_0^\infty(0,1)} G[y]$. Let $\{y_k\}$ be a minimizing sequence. By the homogeneity we can assume that $\int_0^1 y_k''(x)^2\, dx = 1$. The sequence $\{y_k\}$ contains a subsequence converging to y uniformly and weakly in $W_{2,0}^2(0,1)$. The Euler-Lagrange equation for the functional L has the form

$$y^{(4)} - m|y|^{(\beta+1)/(\beta-1)} \operatorname{sign} y = 0,$$

$$y(0) = y'(0) = 0,\ y(1) = y'(1) = 0,\ \int_0^1 y''(x)^2\, dx = 1.$$

Set $q(x) = \lambda |y|^{2/(\beta-1)}$, where λ is such that $\int_0^1 q(x)^\beta\, dx = 1$. Then the problem

$$-y^{(4)} + \lambda q(x) y = 0,\ y(0) = y'(0) = 0,\ y(1) = y'(1) = 0$$

has an eigenvalue m, to which the eigenfunction y corresponds. This eigenvalue is minimal for the considered class of the functions q. So finding the extremal q and λ_1 is reduced to the boundary value problem for an equation of fourth order. The same is true for other values of β and for equations of order $n > 4$.

Let us consider the boundary value problem

$$(-1)^{k_s} y^{(2n)}(x) + \lambda q(x) y(x) = 0$$

on the segment $0 \le x \le 1$, with the boundary conditions

$$y^{(i)}(x_j) = 0,\ i = 0, \dots, k_j - 1,\ 0 = x_1 < x_2 < \cdots < x_s = 1,$$

where $s \le 2n-1$, $k_1 + \cdots + k_s = 2n$. We assume that $q(x) \ge 0$, $\int_0^1 q(x)^\beta = 1$.

We have shown that this problem has positive eigenvalues. Let λ_1 be the minimal of them. Let us show that $\lambda_1 \ge 1$. Indeed, the corresponding eigenfunction y_1 has at least $2n$ zeroes (taking into account their multiplicity.) The function $y^{(2n-1)}$ has at least one zero ξ. Hence

$$y^{(2n-1)}(x) = \int_\xi^1 y^{(2n)}(x)\, dx,$$

and therefore

$$|y^{(2n-1)}(x)| \le \lambda_1 \int_0^1 q(x)\, dx \cdot \max |y(x)| \le \lambda_1 \max |y(x)|.$$

Each function $y'(x), \dots, y^{(n-1)}(x)$ has at least one zero on $(0,1)$. Therefore,

$$y^{(2n-i)}(x) = \int_{\xi_i}^x y^{(2n-i+1)}(t)\, dt,$$

where ξ_i is a zero of $y^{(2n-i)}(x)$. Hence $|y^{(2n-i)}(x)| \le \lambda_1 \max |y(x)|$. In particular, $|y(x)| \le \lambda_1 \max |y(x)|$ and thus $\lambda_1 \ge 1$. □

6.3 Introduction to a Lagrange problem

The considered Lagrange problem consists in finding the extremal values of the following functional:

$$L[Q,y] = \frac{\int_0^1 Q(x) y''(x)^2\, dx}{\int_0^1 y'(x)^2\, dx}$$

under the conditions $y \in H^2(0,1)$, $Q(x)$ is a bounded measurable function,

$$\int_0^1 Q(x)^\alpha\, dx = 1,\ Q(x) \ge 0, \tag{9}$$

$$y(0) = 0,\ y'(0) = 0,\ y(1) = 0,\ y'(1) = 0, \tag{10}$$

where $\alpha \in \mathbf{R}\backslash\{0\}$. It is easy to see that this problem is equivalent to the variational problem on the extremum of the functional

$$F[Q,y] = \frac{\int_0^1 Q(x)y'(x)^2\,dx}{\int_0^1 y(x)^2\,dx},$$

under the conditions (9) and

$$y(0)=0,\ y(1)=0,\ \int_0^1 y(x)\,dx = 0.$$

The Euler-Lagrange equation for the functional L has the form

$$(Q(x)y'')'' + \lambda y'' = 0,\ y(0)=0,\ y'(0)=0,\ y(1)=0,\ y'(1)=0. \tag{11}$$

This problem is very important for applications. For example, it is essential in finding the strongest column of a given volume (the most important values are then $\alpha = 1/2$ or $1/3$) and was considered by many authors (see, for example, [La], [Ke], [KeTa], [OlRa], [BrSe], [Co], [CoOv]). They used methods of functional analysis and of variational calculus, sometimes very complicated.

Let us reproduce some Keller-Tadjbakhsh arguments. Let us suppose that there exists a function $Q_0(x)$ that maximizes the lowest eigenvalue. Let $Q(x,\varepsilon)$ be a family of functions that depend smoothly on ε and such that $Q(x,0) = Q_0(x)$. Assume that λ and y, the lowest eigenvalue and the corresponding eigenfunction with $Q = Q(x,\varepsilon)$, also depend smoothly on ε. Then we may differentiate the equation

$$(Q(x,\varepsilon)y''(x,\varepsilon))'' + \lambda(\varepsilon)y''(x,\varepsilon) = 0$$

with respect to ε to obtain the equation

$$(Q_0(x)z'')'' + \lambda z'' + (Q_1(x)y'')'' + \mu y'' = 0,$$

where $Q_1(x) = \partial Q(x,0)/\partial\varepsilon$, $z(x) = \partial y(x,0)/\partial\varepsilon$, and $\mu = \partial\lambda(0)/\partial\varepsilon$. We multiply the first equation by z, the second by y, subtract one from the other, and integrate the result over $[0,1]$. In virtue of the boundary conditions we have after integrating by parts

$$\int_0^1 ((Q_0(x)y'')''z - (Q_0(x)z'')''y)\,dx = 0,\ \int_0^1 (y''z - z''y)\,dx = 0.$$

Therefore

$$\int_0^1 Q_1(x)y''(x)^2\,dx - \int_0^1 \mu y'(x)^2\,dx = 0.$$

Since λ is the maximal value of $\lambda(\varepsilon)$, we see that $\mu = 0$ and so $\int_0^1 Q_1(x)y''(x)^2\,dx = 0$ and this is true for any function $Q_1(x)$ such that $\int_0^1 Q_0(x)^{\alpha-1}Q_1(x)\,dx = 0$. Thus we have as a necessary condition for a maximum the relation $y''(x)^2 = CQ_0(x)^{\alpha-1}$. This leads to a nonlinear equation for y

$$(|y''|^{2/(\alpha-1)}y'')'' + \lambda y'' = 0,$$

which is integrable. Indeed, if we set $y'' = z$, then $(|z|^{2/(\alpha-1)}z)'' + \lambda z = 0$, and if we set now $z' = P(z)$ we obtain a linear equation of first order for P^2. The weak point of this proof is that the function $\lambda(\varepsilon)$ can be nonsmooth, because the lowest eigenvalue λ_1 can be double.

The authors of [OlRa] and [CoOv] claimed to prove that the result of Keller-Tadjbakhsh [KeTa] is not correct. However, their calculations are erroneous and the value $16\pi^2/3$ found in [KeTa] is optimal.

We will state here another approach, which allows us to say that the indicated solution is really optimal and gives a globally extremal value to the functional L. Let us remark that we shall use the Sobolev type spaces $W_p^l(0,1)$ with $l = 1, 2$ and any real values of $p \neq 0$, which is also interesting outside of the frame of the Lagrange problem. Furthermore, we shall prove the existence of the optimal solution.

The important role in what follows belongs to the functional

$$G[y] = \frac{(\int_0^1 |y'(x)|^p\,dx)^{2/p}}{\int_0^1 y(x)^2\,dx}, \text{ where } p = \frac{2\alpha}{\alpha-1}.$$

Let R_α be the set of bounded measurable functions Q defined on $[0,1]$ satisfying conditions (9).

Let $K_p(a,b)$ for real $p \neq 0$ be the set of nondecreasing real functions y defined on $[a,b]$, absolutely continuous on $[a, b-\varepsilon]$ for any $\varepsilon > 0$, and such that $y(a) \geq 0$, $\int_a^b y'(x)^p\,dx < \infty$, $\int_a^b y(x)^2\,dx < \infty$.

Let $K_p(a,b,c)$ be the set of real functions y defined on $[a,c]$ and such that $y \in K_p(a,b)$, $y(-x) \in K_p(-c,-b)$, $\int_a^b |y'(x)|^p\,dx < \infty$, and $\int_b^c |y'(x)|^p\,dx < \infty$.

Let H be the set of functions y belonging to $H^2(0,1)$ and satisfying conditions (10).

Finally, set

$$m_\alpha = \inf_{Q\in R_\alpha} \inf_{y\in H} L[Q,y], \; M_\alpha = \sup_{Q\in R_\alpha} \inf_{y\in H} L[Q,y].$$

Our aim is to find the values of m_α and M_α and the functions Q, y realizing these extremal values.

6.4 Preliminary estimates

We first estimate m_α, and M_α.

Theorem 12. *Let $\alpha \in \mathbf{R} \setminus 0$. Then*

1. *M_α is finite for $\alpha > -1/2$, $\alpha \neq 0$, and $M_\alpha = \infty$ for $\alpha \le -1/2$;*
2. *$m_\alpha > 0$ for $\alpha \le -1$ and $m_\alpha = 0$ for $\alpha > -1$.*

In order to prove Theorem 12 we need some lemmas.

Lemma 13. *Let a function f be summable on $(0,1)$, $f(x) \ge 0$. Then there are two points a and b such that $0 < a < b < 1$ and a function $y \in C^1[0,1]$ such that $y'(x)$ is absolutely continuous, $y(0) = y(1) = y'(0) = y'(1) = 0$, $y''(x) = f(x)$ if $0 < x < a$ or $b < x < 1$ and $y''(x) = -f(x)$ if $a < x < b$.*

Proof. Let at first $f(x) > 0$. Set $F(x) = \int_0^x f(t)\,dt$, $G(x) = \int_0^x F(t)\,dt$, and

$$y'(x) = \begin{cases} F(x), & \text{if } 0 < x < a, \\ 2F(a) - F(x), & \text{if } a < x < b, \\ F(x) - F(1), & \text{if } b < x < 1, \end{cases}$$

where the values of a and b will be indicated in what follows. This function is continuous at the point $x = a$ and at the point $x = b$ if

$$2F(a) + F(1) = 2F(b). \tag{12}$$

Let

$$y(x) = \begin{cases} G(x), & \text{if } 0 < x < a, \\ 2F(a)(x-a) - G(x) + 2G(a), & \text{if } a < x < b, \\ G(x) - G(1) - (x-1)F(1), & \text{if } b < x < 1. \end{cases}$$

This function is continuous at the point $x = a$, and at the point $x = b$ if

$$2F(a)(b-a) - G(b) + 2G(a) = G(b) - G(1) - (b-1)F(1). \tag{13}$$

Set $H(x) = xF(x) - G(x)$. Then $H'(x) = xf(x) \ge 0$, $H(0) = 0$, and conditions (12), (13) imply that $2H(a) + H(1) = 2H(b)$. The points a and b can be found in the following way. Set

$$K(x) = F(x) - F(1)H(x)/H(1) = H(x)[\frac{F(x)}{H(x)} - \frac{F(1)}{H(1)}].$$

Since

$$(F(x)/H(x))' = -F'(x)G(x)/H(x)^2,$$

we see that $K(x) \ge 0$, $K(0) = 0$, $K(1) = 0$, and $K(a) - K(b) = -1/2K(1) = 0$. Let the function $\rho(t)$ be defined by the equality $F(\rho(t)) - 1/2F(1) = F(t)$.

Then $\rho(0) = \xi$, where ξ is a point such that $F(\xi) = F(1)/2$, $0 < \xi < 1$, and $\rho(\xi) = 1$. The function $S(t) = K(t) - K(\rho(t))$ is such that $S(0) = K(0) - K(\xi) \leq 0$ and $S(\xi) = K(\xi) - K(1) \geq 0$. Therefore, there is a point $a \in (0, \xi)$ such that $S(a) = 0$. If we set $b = \rho(a)$, then we obtain (4) and the equality $K(a) = K(b)$ will also be satisfied.

If $f(x) \geq 0$, then we can construct the function y_ε corresponding to the function $f(x) + \varepsilon$ and pass to the limit, which is easy. □

Lemma 14. *Let p be a negative number. Then for all functions $y \in K_p(0,h)$ the following estimate is valid:*

$$(\int_0^h y'(x)^p\, dx)^{1/p} \leq (\frac{h}{4})^{1/p-3/2}(\int_0^h y(x)^2\, dx)^{1/2}.$$

Proof. Let at first $h = 1$, and $\int_0^1 y'(x)^p\, dx = 1$. Then $\int_0^{1/2} y'(x)^p\, dx < 1$. Let E be the subset of points x in $[0, 1/2]$ such that $y'(x) > 4^{1/p}$ and μ its measure. Then obviously $1 > 4(1/2 - \mu)$, i.e. $\mu > 1/4$. Therefore

$$y(1/2) \geq \int_0^{1/2} y'(x)\, dx > \int_E y'(x)\, dx > 4^{1/p-1}.$$

Since y is increasing, we have $y(x) > 4^{1/p-1}$ for $x > 1/2$. Then $\int_0^1 y(x)\, dx > \int_{1/2}^1 y(x)\, dx > 4^{1/p-3/2}$. Therefore

$$(\int_0^1 y'(x)^p\, dx)^{1/p} \leq 4^{3/2-1/p} \int_0^1 y(x)\, dx \leq 4^{3/2-1/p}(\int_0^1 y^2(x)\, dx)^{1/2}.$$

Let $h \neq 1$ and $y \in K_p(0,h)$. Then $z(x) = y(hx) \in K_p(0,h)$ and we can apply the proved estimate to the function z, so that

$$(\int_0^1 z'(x)^p\, dx)^{1/p} \leq 4^{3/2-1/p}(\int_0^1 z^2(x)\, dx)^{1/2}.$$

Thus

$$(\int_0^h y'(x)^p\, dx)^{1/p} \leq (4/h)^{3/2-1/p}(\int_0^h y^2(x)\, dx)^{1/2}$$

and the proof is complete. □

Lemma 15. *Let p be a real number, $0 < p < 2/3$. Then there exists a constant $C = C(p)$ independent of y and h such that*

$$(\int_0^h y'(x)^p\, dx)^{1/p} \leq C(p)h^{1/p-3/2}(\int_0^h y(x)^2\, dx)^{1/2},\ y \in K_p(0,h).$$

Proof. Let at first $h = 1$, and $\int_0^1 y'(x)^p\, dx = 1$. Then there is a point $t_1 \in (0,1)$ such that $\int_0^{t_1} y'(x)^p\, dx = \int_{t_1}^1 y'(x)^p\, dx = 1/2$. By the Hölder inequality we have

$1/2 \le (\int_0^{t_1} y'(x)\,dx)^p \cdot t_1^{1-p}$. Therefore, $1/2 \le y(t_1)^p \cdot t_1^{1-p} \le y(t_1)^p$ and $y(x) \ge 2^{-1/p}$ for $x \ge t_1$. By the Hölder inequality

$$\begin{aligned} 1/2 &= \int_{t_1}^1 y'(x)^p y^{2p-2} y^{2-2p}\,dx \\ &\le (\int_{t_1}^1 y'(x) y(x)^{2-2/p}\,dx)^p (\int_{t_1}^1 y(x)^2\,dx)^{1-p} \le C_p (\int_0^1 y(x)^2\,dx)^{1-p}, \end{aligned}$$

where $C_p = p^p 2^{(2-3p)/p}/(2-3p)^p$. Thus for $p < 2/3$ we have $\int_0^1 y(x)^2\,dx \ge (2C_p)^{1/(p-1)}$, which gives the result with $C = (2C_p)^{1/2(1-p)}$.

If $\int_0^1 y'(x)^p\,dx = I \neq 1$, then one can take instead of the function $y(x)$ the function $y(x)I^{-1/p}$. If $h \neq 1$, one can apply the obtained inequality to the function $y(xh)$. □

Note that the constants in Lemmas 14 and 15 are not the best possible. The sharp constants are indicated in Lemmas 27 and 32 of Chapter 5.

Lemma 16. *Let $p < 2/3$, $p \neq 0$, and $0 < r \le 1$. Let*

$$m_1 = \sup_{h \in (0,r)} \sup_{y \in K_p(0,h,r)} G[y], \quad \textit{where} \quad G[y] = \frac{(\int_0^r |y'(x)|^p\,dx)^{2/p}}{\int_0^r y(x)^2\,dx}.$$

Then there exists a constant C_1 independent of r such that $m_1 \le C_1 r^{2/p-3}$.

Proof. Let at first $r = 1$. By Lemmas 14 and 15 we have for $y \in K_p(0,h,r)$ the inequalities

$$h^{3-2/p}(\int_0^h |y'(x)|^p\,dx)^{2/p} \le C \int_0^h y(x)^2\,dx,$$
$$(1-h)^{3-2/p}(\int_h^1 |y'(x)|^p\,dx)^{2/p} \le C \int_h^1 y(x)^2\,dx,$$

where the value of C, corresponding to $h = 1$, was found in Lemmas 14 and 15. Let $\int_0^1 y'(x)^p\,dx = 1$ and $\int_0^h y'(x)^p\,dx = a$. Then

$$a^{2/p} h^{3-2/p} + (1-a)^{2/p}(1-h)^{3-2/p} \le C \int_0^1 y(x)^2\,dx.$$

By Lemma 26 of Chapter 5 the function $F(a,h) = a^{2/p}h^{3-2/p} + (1-a)^{2/p}(1-h)^{3-2/p}$ defined in the square $0 < a < 1$, $0 < h < 1$, has a minimal value $1/4$ at the point $a = h = 1/2$. Therefore, $C\int_0^1 y(x)^2\,dx \ge 1/4$, i.e. $G[y] \le 4C$ for all admissible y. In order to obtain the result for an arbitrary r it suffices to substitute the function $y(x)$ by $y(xr)$. □

Proof of Theorem 12.

1. If $\alpha > 1$, then by the Hölder inequality

$$\int_0^1 Q(x)y''(x)^2\,dx \le \Big(\int_0^1 Q(x)^\alpha\,dx\Big)^{1/\alpha}\Big(\int_0^1 |y''(x)|^p\,dx\Big)^{2/p},$$

where $p = 2\alpha/(\alpha-1)$. Set $y_0(x) = x^2(1-x)^2$. Then

$$\inf_{y\in H} L[Q,y] \le \frac{(\int_0^1 |y_0''(x)|^p\,dx)^{2/p}}{\int_0^1 y_0'(x)^2\,dx} = C.$$

Therefore, $M_\alpha \le C$.

Similarly, for $\alpha = 1$ we have

$$\inf_{y\in H} L[Q,y] \le \frac{\max y_0''(x)^2}{\int_0^1 y_0'(x)^2\,dx} = C.$$

2. Let $0 < \alpha < 1$ and Q be a function from the class R_α. According to Lemma 13, we can construct a function $y(x)$ such that $y(x) \ge 0$,

$$y''(x)^2 = [Q(x)+1]^{\alpha-1},\ y(0) = y'(0) = y(1) = y'(1) = 0.$$

By our construction

$$\int_0^1 |y''(x)|^p\,dx = \int_0^1 [Q(x)+1]^\alpha\,dx,$$

where $p = 2\alpha/(\alpha-1)$. Let r be the maximum point of the function $y(x)$. The function y' satisfies the conditions of Lemma 14 on the intervals $[0,r]$ and $[r,1]$. Therefore, $L[Q,y] \le L[Q+1,y] \le C_1 G[y'] \le C$ and thus $M_\alpha \le C$.

3. Let now $0 > \alpha > -1/2$. We will use the same function $y(x)$ as above, in step 2. Let $p = 2\alpha/(\alpha-1)$. Then $0 < p < 2/3$. Using Lemma 15 we obtain that $M_\alpha \le C$.

4. Let $\alpha < -1/2$ and $\varepsilon \in (0, 1/10)$. Set

$$Q(x) = \begin{cases} \varepsilon^{-1/\alpha}(1-\varepsilon)^{1/\alpha}, & \text{if } 0 < x < \varepsilon,\\ (1-\varepsilon)^{-1/\alpha}\varepsilon^{1/\alpha}, & \text{if } \varepsilon < x < 1,\end{cases}$$

so that $\int_0^1 Q(x)^\alpha\,dx = 1$.

Since

$$\begin{aligned}\int_0^1 y'(x)^2\,dx &\le \varepsilon^2\int_0^\varepsilon y''(x)^2\,dx + (1-\varepsilon)^2\int_\varepsilon^1 y''(x)^2\,dx\\ &= (1-\varepsilon)^{-1/\alpha}\varepsilon^{2+1/\alpha}\int_0^\varepsilon Q(x)y''(x)^2\,dx\\ &\quad + (1-\varepsilon)^{2+1/\alpha}\varepsilon^{-1/\alpha}\int_\varepsilon^1 Q(x)y''(x)^2\,dx\\ &\le 2\varepsilon^\gamma\int_0^1 Q(x)y''(x)^2\,dx,\end{aligned}$$

where $\gamma = \min(2 + 1/\alpha, -1/\alpha) > 0$, we obtain that $M_\alpha \geq \varepsilon^{-\gamma}/2$ and therefore $M_\alpha = \infty$.

5. Consider now the case when $\alpha = -1/2$. Set $Q(x) = \max(x^2/\varepsilon^2,\ \delta^2/\varepsilon^2)$, where $\delta = \exp(1 - 1/\varepsilon)$. Observe that

$$\int_0^1 Q(x)^{-1/2}\,dx = \delta \cdot \varepsilon/\delta + \int_\delta^1 \varepsilon/x\,dx = \varepsilon - \varepsilon \ln \delta = 1.$$

On the other hand, the well-known estimate

$$\int_0^1 y'(x)^2\,dx \leq 4 \int_0^1 x^2 y''(x)^2\,dx,$$

valid for all functions $y' \in W_2^1(0,1)$, vanishing at 0, implies that

$$\int_0^1 y'(x)^2\,dx \leq 4\delta^2 \int_0^\delta y''(x)^2\,dx + 4 \int_\delta^1 x^2 y''(x)^2\,dx \leq 4\varepsilon^2 \int_0^1 Q(x) y''(x)^2\,dx.$$

This means that

$$\inf_{y \in H} \frac{\int_0^1 Q(x) y''(x)^2\,dx}{\int_0^1 y'(x)^2\,dx} \geq \frac{1}{4\varepsilon^2},$$

so that $M_\alpha = \infty$.

6. Let $\alpha \leq -1$. By the Hölder inequality

$$\int_0^1 |y''(x)|^p\,dx \leq \Big(\int_0^1 Q(x) y''(x)^2\,dx\Big)^{p/2} \Big(\int_0^1 Q(x)^\alpha\,dx\Big)^{1/(1-\alpha)},$$

where $p = 2\alpha/(\alpha - 1)$. Since $p \geq 1$, the inequality

$$\int_0^1 y'(x)^2\,dx \leq \int_0^1 \Big(\int_0^x |y''(t)|\,dt\Big)^2 dx \leq \Big(\int_0^1 |y''(t)|\,dt\Big)^2 \leq \Big(\int_0^1 |y''(t)|^p\,dt\Big)^{2/p}$$

holds for all functions $y \in H$, and thus $m_\alpha \geq 1$.

7. If $\alpha > 0$, then we can take a function y vanishing in $[0, 1/2]$ and such that $\int_0^1 y'^2\,dx = 1$. Since the function Q can have arbitrarily small values in $[1/2, 1]$, the value of m_α is equal to 0.

8. Let $0 > \alpha > -1$. Let us show that in this case $m_\alpha = 0$. Set the following:

$$y'(x) = \begin{cases} 2x, & \text{if } 0 < x < \varepsilon, \\ 2\varepsilon, & \text{if } \varepsilon < x < 1/2 - \varepsilon, \\ (1 - 2x), & \text{if } 1/2 - \varepsilon < x < 1/2, \end{cases}$$

$$Q(x) = \begin{cases} \varepsilon^{-1/\alpha}(1-\varepsilon)^{1/\alpha}, & \text{if } |x - 1/2| < \varepsilon/4 \\ & \text{or } |x - 1/2| > 1/2 - \varepsilon/4, \\ (1-\varepsilon)^{-1/\alpha}\varepsilon^{1/\alpha} & \text{for other } x. \end{cases}$$

If $y(0) = 0$, the function y is defined for $0 \le x \le 1/2$. Let us now set $y(x) = -y(1-x)$ for $x \in (1/2, 1)$. It is easy to see that

$$\int_0^1 Q(x)^\alpha\, dx = 1; \quad \int_0^1 Q(x) y''(x)^2\, dx = 4\varepsilon^{1-1/\alpha}(1-\varepsilon)^{1/\alpha},$$

$$\int_0^1 y'(x)\, dx = 0, \qquad \int_0^1 y'(x)^2\, dx = 16\varepsilon^3/3 + 4\varepsilon^2(1-\varepsilon).$$

Therefore,

$$m_\alpha \le \frac{\int_0^1 Q(x) y''(x)^2\, dx}{\int_0^1 y'(x)^2\, dx} \le \varepsilon^{-1-1/\alpha},$$

and since $-1 - 1/\alpha > 0$, the value of m_α is equal to zero. □

6.5 Precise results

Now we consider the question of the attainability of the extremal values of the functional L.

Theorem 17. *If $\alpha < -1$, then there exist a function $y \in H$ and a function Q satisfying* (9), *such that*

$$L[Q, y] = m_\alpha = \frac{4(2\alpha+1)}{\alpha}\Big(\frac{\alpha+1}{2\alpha+1}\Big)^{1-1/\alpha} B\big(\frac{1}{2}, \frac{1}{2} + \frac{1}{2\alpha}\big)^2,$$

where B is the Euler beta function.

If $1 > \alpha > -1/2$, $\alpha \ne 0$, then there exist a function $y_0 \in H$ and a function Q, satisfying (9), *such that $\inf_y L[Q, y] = L[Q, y_0] = M_\alpha$.*

Furthermore, if $\alpha \ge 1$, then

$$M_\alpha \le \frac{4(2\alpha+1)}{\alpha}\Big(\frac{1+\alpha}{1+2\alpha}\Big)^{1-1/\alpha} B\big(\frac{1}{2}, \frac{1}{2} + \frac{1}{2\alpha}\big)^2.$$

If $0 < \alpha < 1$, then

$$M_\alpha = 4\frac{(2\alpha+1)}{\alpha}\Big(\frac{1+\alpha}{1+2\alpha}\Big)^{1-1/\alpha} B\big(\frac{1}{2}, \frac{1}{2} + \frac{1}{2\alpha}\big)^2.$$

If $-1/2 < \alpha < 0$, then

$$M_\alpha \le -4\frac{2\alpha+1}{\alpha}\Big(\frac{1+\alpha}{1+2\alpha}\Big)^{1-1/\alpha}\Big(\int_0^\infty \frac{dt}{(1+t^2)^{1/2-1/2\alpha}}\Big)^2.$$

Remark 18. In particular, in the classical Lagrange problem with $\alpha = 1/2$ the optimal function $Q(x)$ is defined on $(0,1)$ in the parametric form as follows:

$$x = (2t + \sin 2t)/4\pi,\ Q(x) = 16\cos^4 t/9;\ 0 \le t \le 2\pi.$$

This function and the value $M_{1/2} = 16\pi^2/3$ have been found by Keller-Tadjbakhsh in [KeTa]. The optimal column has two points at which $Q(x)$ vanishes.

Note also that we do not know the sharp value of M_α if $\alpha > 1$ or $-1/2 < \alpha < 0$ and cannot prove that the optimal functions Q, y_0 do exist in this case.

The proof of Theorem 17 is based on the following lemmas.

Lemma 19. *Let $p > 1$ and K be the class of functions y of the space $W^1_{p,0}(0,1)$ such that $\int_0^1 y(x)\,dx = 0$. Let $m = \inf_{y\in K} G[y]$, where*

$$G[y] = \frac{(\int_0^1 |y'(x)|^p\,dx)^{2/p}}{\int_0^1 y(x)^2\,dx}.$$

Then

$$m = 4\Big(\frac{2p-2}{3p-2}\Big)^{2/p}\Big(3-\frac{2}{p}\Big)B\big(\frac{1}{2}, 1-\frac{1}{p}\big)^2,$$

where B is the Euler function, and there exists a function $y_0 \in K$ such that $G[y_0] = m$. Moreover, $y_0(x) = -y_0(1-x)$ and

$$\begin{aligned} y_0(x) &= y_0(1/4) - c_1|x-1/4|^{p/(p-1)}[1+o(1)],\\ y_0'(x) &= -c_1 p/(p-1)|x-1/4|^{1/(p-1)} sgn(x-1/4)[1+o(1)], \end{aligned}$$

as $\alpha \to 1/4$. **Proof.** It is evident that the number m is finite and is not greater than, for example, $G[y_1]$, where $y_1(x) = 1/4 - |x-1/4|$ for $x \in (0,1/2)$ and $y_1(x) = |x-3/4| - 1/4$ for $x \in (1/2,1)$.

Let $\{y_k\}$ be a minimizing sequence such that

$$\int_0^1 y_k(x)^2\,dx = 1, \int_0^1 y_k(x)\,dx = 0 \quad\text{and}\quad \int_0^1 |y_k'(x)|^p\,dx \to m^{p/2}.$$

This sequence is compact in $L_2(0,1)$ and weakly compact in $W^1_{p,0}(0,1)$ so that there is a subsequence $\{y_{n_k}\}$ converging in $L_2(0,1)$ to $y_0(x)$ and

$$\int_0^1 y_0(x)^2\,dx = 1, \int_0^1 |y_0'(x)|^p\,dx \le m^{p/2}, \int_0^1 y_0(x)\,dx = 0.$$

Since $m^{p/2}$ is the minimal value of integrals $\int_0^1 |y'(x)|^p\,dx$, we have in fact the equality: $\int_0^1 |y_0'(x)|^p\,dx = m^{p/2}$, and the function y_0 is extremal. Since the integral of y_0 vanishes, this function has at least one zero in $(0,1)$. We can reconstruct the function y_0 without changing the values of the integrals $\int_0^1 y_0(x)^2\,dx$, $\int_0^1 y_0'(x)^2\,dx$ and $\int_0^1 y_0(x)\,dx$ in such a way that it will be positive on $(0,x_0)$ and negative on $(x_0,1)$, where $0 < x_0 < 1$. To do this we shift all intervals on which $y(x) > 0$ to the left, not changing the values of y on them.

The functional $G[y]$ is differentiable since $p>1$ and therefore the function y_0 satisfies the Euler-Lagrange equation

$$(|y_0'|^{p-2}y_0')' + m^{p/2}y_0 + m_1 = 0,\ y_0(0) = y_0(1) = 0,$$

which implies that

$$(p-1)|y_0'|^{p-1}y_0'' + m^{p/2}y_0y_0' + m_1y_0' = 0.$$

Therefore

$$(p-1)|y_0'|^p/p + m^{p/2}y_0^2/2 + m_1y_0 = C.$$

Integrating this equality over $(0,1)$, we obtain that $C = m^{p/2}[(p-1)/p+1/2]$.

Moreover, $y_0'(0) = -y_0'(x_0) = y_0'(1) = (Cp/(p-1))^{1/p}$ and $y_0(0) = y_0(x_0) = y_0(1) = 0$. Therefore, $y_0(x) = y_0(x_0-x)$ for $0 \le x \le x_0$ and $y_0(x) = y_0(1+x_0-x)$. Set $z(x) = m^{p/2}y_0 + m_1$. Then

$$(|z'|^{p-2}z')' + m^{p(p-1)/2}z = 0.$$

All solutions of this equation are oscillating periodic functions, with the distance between zeroes equal to half of the period, and odd with respect to each zero. Therefore, the mean value in the period is equal to zero. Since $z(0) = z(1),\ z'(0) = z'(1)$, the mean value of z on $(0,1)$ is equal to zero. However, then we have $m_1 = \int_0^1 z(x)\,dx - m^{p/2}\int_0^1 y_0(x)\,dx = 0$.

Therefore, $x_0 = 1/2$ and $y_0(x) = -y_0(1-x)$. The maximal value M of y_0 is defined from the equation

$$(p-1)|y_0'|^p/p + m^{p/2}y_0^2/2 = m^{p/2}[(p-1)/p+1/2], \tag{14}$$

so that $M^2 = 3-2/p$. Integrating, we obtain that for $0 \le x \le 1/4$

$$\int_0^{y_0} \frac{dz}{[3p-2-pz^2]^{1/p}} = \frac{xm^{1/2}}{[2(p-1)]^{1/p}}.$$

In particular, $y_0(1/4) = M$ and

$$\int_0^{M} \frac{dz}{[3p-2-pz^2]^{1/p}} = \frac{m^{1/2}}{4[2(p-1)]^{1/p}}.$$

Changing the variable z to $t(3-2/p)^{1/2}$, we see that

$$\int_0^1 \frac{dt}{(1-t^2)^{1/p}} = \frac{m^{1/2}}{4}\Big(\frac{3p-2}{2p-2}\Big)^{1/p}\Big(\frac{p}{3p-2}\Big)^{1/2},$$

or

$$m^{1/2} = 4\Big(\frac{2p-2}{3p-2}\Big)^{1/p}\Big(3-\frac{2}{p}\Big)^{1/2}\int_0^1 \frac{dt}{(1-t^2)^{1/p}}.$$

Observe that

$$\int_0^1 \frac{dt}{(1-t^2)^{1/p}} = \frac{1}{2}B(\frac{1}{2}, 1-\frac{1}{p}),$$

where B is the Euler function. Thus

$$m^{1/2} = 2\Big(\frac{2p-2}{3p-2}\Big)^{1/p}(3-\frac{2}{p})^{1/2}B(\frac{1}{2}, 1-\frac{1}{p}).$$

On the other hand

$$\int_{y_0}^M \frac{dz}{[3p-2-pz^2]^{1/p}} = (x-1/4)\frac{m^{1/2}}{[2(p-1)]^{1/p}},$$

so that $y_0(x) = M + A(1/4-x)^\gamma[1+o(1)]$ as $x \to 1/4-0$ with $\gamma = p/(p-1)$. □

Corollary 20. *If* $p = 1$, *then* $m = 16$.

Proof. For all $y \in K$, and $p > 1$ we have

$$m\int_0^1 y(x)^2\,dx \le (\int_0^1 |y'(x)|^p\,dx)^{2/p},$$

and $\lim_{p\to 1+0} m = 16$. Therefore, $m \ge 16$. On the other hand, setting $y(x) = 1$ for $\varepsilon < x < 1/2-\varepsilon$, $y(x) = -1$ for $1/2+\varepsilon < x < 1-\varepsilon$, $y(x) = x/\varepsilon$ for $0 < x < \varepsilon$, $y(x) = (1/2-x)/\varepsilon$ for $1/2-\varepsilon < x < 1/2+\varepsilon$, and $y(x) = (x-1)/\varepsilon$ for $1-\varepsilon < x < 1$, we can see that $\int_0^1 y(x)^2\,dx = 1-8\varepsilon/3$, $\int_0^1 |y'(x)|\,dx = 4$, and so $m = 16$. □

In the same way one can prove the following.

Corollary 21. *If* $p = \infty$, *then* $m = \lim_{p\to\infty} m(p) = 48$. *The estimate is realized by the function* y_1 *equal to* $1/4 - |x-1/4|$ *for* $0 < x < 1/2$ *and to* $|x-3/4| - 1/4$ *for* $1/2 < x < 1$.

Lemma 22. *Let* $-1/2 < \alpha < 1$, $\alpha \ne 0$ *and* $z(x) = y_0(x)$ *for* $0 < x < 1/2$, $z(x) = -y_0(1-x)$ *for* $1/2 < x < 1$, *where* y_0 *is the function found in Lemmas* 28 *and* 33 *of Chapter* 5 *for* $l = 1/2$ *and* $p = 2\alpha/(\alpha-1)$. *Set* $Q(x) = |z'(x)|^{2/(\alpha-1)}$. *Then*

$$\begin{aligned}
&Q(x) = Q(1-x) && \text{for } 0 < x < 1,\\
&Q(x) = Q(1/2-x) && \text{for } 0 < x < 1/2,\\
&Q(x) = c|x-1/4|^\gamma[1+o(1)] && \text{as } x \to 1/4,\\
&Q(x) = c|x-3/4|^\gamma[1+o(1)] && \text{as } x \to 3/4,
\end{aligned}$$

where $\gamma = 2/(\alpha+1) \in]1,2[$ *if* $\alpha > 0$ *and* $\gamma = 2$ *if* $\alpha < 0$.

Proof. See Corollaries 29 and 34 of Chapter 5. □

Lemma 23. *Let $Q(x)$ be a smooth positive function on $[0, d[$, such that*

$$\lim_{x \to d} Q(x)(d-x)^{-\gamma} = a \neq 0,$$

where $1 < \gamma \leq 2$. Let $y(x)$ be a solution of the equation

$$Q(x)y''(x) + my(x) = 0, \ 0 < x < d,$$

such that $\int_0^d Q(x)y''(x)^2\,dx < \infty$. Then $\lim_{x\to d} y(x) = 0$, $\lim_{x\to d} y(x)y'(x) = 0$. Moreover, we have as $x \to d-$

$$\begin{aligned} y(x) &= C[d - x + c_1(d-x)^{3-\gamma}(1+o(1))], \\ y'(x) &= -C[1 + c_1(3-\gamma)(d-x)^{2-\gamma}(1+o(1))], \\ y''(x) &= Cc_1(3-\gamma)(2-\gamma)(d-x)^{1-\gamma}(1+o(1)) \end{aligned}$$

if $1 < \gamma < 2$ and

$$\begin{aligned} y(x) &= (d-x)^{\rho}(C + o(1)), \\ y'(x) &= (d-x)^{\rho-1}(-\rho C + o(1)), \\ y''(x) &= (d-x)^{\rho-2}(\rho(\rho-1)C + o(1)), \end{aligned}$$

if $\gamma = 2$ with $\rho > 1/2$.

Proof. Set $I[y] = \int_0^d Q(x)y''(x)^2\,dx$.
If $1 < \gamma < 2$, then by solving the Cauchy problem we can find two linearly independent solutions $y(x)$ and $z(x)$ such that as $x \to d-$

$$\begin{aligned} y(x) &= d - x + c_1(d-x)^{3-\gamma} + \dots, \ c_1 = \frac{m}{a(3-\gamma)(2-\gamma)}; \\ z(x) &= 1 + c_2(d-x)^{2-\gamma} + \dots, \ c_2 = \frac{m}{a(2-\gamma)(1-\gamma)}. \end{aligned}$$

However, $z''(x) = (d-x)^{-\gamma}(m/a + o(1))$ and the integral $I[z] = \int_0^d Q(x)z''(x)^2\,dx$ is divergent. So the solution with a finite value of I is proportional to y and vanishes at $x = d$.

If $\gamma = 2$, the corresponding solutions have the form

$$y(x) = (d-x)^{\kappa_1}(1+o(1)), \ z(x) = (d-x)^{\kappa_2}(1+o(1)),$$

where κ_j are the different roots of the characteristic equation $a\kappa(\kappa-1)+m = 0$ so that $\kappa_{1,2} = \frac{1}{2} \pm \sqrt{\frac{1}{4} - \frac{m}{a}}$. If $4m > a$, then both roots are complex and $\operatorname{Re} \kappa_{1,2} = \frac{1}{2}$. The integrals $I[y]$ and $I[z]$ are divergent. If $4m < a$, then both roots are real, the integral $I[y]$ is finite, but $I[z]$ is divergent. We have $y(d) = 0$.

At last, if $\kappa_1 = \kappa_2$, i.e. $a = 4m$, the solutions have the form

$$y(x) = (d-x)^{1/2}[1+o(1)], \ z(x) = (d-x)^{1/2}\ln(d-x)[1+o(1)],$$

so that both integrals $I[y]$ and $I[z]$ are divergent.

Therefore, if $\gamma = 2$ and $I[y]$ is finite, then $4m < a$ and the solution vanishes at $x = d$. □

Lemma 24. *Let $0 < \alpha < 1$ and $Q(x)$ be the function found in Lemma 22. Let*

$$m_1 = \inf_{y \in H_0^2(0,1)} \frac{\int_0^1 Q(x) y''(x)^2 \, dx}{\int_0^1 y'(x)^2 \, dx}.$$

Then $m_1 = M_\alpha$, where M_α was indicated in Theorem 17. The minimal value is attained on the function w that is equal to $\int_0^x z(t)\, dt$.

Proof. Consider a minimizing sequence $y_k(x)$ such that $\int_0^1 y_k'(x)^2 \, dx = 1$. The integrals $\int_0^{1/4-\varepsilon} y_k''(x)^2 \, dx$, $\int_{1/4+\varepsilon}^{3/4-\varepsilon} y_k''(x)^2 \, dx$, and $\int_{3/4+\varepsilon}^1 y_k''(x)^2 \, dx$ are bounded and one can choose a subsequence converging almost everywhere in $(0, 1/4-\varepsilon)$, $(1/4+\varepsilon, 3/4-\varepsilon)$, and $(3/4+\varepsilon, 1)$ in H^1 and weakly in $H^2(0, 1/4-\varepsilon)$, $H^2(1/4+\varepsilon, 3/4-\varepsilon)$, and $H^2(3/4+\varepsilon, 1)$. Using diagonalization, one can find a subsequence converging almost everywhere in $(0,1)$ to $y_1(x)$.

Let us show that

$$\int_{1/4-\varepsilon}^{1/4+\varepsilon} y_k'(x)^2 \, dx + \int_{3/4-\varepsilon}^{3/4+\varepsilon} y_k'(x)^2 \, dx \le C\varepsilon^{2-\gamma} \tag{15}$$

with a constant C independent of ε and k. Indeed, using the equality $y_k'(1/4-\varepsilon) = \int_0^{1/4-\varepsilon} y_k''(x)\, dx$, we see that

$$y_k'(1/4-\varepsilon)^2 \le \int_0^{1/4-\varepsilon} Q(x) y_k''(x)^2 \, dx \int_0^{1/4-\varepsilon} Q(x)^{-1} \, dx \le C_1 \varepsilon^{1-\gamma}.$$

Since $\int_{1/4+\varepsilon}^{3/4-\varepsilon} y_k'(x)^2 \, dx < 1$, there exists a $\theta_k \in]1/4+\varepsilon, 3/4-\varepsilon[$ such that $y_k'(\theta_k)^2 \le 2$ and

$$y_k'(1/4+\varepsilon)^2 \le 4 + 2\int_{\theta_k}^{1/4+\varepsilon} Q(x) y_k''(x)^2 \, dx \int_{\theta_k}^{1/4+\varepsilon} Q(x)^{-1} \, dx \le C_2 \varepsilon^{1-\gamma}.$$

Analogously, we have $y_k'(3/4-\varepsilon)^2 \le C_2\varepsilon^{1-\gamma}$, $y_k'(3/4+\varepsilon)^2 \le C_1 \varepsilon^{1-\gamma}$. If $1/4-\varepsilon < x < 1/4$, then

$$\begin{aligned} y_k'(x)^2 &\le 2y_k'(1/4-\varepsilon)^2 + 2\int_{1/4-\varepsilon}^x Q(x) y_k''(x)^2 \, dx \int_{1/4-\varepsilon}^x Q(x)^{-1} \, dx \\ &\le C_3 (1/4 - x)^{1-\gamma} \end{aligned}$$

and therefore, $\int_{1/4-\varepsilon}^{1/4} y_k'(x)^2 \, dx \le C_3 \varepsilon^{2-\gamma}$. So (15) is valid. Since we can assume that $y_k'(x)$ converges uniformly to $y_1'(x)$ outside of the ε-neighborhood of the points $1/4$ and $3/4$, we see that

$$\int_0^1 y_1'(x)^2 \, dx = 1, \ \int_0^1 Q(x) y_1''(x)^2 \, dx \le m_1.$$

But m_1 is the minimal value of the latter integral. Therefore, $\int_0^1 Q(x)y_1''(x)^2\,dx = m_1$. The function y_1 satisfies the equation

$$(Q(x)y_1''(x))'' + m_1 y_1''(x) = 0,$$

$y_1(0) = 0,\ y_1'(0) = 0,\ y_1(1) = 0,\ y_1'(1) = 0$. The function $z_1(x) = y_1(x) + y_1(1-x)$ is also minimizing, if it does not vanish identically.

If $z_1(x) \not\equiv 0$, then it is even and

$$Q(x)z_1''(x) + m_1 z_1 = C.$$

Using Lemma 23 we see that $C = m_1 z_1(1/4) = m_1 z_1(3/4)$. Set $u(x) = z_1(x) - z_1(1/4)$. Then

$$Q(x)u'' + m_1 u = 0,\ u(1/4) = u(3/4) = 0.$$

On the other hand, if $v(x) = w(x) - w(1/4)$, where $w(x) = \int_0^x z(t)\,dt$ and z is the function indicated in Lemma 22, then

$$Q(x)v'' + mv = 0,\ v(1/4) = v(3/4) = 0$$

and $v > 0$ in $(1/4, 3/4)$. If u vanishes in $(1/4, 3/4)$, then by the Sturm theorem, $m_1 > m$, which is impossible. If u does not vanish in $(1/4, 3/4)$, then we obtain using Lemma 23 that

$$(m_1 - m)\int_{1/4}^{3/4} \frac{uv}{Q}\,dx = (u'v - uv')|_{x=1/4}^{x=3/4} = 0$$

and $m_1 = m$.

If $z_1(x) \equiv 0$, then y_1 is odd and

$$Q(x)y_1''(x) + m_1 y_1 = A(x - 1/2),$$

where A is a constant. Lemma 23 implies that $A = -4m_1 y_1(1/4)$. Setting $u(x) = y_1(x) - A(x - 1/2)/m_1$, we obtain that

$$Qu'' + m_1 u = 0.$$

Moreover, $u(1/4) = u(1/2) = u(3/4) = 0$. Applying once again the Sturm theorem, we see that $m_1 > m$, which is impossible. □

Proof of Theorem 17. Let Q be an arbitrary positive function, satisfying (8).

Let at first $\alpha \le -1$. Then by the Hölder inequality

$$\int_0^1 Q(x)|y''(x)|^2\,dx \ge \Big(\int_0^1 |y''(x)|^p\,dx\Big)^{2/p}\Big(\int_0^1 Q(x)^\alpha\,dx\Big)^{1/\alpha},$$

where $p = 2\alpha/(\alpha - 1)$, $2 > p \geq 1$. Therefore, $L[Q, y] \geq G[y']$. Lemma 19 implies the existence of the function $y_0 \in W^2_{p,0}(0, 1)$ for $p > 1$ satisfying the conditions

$$\int_0^1 |y_0''(x)|^p \, dx = 1, \;\; G[y_0'] \leq G[y']$$

for all $y \in W^2_{2,0}(0, 1)$ such that $\int_0^1 |y''(x)|^p \, dx = 1$. The value of m is also indicated in Lemma 19. Therefore, we have $m = m_\alpha = L[Q, y_0]$, if $Q(x) = |y_0''|^{2/(\alpha-1)}$. If $p = 1$, the optimal functions Q and y do not exist, but the value of m was indicated in Corollary 20.

Now let $\alpha > -1/2, \alpha \neq 0$. Then by Lemma 13 there exists a function $y_Q(x)$ such that $Q(x) = |y_Q''|^{2/(\alpha-1)}$, so that $L[Q, y_Q] = G[y_Q']$ and therefore, $M_\alpha \leq m$. On the other hand, we have the equality $L[Q_0, y_0] = G[y_0']$, if $Q_0(x) = |y_0''(x)|^{2/(\alpha-1)}$ and y_0 is equal to $\int_0^x z(t) \, dt$, where z is the function indicated in Lemma 22. If $0 < \alpha < 1$, then by Lemma 24 we have

$$m = \inf_{y \in H_0^2} \frac{\int_0^1 Q_0(x) y''(x)^2 \, dx}{\int_0^1 y'(x)^2 \, dx} = L[Q_0, y_0].$$

The proof is complete. □

Chapter 7

Eigenfunctions of Elliptic Operators in Bounded Domains

We are going to study estimates of eigenfunctions and eigenvalues of elliptic operators. Note that some estimates have been considered before, in Chapter 4.

For example, consider the boundary value problem:

$$\sum_{|\alpha|\le m, |\beta|\le m} D^\alpha a_{\alpha\beta}(x) D^\beta u = \lambda u$$

in the space $W^m_{2,0}(\Omega)$. A number λ is called an *eigenvalue*, if there is a solution $u(x) \not\equiv 0$. This solution is called an *eigenfunction.* Let $a_{\alpha\beta}(x)$ be measurable functions in Ω, such that $a_{\alpha\beta}(x) = \overline{a_{\beta\alpha}(x)}$ for $|\alpha| = |\beta| = m$,

$$|a_{\alpha\beta}| \le M, \sum_{|\alpha|=|\beta|=m} a_{\alpha\beta}(x)\xi^\alpha\xi^\beta \ge \gamma \sum_{|\alpha|=m} |\xi^\alpha|^2, \ \xi \in \mathbf{R}^n, \ \gamma > 0.$$

Theorem 3 of Chapter 3 and theorems of Chapter 2 imply that there is a $\delta > 0$ such that $|\lambda| > 1$ if $\operatorname{diam}\Omega < \delta$. Now we are going to study the problem in more detail.

7.1 On the Dirichlet problem for strongly elliptic equations

Consider the equation

$$Lu \equiv \sum_{|\alpha|\le m, |\beta|\le m} D^\alpha a_{\alpha\beta}(x) D^\beta u + \sum_{|\alpha|\le m} a_\alpha(x) D^\alpha u = F(x), \qquad (1)$$

where $m \geq 1$,

$$\mathrm{Re} \sum_{|\alpha|=|\beta|=m} a_{\alpha\beta}(x)\xi^{\alpha+\beta} \neq 0$$

for all $x \in \mathbf{R}^n$, $\xi \in \mathbf{R}^n \setminus 0$, $a_{\alpha\beta}(x)$, $a_\alpha(x)$ are bounded measurable in a domain $\Omega \subset \mathbf{R}^n$ functions, $a_{\alpha\beta}(x) = \overline{a_{\beta\alpha}(x)}$ for $|\alpha| = |\beta| = m$, $F(x) \in L_2(\Omega)$.

We consider the problem of the existence and the uniqueness of a generalized solution u to equation (1) from the space $W_{2,0}^m(\Omega)$, assuming that the domain Ω is bounded. In Section 3.1, we proved that this problem is Fredholmian, i.e. the uniqueness of the solution implies the solvability of the problem for any $F \in L_2(\Omega)$.

It is well known that in the general case the uniqueness has no place. For instance, the equation $y'' + y = 0$ has a solution $y = \sin\ x$ in the space $W_{2,0}^1(\Omega)$, where $\Omega = \{x:\ 0 < x < \pi\}$.

We will find some sufficient conditions for uniqueness of the solution to equation (1) in the space $W_{2,0}^m(\Omega)$. Let

$$\sum_{|\alpha|\leq m, |\beta|\leq m} |a_{\alpha\beta}| + \sum_{|\gamma|\leq m} |a_\gamma| \leq M, \ |\mathrm{Re} \sum_{|\alpha|=|\beta|=m} a_{\alpha\beta}(x)\xi^{\alpha+\beta}| \geq \gamma_0|\xi|^{2m},$$

$\gamma_0 > 0$, and $a_{\alpha\beta}(x)$ be uniformly continuous in Ω. Then the Gårding inequality (see Theorem 3 of Chapter 3):

$$|\mathrm{Re} \int_\Omega Lu(x) \cdot \overline{u(x)}\, dx| \geq \gamma \sum_{|\alpha|=m} \int_\Omega |D^\alpha u|^2\, dx - \gamma_1 \int_\Omega |u|^2\, dx, \ \forall u \in W_{2,0}^m(\Omega) \quad (2)$$

holds. Here the constants $\gamma > 0, \gamma_1$ depend only on M, γ_0 and on the module of continuity of the coefficients $a_{\alpha\beta}$ with $|\alpha| = |\beta| = m$.

Theorem 1. *There exists a positive number μ depending only on γ/γ_1, M such that if* $\mathrm{meas}\,\Omega < \mu$, *then equation* (1) *has a unique solution in the class* $W_{2,0}^m(\Omega)$. *If*

$$|\mathrm{Re} \int_\Omega Lu(x) \cdot \overline{u(x)}\, dx| \geq \gamma \sum_{|\alpha|=m} \int_\Omega |D^\alpha u|^2\, dx, \ u \in W_{2,0}^m(\Omega), \quad (3)$$

where $\gamma > 0$, then each eigenvalue λ of the operator L satisfies the inequality $|\mathrm{Re}\ \lambda| \geq (\mathrm{meas}\,\Omega)^{-2m/n}$.

Proof. Let $u(x) \in W_{2,0}^m(\Omega)$ be a generalized solution to the equation $Lu = 0$. Using (2) we obtain that

$$\int_\Omega \sum_{|\alpha|=m} |D^\alpha u|^2\, dx \leq \frac{\gamma_1}{\gamma} \int_\Omega |u|^2\, dx.$$

Now we are going to show that the latter inequality is true in a domain Ω of small measure, only if $u = 0$.

The Sobolev Theorem 57 of Chapter 2 implies that for $w(x) \in C_0^\infty(\mathbf{R}^n)$ the inequality

$$\int_{\mathbf{R}^n} |w|^2\, dx \le C_1 \Big(\int_{\mathbf{R}^n} \sum_{|\alpha|=m} |D^\alpha w|^p\, dx\Big)^{2/p}$$

is valid with $p = 2n/(n+2m) < 2$.

If $\operatorname{supp} w \subset \Omega$, then from the Hölder inequality it follows that

$$\int_\Omega |D^\alpha w|^p\, dx \le C_2 \Big(\int_\Omega |D^\alpha w|^2\, dx\Big)^{p/2} \cdot (\operatorname{meas}\Omega)^{1-p/2},$$

and therefore

$$\int_\Omega |w|^2\, dx \le C_3 (\operatorname{meas}\Omega)^{2m/n} \int_\Omega \sum_{|\alpha|=m} |D^\alpha w|^2\, dx \tag{4}$$

for any $w(x) \in C_0^\infty(\Omega)$. Setting $w = u$ we see that $\gamma/\gamma_1 \le C_3(\operatorname{meas}\Omega)^{2m/n}$, if $u \not\equiv 0$. Thus the solution is unique, if $(\operatorname{meas}\Omega)^{2m/n} < \gamma/(C_3\gamma_1)$. Its existence follows now from the Fredholm theorem.

If $Lu = \lambda u$ and (3) is valid, then

$$|\operatorname{Re}\lambda| \int_\Omega |u(x)|^2\, dx = |\operatorname{Re}(Lu,u)| \ge \gamma \int_\Omega \sum_{|\alpha|=m} |D^\alpha u|^2\, dx.$$

As we have seen, for any $u \in W_{2,0}^m(\Omega)$ inequality (4)

$$\int_\Omega |u|^2\, dx \le C_3 (\operatorname{meas}\Omega)^{2m/n} \int_\Omega \sum_{|\alpha|=m} |D^\alpha u|^2\, dx.$$

holds with a constant C_3 independent of Ω. Therefore, $C_3(\operatorname{meas}\Omega)^{2m/n} |\operatorname{Re}\lambda| \ge \gamma$ and $|\operatorname{Re}\lambda| \ge C_4 (\operatorname{meas}\Omega)^{-2m/n}$. □

Lemma 2. *If Q is a cube in $\mathbf{R}^n$ with size h and $2m > n$, then*

$$\int_Q |u(x)|^2\, dx \le C \sum_{1\le|\alpha|\le m} \int_Q h^{2|\alpha|} |D^\alpha u(x)|^2\, dx$$

for all $u \in C^\infty(Q)$ vanishing at a point x_0 of Q.

Proof. Let at first $h = 1$. The Poincaré inequality

$$\int_Q |u(x) - \overline{u}|^2\, dx \le C \int_Q |\nabla u(x)|^2\, dx, \ \overline{u} = h^{-n} \int_Q u(x)\, dx,$$

and the Sobolev theorem on embedding of the space $W_2^m(Q)$ in the space of continuous functions imply that

$$|u(x) - \overline{u}|^2 \le C_1 \sum_{1\le|\alpha|\le 2m} \int_Q |D^\alpha u|^2\, dx.$$

Setting here $x = x_0$, we obtain the inequality

$$|\overline{u}|^2 \leq C_1 \sum_{1\leq|\alpha|\leq 2m} \int_Q |D^\alpha u|^2\,dx.$$

Therefore

$$\int_Q |u(x)|^2\,dx \leq 2C_1 \sum_{1\leq|\alpha|\leq 2m} \int_Q |D^\alpha u|^2\,dx.$$

So we have obtained our inequality for $h = 1$. In the general case it suffices to substitute $x = hx'$. □

Recall that the *inner radius* of a domain is the least upper bound of radii of balls lying in the domain.

Theorem 3. *If $2m > n$, then there is a number $h_0 > 0$ such that in each domain Ω in $\mathbf{R}^n$, whose inner radius ρ is less than h_0, the equation $Lu = 0$ has no nontrivial solutions from $W_{2,0}^m(\Omega)$. If condition (3) holds, then all eigenvalues λ satisfy the inequality $|\operatorname{Re}\lambda| \geq C\rho^{-2m}$, $C > 0$.*

Proof. Set $u(x) = 0$ in $\mathbf{R}^n \setminus \Omega$. Let us represent the space $\mathbf{R}^n$ as a sum of closed cubes Q with nonintersecting interiors and their edges equal to h. If $\rho < h\sqrt{n}/2$, then u vanishes at a point of Q and by Lemma 2 we have

$$\int_Q |u(x)|^2\,dx \leq C_1 \sum_{0<|\alpha|\leq m} \int_Q h^{2|\alpha|}|D^\alpha u|^2\,dx,$$

where Q is a cube of this partition. Summing over all Q we see that

$$\int_\Omega |u(x)|^2\,dx \leq C_1 \sum_{0<|\alpha|\leq m} \int_\Omega h^{2|\alpha|}|D^\alpha u|^2\,dx.$$

Since for any $\varepsilon > 0$ the inequality

$$\sum_{0<|\alpha|\leq m-1} \int_\Omega h^{2|\alpha|}|D^\alpha u|^2\,dx \leq \varepsilon \int_\Omega |u(x)|^2\,dx + C_\varepsilon \sum_{|\alpha|=m} \int_\Omega h^{2m}|D^\alpha u|^2\,dx$$

holds, we have

$$\int_\Omega |u(x)|^2\,dx \leq C_2 \sum_{|\alpha|=m} \int_\Omega h^{2m}|D^\alpha u|^2\,dx.$$

If $u(x) \in W_{2,0}^m(\Omega)$ is a solution to the equation $Lu = 0$, then by (2)

$$\int_\Omega \sum_{|\alpha|=m} |D^\alpha u|^2\,dx \leq C_3 \int_\Omega |u|^2\,dx.$$

If $h^{2m} < \min(1, 1/C_2C_3) = h_0^{2m}$, then these inequalities imply that $u \equiv 0$.

On the other hand, if $Lu = \lambda u$, $u \in W^m_{2,0}(\Omega)$, and (3) holds, then

$$|\operatorname{Re}\lambda| \int_\Omega |u|^2\,dx = |\operatorname{Re}(Lu,u)| \geq \gamma \int_\Omega \sum_{|\alpha|=m} |D^\alpha u|^2\,dx \geq \frac{\gamma}{C_2 h^{2m}} \int_\Omega |u|^2\,dx$$

and $|\operatorname{Re}\lambda| \geq \gamma h^{-2m}/C_2$. The proof is complete. □

Note that the condition $n < 2m$ is essential, as is seen from the following theorem.

Theorem 4. *Let $2m \leq n$, L be a symmetric elliptic operator of order $2m$, $L(x,D)1 = 0$, and condition (3) hold. Then for any $\varepsilon > 0$ there exists a domain Ω with its inner radius smaller than ε and such that the equation $Lu = \lambda u$ has a nontrivial solution $u \in W^m_{2,0}(\Omega)$ for some λ with $|\lambda| < 1$.*

Proof. In order to show this, consider the function $v_R(x) = (R^2 - |x|^2)^m$. If R is sufficiently large, then a simple calculation shows that

$$\left|\int_{|x|<R} Lv_R \cdot v_R\,dx\right| \leq \frac{1}{2}\int_{|x|<R} v_R^2\,dx.$$

Let us fix such R.

Let us also fix a finite set of points $P_1,\dots,P_N$ in $K^0_R = \{x : |x| < R\}$, such that the inner radius of the domain $K^0_R \setminus \cup_{i=1}^N P_i$ is less than ε. For any small $\delta > 0$, $\eta > 0$ there is a function $g(x)$ such that $g(x) \equiv 1$ for $|x| > \eta$, $g(x) \equiv 0$ in a neighborhood of the origin, $|g(x)| \leq 1$, and

$$\sum_{1\leq|\alpha|\leq m} \int_{\mathbf{R}^n} |D^\alpha g(x)|^2\,dx \leq \delta.$$

For example, one can take as g the function $g(x) = g_0(\ln|x|/|\ln\eta|)$, where $g_0(t) \in C^\infty(\mathbf{R}^1)$ and $g_0(t) = 1$ for $t > -2$, $g_0(t) = 0$ for $t < -3$. It is easy to see that $g(x) = 1$ for $|x| > \eta^2$ and $g(x) = 0$ for $|x| < \eta^3$. Moreover, $|D^\alpha g(x)| \leq C_\alpha |x|^{-|\alpha|}/|\ln\eta|$ for $|\alpha| > 0$. Therefore, for $0 < |\alpha| \leq m \leq n/2$ we have

$$\int_{\mathbf{R}^n} |D^\alpha g(x)|^2\,dx \leq \frac{C_\alpha^2 \omega_n}{|\ln\eta|^2}\int_{\eta^3<|x|<\eta^2} \frac{|x|^{n-1}d|x|}{|x|^{2|\alpha|}} \leq \frac{C'_\alpha}{|\ln\eta|^2}\int_{\eta^3<|x|<\eta^2}\frac{d|x|}{|x|} \to 0$$

as $\eta \to 0$.

If η is small enough, then the balls $K^{P_i}_{\eta^3}$ do not mutually intersect. Let $w(x) = v_R(x)\Pi_{i=1}^N g(x - P_i)$ in the domain $\Omega = K_R \setminus \cup_{i=1}^N K^{P_i}_{\eta^3}$. We have

$$\left|\int_\Omega Lw \cdot w\,dx\right| \leq \frac{1}{2}\int_{|x|<R} w(x)^2\,dx + O\left(\frac{1}{|\ln\eta|^2}\right) = \frac{1}{2}\int_{|x|<R} w^2\,dx + o(1).$$

If η is small enough, then $|\int_\Omega Lw \cdot w\,dx| < \int_\Omega |w|^2\,dx$. The variational principle implies that there exists an eigenvalue of the problem $Lu = \lambda u$, $u \in W^m_{2,0}(\Omega)$ such that $|\lambda| < 1$. □

Theorem 5. *Let $2m = n$, and a domain $\Omega \subset \mathbf{R}^n$ be bounded and simply connected. There is a $\rho_0 > 0$ such that if the inner radius of Ω is less than ρ_0, $Lu = 0$, and $u(x) \in W^m_{2,0}(\Omega)$, then $u(x) \equiv 0$. If condition (3) holds, then all eigenvalues λ satisfy the relation $|\operatorname{Re}\lambda| \geq C\rho^{-2m}$, where ρ is the inner radius of Ω.*

Proof. Let us represent the space $\mathbf{R}^n$ as a sum of closed cubes with edges h, whose interiors do not intersect. Let Q be one of them. If $\rho = h/5$, then in the concentric cube Q' with the edge equal to $h/2$ there is a point y from $\mathbf{R}^n \setminus \Omega$. Since the complement of Ω is connected, there is a curve connecting y with some point of ∂Q. Let the center of Q be the origin. Then there is an i $(1 \leq i \leq n)$ such that for any ξ, satisfying the inequalities $h/4 < \xi < h/2$ or $-h/2 < \xi < -h/4$, the section Q_ξ of the cube Q' by the plane $x_i = \xi$ contains a point y_ξ from $\mathbf{R}^n \setminus \Omega$. Let $i = n$, $\frac{h}{4} < \xi < \frac{h}{2}$. Then by Lemma 2

$$\int_{Q_\xi} |u(x)|^2\, dx_1 \dots dx_{n-1} \leq \sum_{0<|\alpha|\leq m} C|h|^{2|\alpha|} \int_{Q_\xi} |D^\alpha u|^2\, dx_1 \dots dx_{n-1}.$$

Integrate both sides of this inequality with respect to ξ from $h/4$ to $h/2$. Then we have

$$\int_{\substack{|x_i|<1/2, i=1,\dots,n-1\\ h/4<x_n<h/2}} |u(x)|^2\, dx \leq \sum_{0<|\alpha|\leq m} C_1 \int_{\substack{|x_i|<1/2, i=1,\dots,n-1\\ h/4<x_n<h/2}} h^{2|\alpha|}|D^\alpha u|^2\, dx.$$

This inequality implies that there is a $\overline{\xi}$, $h/4 < \overline{\xi} < h/2$ such that

$$\int_{\substack{|x_i|<1/2, i=1,\dots,n-1\\ x_n=\overline{\xi}}} |u(x)|^2\, dx_1 \dots dx_{n-1} \leq \frac{4}{h} \int_{\substack{|x_i|<1/2, i=1,\dots,n-1\\ h/4<x_n<h/2}} |u(x)|^2\, dx$$
$$\leq \frac{4C_1}{h} \sum_{0\leq|\alpha|\leq m} \int_Q h^{2|\alpha|}|D^\alpha u(x)|^2\, dx.$$

Since for $-h \leq 2x_n \leq h$ we have

$$|u(x_1, \dots, x_{n-1}, x_n)|^2 \leq 2|u(x_1, \dots, x_{n-1}, \overline{\xi})|^2 + 2h \int_{\overline{\xi}}^{x_n} |\frac{\partial u}{\partial x_n}|^2\, dx_n,$$

it follows that

$$\int_Q |u(x)|^2\, dx \leq C_2 \sum_{0<|\alpha|\leq m} h^{2|\alpha|} \int_Q |D^\alpha u|^2\, dx.$$

Summing over all cubes Q, we obtain

$$\int_\Omega |u(x)|^2\, dx \leq C_2 \sum_{0<|\alpha|\leq m} h^{2|\alpha|} \int_\Omega |D^\alpha u|^2\, dx.$$

If h is sufficiently small, $Lu = 0$, and $u(x) \in W^m_{2,0}(\Omega)$, then this inequality and (3) imply that $u \equiv 0$.

On the other hand, if $Lu = \lambda u$, then

$$|\operatorname{Re}\lambda| \int_\Omega |u(x)|^2\,dx = |\operatorname{Re}(Lu,u)| \geq \gamma \int_\Omega \sum_{|\alpha|=m} |D^\alpha u|^2\,dx.$$

We can use the interpolatory inequality

$$\sum_{0<|\alpha|\leq m-1} \int_\Omega h^{2|\alpha|}|D^\alpha u|^2\,dx \leq \varepsilon \int_\Omega |u(x)|^2\,dx + C_\varepsilon \sum_{|\alpha|=m} \int_\Omega h^{2m}|D^\alpha u|^2\,dx.$$

Therefore

$$\int_\Omega |u(x)|^2\,dx \leq C_3 \sum_{|\alpha|=m} \int_\Omega h^{2m}|D^\alpha u|^2\,dx$$

and $|\operatorname{Re}\lambda| \geq \gamma h^{-2m}/C_3$. □

7.2 Estimates of eigenfunctions of strongly elliptic operators

In this section we state some estimates of eigenfunctions in the case $n = 2m$.

Theorem 6. *Let* $n = 2m$, $u(x) \in W^m_2(K^0_1)$, $K^0_1 = \{x :\ x \in \mathbf{R}^n,\ |x| < 1\}$, *and* $u(x)$ *be a generalized solution of the equation*

$$L_0 u \equiv \sum_{|\alpha|=|\beta|=m} D^\alpha(a_{\alpha\beta}(x)D^\beta u) = 0,$$

where $a_{\alpha\beta}(x)$ *are real and*

$$\sum_{|\alpha|=|\beta|=m} a_{\alpha\beta}(x)\xi_\beta\xi_\alpha \geq \gamma \sum_{|\alpha|=m} |\xi_\alpha|^2, \quad \sum_{|\alpha|=|\beta|=m} |a_{\alpha\beta}(x)| \leq M,$$

$x \in K^0_1$, $\gamma = \text{const} > 0$. *Then there is an* A, *depending on* γ, M, n, m *such that for* $|x| < 1/2$

$$|u(x)| \leq A\|u(x)\|_{W^m_2(K^0_1)}.$$

Proof. Let x_0 be a point such that $|x_0| < 1/2$, $K^{x_0}_{\eta/2}$ be a ball of the radius $\eta/2$ with its center at the point x_0, and $0 < \eta < 1/2$. The definition of the generalized solution implies that

$$\sum_{|\alpha|=|\beta|=m} \int_{|x-x_0|<\eta} a_{\alpha\beta}(x)D^\beta u D^\alpha(\theta u)\,dx = 0,$$

where $\theta(x) \equiv 1$ for $|x - x_0| < \eta/2$, $\theta(x) \in C_0^\infty(K_\eta^{x_0})$. Then we have

$$\int_{|x-x_0|<\eta/2} \sum_{|\alpha|=m} |D^\alpha u|^2 \, dx \le C \sum_{|\alpha|=|\beta|=m} \int_{\eta/2<|x-x_0|<\eta} |D^\beta u| \, |D^\alpha(\theta u)| \, dx.$$

Therefore

$$\int_{|x-x_0|<\eta/2} \sum_{|\alpha|=m} |D^\alpha u|^2 \, dx \le C \int_{\eta/2<|x-x_0|<\eta} \sum_{|\alpha|\le m} \eta^{2(m-|\alpha|)} |D^\alpha u|^2 \, dx.$$

If $\int_{K_\eta^{x_0} \setminus K_{\eta/2}^{x_0}} u x^\alpha \, dx = 0$ for all α with $|\alpha| < m$, then we can use the Poincaré inequality in the form

$$\int_{\eta/2<|x-x_0|<\eta} \sum_{|\alpha|\le m} \eta^{-2|\alpha|} |D^\alpha u|^2 \, dx \le C\eta^{-2m} \int_{\eta/2<|x-x_0|<\eta} \sum_{|\alpha|\le m} |D^\alpha u|^2 \, dx,$$

which implies the inequality

$$\int_{K_{\eta/2}^{x_0}} \sum_{|\alpha|=m} |D^\alpha u|^2 \, dx \le C_1 \int_{K_\eta^{x_0} \setminus K_{\eta/2}^{x_0}} \sum_{|\alpha|=m} |D^\alpha u|^2 \, dx$$

with a constant C_1 independent of η. In the general case the function u can be represented in the form $u = u_1 + P(x)$, where P is a polynomial of degree less than m and is chosen from the condition

$$\int_{K_\eta^{x_0} \setminus K_{\eta/2}^{x_0}} u_1 x^\alpha \, dx = 0 \text{ for } |\alpha| < m.$$

Since P is a solution to the equation $LP = 0$, we can always assume that this condition is satisfied and the estimate

$$\int_{K_{\eta/2}^{x_0}} \sum_{|\alpha|=m} |D^\alpha u|^2 \, dx \le c_0 \int_{K_\eta^{x_0}} \sum_{|\alpha|=m} |D^\alpha u|^2 \, dx$$

holds with $c_0 = C_1/(1 + C_1) < 1$. This estimate implies that

$$\int_{K_{\eta/4}^{x_0}} \sum_{|\alpha|=m} |D^\alpha u|^2 \, dx \le c_0^2 \int_{K_\eta^{x_0}} \sum_{|\alpha|=m} |D^\alpha u|^2 \, dx$$

and in general

$$\int_{K_{\eta/2^m}^{x_0}} \sum_{|\alpha|=m} |D^\alpha u|^2 \, dx \le c_0^m \int_{K_\eta^{x_0}} \sum_{|\alpha|=m} |D^\alpha u|^2 \, dx,$$

so that

$$\int_{K_\varepsilon^{x_0}} \sum_{|\alpha|=m} |D^\alpha u|^2 \, dx \le C\varepsilon^a \int_\Omega \sum_{|\alpha|=m} |D^\alpha u|^2 \, dx$$

for all $\varepsilon < 1/4$, $a = -\log_2 c_0$, and some $C > 0$ not depending on u.

Now Theorem 6 follows from Theorem 95 of Chapter 2. □

Theorem 7. *Let $u(x)$ be a generalized solution to the equation $L_0 u = f(x)$ in the unit ball K_1^0, where the operator L_0 satisfies the conditions of Theorem 6, $n = 2m$, and $\int_{K_1^0} |f(x)|^2\, dx < \infty$. There exists a constant A_1 such that for $|x_0| < 3/4$*

$$|u(x_0)|^2 \le A_1\Big[\int_{K_1^0} |u(x)|^2\, dx + \int_{K_1^0} |f(x)|^2\, dx\Big],$$

and A_1 does not depend on u, f.

Proof. Consider $K_\delta^{x_0} = \{x : \ |x - x_0| < \delta\}$, $\delta < 1/4$, $|x_0| < 1/4$. Let $u_\delta(x)$ be a generalized solution of the equation $L_0 u_\delta = f$ from $W_{2,0}^m(K_\delta^{x_0})$. Then we have

$$\sum_{|\alpha|=|\beta|=m} \int_{K_\delta^{x_0}} (-1)^{|\alpha|} a_{\alpha\beta}(x) D^\beta u_\delta(x) \overline{D^\alpha u_\delta(x)}\, dx = \int_{K_\delta^{x_0}} f(x)\overline{u_\delta(x)}\, dx.$$

As in the proof of Theorem 6 hence and from the Cauchy inequality it follows that

$$C_1 \int_{K_\delta^{x_0}} \sum_{|\alpha|=m} |D^\alpha u_\delta(x)|^2\, dx \le \varepsilon^2 \int_{K_\delta^{x_0}} |f(x)|^2\, dx + \varepsilon^{-2} \int_{K_\delta^{x_0}} |u_\delta(x)|^2\, dx.$$

Estimating the second term on the right-hand side with the help of the Friedrichs inequality we obtain that

$$C_1 \int_{K_\delta^{x_0}} \sum_{|\alpha|=m} |D^\alpha u_\delta|^2\, dx \le \varepsilon^2 \int_{K_\delta^{x_0}} |f|^2\, dx + C\varepsilon^{-2}\delta^{2m} \sum_{|\alpha|=m} \int_{K_\delta^{x_0}} |D^\alpha u_\delta|^2\, dx.$$

If $C\varepsilon^{-2}\delta^{2m} = \frac{1}{2}C_1$, then we have

$$\sum_{|\alpha|=m} \int_{K_\delta^{x_0}} |D^\alpha u_\delta|^2\, dx \le C_2 \delta^{2m} \int_{K_\delta^{x_0}} |f|^2\, dx.$$

Set $u - u_\delta = v$. The function v satisfies the equation $L_0 v = 0$ in $K_\delta^{x_0}$.

Repeating the proof of Theorem 6 we can see that

$$\sum_{|\alpha|=m} \int_{K_\eta^{x_0}} |D^\alpha v|^2\, dx \le C_3 \Big(\frac{\eta}{\delta}\Big)^a \int_{K_\delta^{x_0}} \sum_{|\alpha|=m} |D^\alpha v|^2\, dx$$

with some $a > 0$. Therefore for $\eta \le \delta/2$ we have

$$\sum_{|\alpha|=m} \int_{K_\eta^{x_0}} |D^\alpha u|^2\, dx \le C_4 \int_{K_\delta^{x_0}} \Big(\delta^{2m}|f|^2 + \Big(\frac{\eta}{\delta}\Big)^a \sum_{|\alpha|=m} |D^\alpha u|^2\Big)\, dx.$$

Set $\eta = \delta^2$. Then

$$\sum_{|\alpha|=m} \int_{K_\eta^{x_0}} |D^\alpha u|^2\, dx \le C_5 (\eta^{a/2} + \eta^m),$$

where

$$C_5 = C_4[\int_{K_1^0} |f|^2 \, dx + \int_{K_{3/4}^0} \sum_{|\alpha|=m} |D^\alpha u|^2 \, dx] \leq C_4 \int_{K_1^0} |f|^2 \, dx + C_6 \int_{K_1^0} |u|^2 \, dx,$$

if $\delta \leq 1/4$. Furthermore, applying Theorem 95 of Chapter 2, we have

$$|u(x_0)|^2 \leq A_1[\int_{K_1^0} |u|^2 \, dx + \int_{K_1^0} |f|^2 \, dx]$$

and the proof is complete. □

Theorem 8. *Let $L_0 u = \lambda u$, $u \in W_{2,0}^m(\Omega)$, $2m = n$, and let the operator L_0 satisfy the conditions of Theorem 6. Let $\int_\Omega |u(x)|^2 \, dx = 1$. Then for any $\beta > 0$ and for $x \in \Omega$ such that $dist(x, \partial\Omega) \geq \beta$ we have the estimate $|u(x)|^2 \leq C_\beta(|\lambda| + 1)$.*

Proof. Let x_0 be a point of Ω and $\text{dist}(x_0, \partial\Omega) > \beta$. Set $y = (|\lambda|^{1/2m} + \beta^{-1})(x - x_0)$. Then the function $v(y) = u(x(y))$ satisfies the equation

$$L_0 v + \frac{\lambda}{(|\lambda|^{1/2m} + \beta^{-1})^{2m}} v = 0$$

in K_1^0, $\int_{K_1^0} v^2 \, dy \leq (|\lambda|^{1/2m} + \beta^{-1})^n$. By Theorem 7 we have $|u(x_0)|^2 \leq C_\beta[1 + |\lambda|^{n/2m}]$ and the proof is complete. □

7.3 Equations of second order

We consider the important special case of elliptic equations of second order.

Theorem 9. *Let u be a generalized solution of the equation*

$$L_1 u + \lambda u \equiv \sum_{i,j=1}^n \frac{\partial}{\partial x_i} a_{ij}(x) \frac{\partial u}{\partial x_j} + \lambda u = 0$$

in a domain $\Omega \subset \mathbf{R}^n$ and $u = 0$ on $\partial\Omega$. Let $a_{ij}(x)$ be bounded measurable real-valued functions and the operator L_1 be elliptic, i.e. $\gamma^2 \sum_{i,j=1}^n a_{ij}(x)\xi_i\xi_j \geq |\xi|^2$. Then the estimate

$$|u(x)|^2 \leq A^2 |\lambda|^{n/2} \int_\Omega u^2 \, dx$$

is valid for all $x \in \Omega$ with a constant A depending on γ and n only.

Proof. 1. Assume at first that $n > 2$, the surface $\partial\Omega$ is smooth, and $a_{ij}(x) \in C^\infty(\overline{\Omega})$. Then $u \in C^\infty(\overline{\Omega})$. We can assume that u is real. Let $\int_\Omega u^2 \, dx = 1$ and $0 < \lambda \leq 1$. Multiplying the equation by u and integrating over Ω we obtain that $\int_\Omega \sum_{i=1}^n (\frac{\partial u}{\partial x_i})^2 \, dx \leq \gamma^2$. Using Theorem 57 of Chapter 2 we get that

$$(\int_\Omega |u|^{2n/(n-2)} \, dx)^{(n-2)/2n} \leq \gamma C_0. \tag{5}$$

Now multiply the equation by $|u|^{p-1}\text{sign}\, u,\ p \geq 2$. Integrating by parts we obtain that $(p-1)\gamma^{-2}\int_\Omega |u|^{p-2}|\nabla u|^2\,dx \leq \int_\Omega |u|^p\,dx$, i.e.

$$(p-1)\gamma^{-2}\frac{4}{p^2}\int_\Omega |\nabla(u^{p/2})|^2\,dx \leq \int_\Omega |u|^p\,dx.$$

Now Theorem 57 of Chapter 2 yields that

$$\Big(\int_\Omega |u|^{pn/(n-2)}\,dx\Big)^{(n-2)/n} \leq C_0^2\int_\Omega |\nabla(u^{p/2})|^2\,dx \leq \frac{C_0^2 p^2\gamma^2}{4(p-1)}\int_\Omega |u|^p\,dx.$$

Therefore

$$\|u\|_{L_{np/(n-2)}} \leq \Big(\frac{C_0^2\gamma^2 p^2}{4(p-1)}\Big)^{1/p}\Big(\int_\Omega |u|^p\,dx\Big)^{1/p}.$$

Let $p_1 = 2,\ p_{k+1} = p_k \cdot n/(n-2)$, i.e. $p_k = 2(\frac{n}{n-2})^{k-1},\ k = 1, 2, \ldots$.

The obtained estimate implies that

$$\|u\|_{L_{p_{k+1}}} \leq \Big(\frac{C_0^2\gamma^2 p_k^2}{4(p_k-1)}\Big)^{1/p_k}\|u\|_{L_{p_k}}$$

$$\text{or}\quad \|u\|_{L_{p_{k+1}}} \leq \Big[\frac{C_0^2\gamma^2 p_k^2}{4(p_k-1)}\Big]^{1/p_k}\cdot\Big[\frac{C_0^2\gamma^2 p_k^2}{4(p_{k-1}-1)}\Big]^{1/p_{k-1}}\cdot\ \cdots\ \cdot\Big[\frac{C_0^2\gamma^2 p_1^2}{4(p_1-1)}\Big]^{1/p_1}.$$

The series

$$\sum_{k=1}^\infty \frac{1}{p_k}\ln\Big[\frac{C_0^2\gamma^2 p_k^2}{4(p_k-1)}\Big] = \sum_{k=1}^\infty a_k$$

is converging since $a_k \sim h^{-k}\ln[Ch^k],\ h > 1$. Therefore, we have $\|u\|_{L_{p_k}} \leq A$ with A independent of p_k. Thus $\max|u(x)| \leq A$.

2. Now let $a_{ij}(x) \in C^\infty$, but Ω be an arbitrary domain. Consider a sequence Ω_k of domains with smooth boundaries $\partial\Omega_k$, such that $\overline{\Omega}_k \subset\subset \Omega_{k+1}$ and $\cup_{k=1}^\infty \Omega_k = \Omega$. Let $0 < \lambda_1 < \lambda_2 \leq \cdots \leq \lambda_N$ be eigenvalues of this problem such that $\lambda_N \leq 1$.

Consider also the problem

$$\sum_{i,j=1}^n \frac{\partial}{\partial x_i}a_{ij}(x)\frac{\partial u}{\partial x_j} + \lambda u = 0,\ x \in \Omega_k,\ u \in W_{2,0}^1(\Omega_k).$$

Its eigenvalues λ_{jk} are such that $0 < \lambda_{1k} < \lambda_{2k} \leq \cdots \leq \lambda_{Nk}$ and $\lambda_{jk} \geq \lambda_j$, as follows from the variational principle. (See Theorem 9 of Chapter 4.) We write here each λ taking into account its multiplicity and continue u by zero values in $\Omega \setminus \Omega_k$. Let

$$L_1 u_{lk} + \lambda_{lk}u_{lk} = 0,\ \|u_{lk}\|_{L_2(\Omega_k)} = 1.$$

Using the equation we see that $\|u_{lk}\|_{W_2^1(\Omega)} \leq B,\ B = \text{const}$ for $l \leq N,\ k = 1, 2, \ldots$. Therefore, we can use the compactness of embeddings and assume that the sequence u_{lk} converges in $L_2(\Omega)$ and weakly in $W_2^1(\Omega)$ to $u_{l\infty}$. The sequence

λ_{lk} is also converging to some value $\lambda_{l\infty}$. The function $u_{l\infty}$ is an eigenfunction of the boundary problem in Ω, corresponding to $\lambda = \lambda_{l\infty}$. Thus each $\lambda_{l\infty}$ coincides with some λ_s, where $l \leq N$, $s \leq N$.

Let us show that inversely each λ_q with $q \leq N$ coincides with some $\lambda_{l\infty}$ with $l \leq N$. Observe at first that for any $\varepsilon > 0$ we have $\lambda_{qk} < \lambda_q + \varepsilon$, if k is sufficiently big. Indeed, one can find a set of linearly independent functions $v_1, \ldots, v_q \in C_0^\infty(\Omega)$ such that, for any linear combination v of the functions $v_1, \ldots, v_q$,

$$-\sum_{i,j=1}^{n} \int_\Omega a_{ij}(x) \frac{\partial v}{\partial x_i} \frac{\partial v}{\partial x_j}\, dx + (\lambda_q + \varepsilon) \int_\Omega v^2\, dx \geq 0.$$

This inequality and the variational principle imply that $\lambda_{qk} < \lambda_q + \varepsilon$ if k is so big that supp $v_j \subset \Omega_k$, $j = 1, \ldots, q$. Since $\lambda_{qk} \geq \lambda_q$, we can see that $\lim_{k\to\infty} \lambda_{qk} = \lambda_q$. This is true for all q.

All functions u_{jk} are bounded in $W_2^1(\Omega)$ uniformly in $j \leq N$ and k. Since a subsequence u_{jn_k} converges almost everywhere to u_j as $k \to \infty$, we obtain that $|u_j| \leq A$, where A is the constant from the first part of the proof.

3. Now let $\partial\Omega$ be an arbitrary surface and $a_{ij}(x)$ be measurable functions. Using averaging one can construct smooth functions $a_{ijs}(x)$ such that $\|a_{ij}(x) - a_{ijs}(x)\|_{L_2(\Omega)} \to 0$ as $s \to \infty$. Moreover, $a_{ijs}(x) \to a_{ij}(x)$ almost everywhere and

$$\gamma_1^{-2}|\xi|^2 \geq \sum_1^n a_{ijs}(x)\xi_i\xi_j \geq \gamma^{-2}|\xi|^2,$$

uniformly in s. Consider the problem

$$L_{1s}u + \lambda u \equiv \sum_{i,j=1}^{n} \frac{\partial}{\partial x_i} a_{ijs} \frac{\partial u}{\partial x_j} + \lambda u = 0$$

in Ω, $u \in W_{2,0}^1(\Omega)$. Let $\lambda_{1s} < \lambda_{2s} \leq \cdots \leq \lambda_{js} \leq \cdots$ be its eigenvalues, and u_{js} be an orthonormed system of the eigenfunctions. Every sequence λ_{ls}, $s = 1, 2, \ldots$, is bounded. For example, all λ_{ls} are bounded above by the lth eigenvalue of the problem

$$\gamma_1^{-2}\Delta u + \lambda u = 0 \text{ in } K,\ u|_{\partial K} = 0,$$

where K is a ball lying in Ω. Every sequence u_{ls}, $s = 1, \ldots$, is bounded in $W_{2,0}^1(\Omega)$ since

$$\sum_1^n \int_\Omega a_{ijs} u_{ls_{x_j}} u_{ls_{x_i}}\, dx = \lambda_{ls} \int u_{ls}^2\, dx.$$

The compactness of embedding $W_2^1(\Omega) \subset L_2(\Omega)$ implies the existence of a subsequence s_k such that $\lambda_{ls_k} \to \lambda_{l\infty}$, $u_{ls_k} \to u_{l\infty}$ in $L_2(\Omega)$, and $u_{ls_k} \to u_{l\infty}$ weakly in

$W_2^1(\Omega)$. Let us show that $u_{l\infty}$ satisfies the equation $L_1 u_{l\infty} + \lambda_{l\infty} u_{l\infty} = 0$. To this end we consider the identity

$$\int_\Omega \sum_1^n a_{ijs}(u_{ls})_{x_i} \psi_{x_j}\, dx = \lambda_{ls} \int u_{ls} \psi\, dx,$$

which is valid for all functions $\psi(x) \in C_0^\infty(\Omega)$. This identity implies that

$$\sum_{i,j=1}^n \int_\Omega (a_{ijs} - a_{ij})(u_{ls})_{x_i} \psi_{x_j}\, dx + \sum_{i,j=1}^n \int_\Omega a_{ij}(u_{ls})_{x_i} \psi_{x_j}\, dx = \lambda_{ls} \int_\Omega u_{ls} \psi\, dx.$$

The term on the right-hand side tends to $\lambda_{l\infty} \int_\Omega u_{l\infty} \psi\, dx$, since u_{ls} converges to $u_{l\infty}$ in the space $L_2(\Omega)$ as $s \to \infty$. The second sum on the left-hand side tends to $\sum_{i,j=1}^n \int_\Omega a_{ij}(x)(u_{l\infty})_{x_i} \psi_{x_j}\, dx$ since $\frac{\partial}{\partial x_i} u_{ls}$ tends to $\frac{\partial}{\partial x_i} u_{l\infty}$ weakly in $L_2(\Omega)$. The first sum on the left-hand side is estimated by $C \sum_{i,j=1}^n (\int_\Omega |a_{ijs} - a_{ij}|^2\, dx)^{1/2}$, since a weakly converging sequence is bounded, and tends to 0 as $s \to \infty$. Therefore, one can pass to the limit as $s \to \infty$ and see that

$$\sum_{i,j=1}^n \int_\Omega a_{ij}(x) \frac{\partial}{\partial x_i} u_{l\infty}(x) \frac{\partial}{\partial x_j} \psi(x)\, dx = \lambda_{l\infty} \int_\Omega u_{l\infty}(x) \psi(x)\, dx$$

for all ψ from $C_0^\infty(\Omega)$. Since the space $C_0^\infty(\Omega)$ is dense in $W_{2,0}^1(\Omega)$, the relation is valid for all $\psi(x) \in W_{2,0}^1(\Omega)$. Therefore, $u_{l\infty}$ is an eigenfunction corresponding to $\lambda = \lambda_{l\infty}$. We will show now that each λ_l coincides with some $\lambda_{l\infty}$. Let N_p be the number of λ_l such that $\lambda_l < p$. We show that for s big enough the number N_{ps} of eigenvalues λ_{ks}, which are smaller than p, coincides with N_p. If $u_1, \ldots, u_l$ are eigenfunctions of the equation $L_1 u_j + \lambda_j u_j = 0$, corresponding to $\lambda_1, \ldots, \lambda_l$, then

$$\sum_{i,j=1}^n \int_\Omega a_{ij}(x) u_{x_i}(x) u_{x_j}(x)\, dx - p \int_\Omega u^2(x)\, dx \le 0,$$

where u is an arbitrary linear combination of $u_1, \ldots, u_l$. Since $u_i \in W_{2,0}^1(\Omega)$, then there is a set of functions $u_{1,0}, \ldots, u_{l,0}$ from $C_0^\infty(\Omega)$ that are linearly independent and such that the relation is true for all linear combinations of $u_{i,0}$. Hence by the variational principle we get that the eigenvalues of the equation $L_{1s} u + \lambda u = 0$ coincide with $\lambda_{i\infty}$, $i = 1, \ldots$. Thus the estimate $|u(x)| \le A$ is true for all λ such that $\lambda \le 1$.

4. Let a function $u(x) \in W_{2,0}^1(\Omega)$ satisfy the equation $\sum_{i,j=1}^n \frac{\partial}{\partial x_i} a_{ij}(x) \frac{\partial u}{\partial x_j} + \lambda u = 0$, where

$$\gamma_1 |\xi|^2 \le \sum_1^n a_{ij} \xi_i \xi_j \le \gamma_2 |\xi|^2, \ n > 2.$$

Then $|u(x)|^2 \le A^2\lambda^{n/2}\int_\Omega u^2\,dx$. This estimate follows from the proof if one makes a change of variables $y = \sqrt{\lambda}x$. The constant A in this inequality depends only on γ_1, γ_2 and is independent of Ω.

We have proved Theorem 9 for $n > 2$. If $n = 2$, we need only a small modification since instead of (5) we can use the inequality

$$\left(\int_\Omega |u|^p\,dx\right)^{2/p} \le \gamma C_0,$$

which is true for any $p > 1$. □

7.4 Estimates of eigenfunctions of operator pencils

Let

$$L(x, D, \lambda) = \sum_{|\alpha|\le m} a_\alpha(x,\lambda)D^\alpha, \ x \in \Omega$$

be a family of elliptic differential operators in Ω and

$$B_j(x, D, \lambda) = \sum_{|\alpha|\le m_j} b_{\alpha j}(x,\lambda)D^\alpha, \ x \in \partial\Omega, \ j = 1, 2, \dots, m/2$$

be another family of differential operators defined on $\partial\Omega$, i.e. on the boundary of the domain Ω, which is supposed to be infinitely differentiable. The coefficients $a_\alpha(x,\lambda)$, $b_{\alpha j}(x,\lambda)$ of these operators are also infinitely differentiable in x and are polynomials in λ of degrees $m - |\alpha|$, $m_j - |\alpha|$, respectively. Set

$$\mathcal{H} = W_2^m(\Omega) \times \prod_{j=1}^{m/2} H_2^{m-m_j-1/2}(\partial\Omega).$$

Consider the family $\mathcal{L}(\lambda)$ of the operators

$$u \mapsto (Lu|_\Omega, B_1u|_{\partial\Omega}, \dots, B_{m/2}u|_{\partial\Omega}),$$

which are bounded as operators from the space $W_2^m(\Omega)$ to $\mathcal{H}$. A point λ is called *regular* if the operator $\mathcal{L}(\lambda)$ has a bounded inverse. A point λ_0 is called an *eigenvalue of the pencil* $\mathcal{L}(\lambda)$, if the boundary value problem

$$Lu = 0 \text{ for } x \in \Omega, \ B_ju = 0 \text{ for } x \in \partial\Omega, \ j = 1, \dots, m/2$$

has a nontrivial solution. This solution is called an *eigenfunction* of the pencil $\mathcal{L}$. We will assume that the operators L, B_j satisfy the Lopatinsky condition.

Theorem 10. *If $u(x)$ is an eigenfunction of the pencil $\mathcal{L}$ corresponding to an eigenvalue λ, then*

$$|u(x)|^2 \le C(1 + |\lambda|)^n \int_\Omega |u(x)|^2\,dx,$$

where the constant C is independent of u and λ.

Proof. It suffices to verify this estimate for $|\lambda| > 1$, since for other values it follows from Corollary 13 of Chapter 3. The constant C in that Corollary depends on the maximal values of the derivatives of the coefficients of L and B_j and therefore it can be chosen for all small $|\lambda|$ simultaneously.

Let $x_0 \in \partial\Omega$ and $|\lambda| > 1$. Let us change the variables $|\lambda|(x - x_0) = y$. As a result we obtain that $u(x(y))$ is a solution of the equation

$$|\lambda|^{-m} L(y|\lambda|^{-1} + x_0, |\lambda| D_y, \lambda)u = 0, \; |y| < 1, \; y \in \Omega',$$

which satisfies the boundary conditions

$$|\lambda|^{-m_j} B_j(y|\lambda|^{-1} + x_0, |\lambda| D_y, \lambda)u = 0$$

on the boundary of Ω', which is the image of Ω under this transformation. Using again the estimate of Corollary 13 of Chapter 3, we obtain that

$$|u(x(y))|^2 \leq C \int_{\Omega'} |u(x)|^2 \, dy, \; |y| < 1/2, \; y \in \Omega'.$$

Returning to the variable x, we get the estimate $|u(x)|^2 \leq C|\lambda|^n \int_\Omega |u(x)|^2 \, dx$ for $|x - x_0| \leq \frac{1}{2}|\lambda|^{-1}$.

If $x_0 \in \Omega$, $\rho(x_0, \partial\Omega) > \frac{1}{2}|\lambda|^{-1}$, and $|\lambda| > 1$, then applying the transformation $y = |\lambda|(x - x_0)/3$, we obtain the same estimate. Thus the result is valid for all complex λ with $|\lambda| > 1$. □

Theorem 11. *The dimension of the space of eigenfunctions of the pencil $\mathcal{L}(\lambda)$, corresponding to a fixed value $\lambda = \lambda_0$, is finite.*

Proof. Consider the space of the eigenfunctions of the operator $\mathcal{L}(\lambda_0)$. This is a linear space $\mathcal{N}$, $\mathcal{N} \subset L_2(\Omega)$. Let $u(x) \in \mathcal{N}$, $\int u^2 \, dx = 1$. The estimate of Theorem 12 of Chapter 3 yields that $\|u\|_{W_2^m} \leq C$. The compactness of embedding $W_2^m(\Omega) \subset L_2(\Omega)$ implies that the unit sphere in $\mathcal{N}$ is compact. Therefore, the space $\mathcal{N}$ is finite dimensional. □

Theorem 12. *Let λ_0 be a regular point of the pencil $\mathcal{L}(\lambda)$ and G be a domain in the complex plane, containing λ_0 and consisting of regular points. If $\lambda' \in \partial G$, then either λ' is regular or λ' is an eigenvalue of the pencil $\mathcal{L}(\lambda)$.*

Proof. There are two possibilities: either $\overline{\lim}_{\substack{\lambda \to \lambda' \\ \lambda \in G}} \|\mathcal{L}^{-1}(\lambda)\| = \infty$ or

$$\overline{\lim}_{\substack{\lambda \to \lambda' \\ \lambda \in G}} \|\mathcal{L}^{-1}(\lambda)\| < \infty.$$

In the first case there is a sequence $\lambda_k \to \lambda'$, $\lambda_k \in G$, such that $u_k \in W_2^m(\Omega)$ and $\|u_k\|_{W_2^m(\Omega)} = 1$, $\mathcal{L}(\lambda_k)u_k \to 0$ in $\mathcal{H}$. Theorem 11 of Chapter 3 implies that

$$\|u_{k+l} - u_k\|_{W_2^m(\Omega)} \leq C(\|\mathcal{L}(\lambda')u_{k+l} - \mathcal{L}(\lambda')u_k\|_{L_2(\Omega)} + \|u_{k+l} - u_k\|_{L_2(\Omega)}).$$

One can find a subsequence $u_{k'}$ of the sequence u_k converging in $L_2(\Omega)$ to an element $v \in L_2(\Omega)$. The above estimate implies that $u_{k'}$ is converging in $W_2^m(\Omega)$ to v and $\|v\|_{W_2^m(\Omega)} = 1$. Since $\mathcal{L}(\lambda_k)u_k \to 0$ in $\mathcal{H}$, we have $\mathcal{L}(\lambda')v = 0$. Therefore, v is an eigenfunction of $\mathcal{L}$ with $\lambda = \lambda'$.

Now let

$$\overline{\lim}_{\substack{\lambda\to\lambda' \\ \lambda\in G}} \|\mathcal{L}^{-1}(\lambda)\| < \infty.$$

If $[f,\phi] \in \mathcal{H}$ and $\lambda \in G$, then there is a unique element $v_\lambda \in W_2^m(\Omega)$, such that $\mathcal{L}(\lambda)v_\lambda = [f,\phi]$. Furthermore, then $\|v_\lambda\|_{W_2^m(\Omega)} \le C\|[f,\phi]\|_{\mathcal{H}}$, $v_\lambda \to v$ in $W_2^m(\Omega)$, when $\lambda \to \lambda'$, $\lambda \in G$, v is a solution of the problem $\mathcal{L}v = [f,\phi]$ and $\|v\|_{W_2^m(\Omega)} \le C\|[f,\phi]\|_{\mathcal{H}}$. The operator $\mathcal{L}(\lambda')$ is invertible, since the equalities $\mathcal{L}(\lambda')u = 0$, $\|u\|_{W_2^m(\Omega)} = 1$ imply that $\|\mathcal{L}(\lambda)u\| \to 0$ in $\mathcal{H}$ as $\lambda \in G$, $\lambda \to \lambda'$, which is impossible because the operator $\|\mathcal{L}^{-1}(\lambda)\|$ is bounded in $G \cap U$, where U is a neighborhood of λ'. □

7.5 The method of stationary phase

Let $f, S \in C^\infty(R)$, S be a real-valued function, and $\lambda \in R$. Consider the integral

$$I(\lambda) = \int_a^b f(x)e^{i\lambda S(x)}\,dx.$$

We are interested in studying the behavior of $I(\lambda)$ as $\lambda \to +\infty$.

Lemma 13. *If $f \in C_0^\infty(a,b)$ and $S'(x) \neq 0$ in (a,b), then*

$$I(\lambda) = O(\lambda^{-\infty}),\ I^{(j)}(\lambda) = O(\lambda^{-\infty})$$

as $\lambda \to \infty$, $j = 1,2,\ldots$.

Proof. It is evident that

$$I(\lambda) = \int_a^b f(x)\frac{1}{i\lambda S'(x)}\frac{d}{dx}e^{i\lambda S(x)}\,dx = -\frac{1}{\lambda}\int_a^b f_1(x)e^{i\lambda S(x)}\,dx,$$

where $f_1(x) = (f(x)/iS'(x))' \in C_0^\infty(a,b)$. We can repeat this and show that $I(\lambda) = \frac{1}{\lambda^2}\int_a^b f_2(x)e^{i\lambda S(x)}\,dx$, where $f_2(x) = (f_1(x)/iS'(x))'$, and in general for any N we have $I(\lambda) = \frac{1}{\lambda^N}\int_a^b f_N(x)e^{i\lambda S(x)}\,dx$, where $f_N(x) \in C_0^\infty(a,b)$. This means that $|I(\lambda)| \le C_N\lambda^{-N}$ as $\lambda \to \infty$. The same consideration is valid for the integrals $I^{(j)}(\lambda)$ with $j = 1,2\ldots$. □

A point $x_0 \in R^1$ such that $S'(x_0) = 0$ is called *stationary*. It is called *nondegenerate*, if $S''(x_0) \neq 0$.

Lemma 14. *If there is only one stationary point x_0 on (a,b) that is nondegenerate, then*

$$I(\lambda) \sim e^{i\lambda S(x_0)}\sum_{k=0}^{\infty} a_k\lambda^{-1/2-k},$$

where

$$a_0 = f(x_0)\sqrt{\frac{2\pi i}{S''(x_0)}} = f(x_0)\sqrt{\frac{2\pi}{|S''(x_0)|}}e^{i\pi/4 \operatorname{sign} S''(x_0)},$$

and all coefficients a_k with $k \geq 1$ are expressed in terms of the values of the functions f, S and of their derivatives of orders $\leq 2k$ at the point $x = x_0$. This asymptotic representation can be differentiated with respect to λ.

Proof. Let $S''(x_0) > 0$ and let $h(x) \in C_0^\infty(a,b)$, $h(x) = 1$ for $|x - x_0| < \delta/2$, $h(x) = 0$ for $|x - x_0| > \delta$ where δ is small enough. By Lemma 13

$$\frac{d^j}{d\lambda^j}\int_a^b [1 - h(x)]f(x)e^{i\lambda S(x)}\,dx = O(\lambda^{-\infty})$$

for $j = 0, 1, \ldots$, so that

$$I(\lambda) = \int_a^b h(x)f(x)e^{i\lambda S(x)}\,dx + O(\lambda^{-\infty}).$$

Let $\delta > 0$ be so small that the function

$$\psi(x) = S(x) - S(x_0) = \frac{S''(x_0)}{2}(x - x_0)^2 + O(|x - x_0|^2)$$

can be transformed in t^2 for $|x - x_0| \leq \delta$ with the help of a smooth nondegenerate transformation $x = x(t)$, so that $\psi(x(t)) = t^2$, $x'(t) \geq 0$. Then

$$I(\lambda) = e^{i\lambda S(x_0)}\int_{-\infty}^{\infty} h(x(t))f(x(t))x'(t)e^{i\lambda t^2}\,dt + O(\lambda^{-\infty})$$

and $x'(0) = \sqrt{2/S''(x_0)}$. Set

$$q(t) = h(x(t))f(x(t))x'(t) + h(x(-t))f(x(-t))x'(-t).$$

Then

$$I(\lambda) = e^{i\lambda S(x_0)}\int_0^\infty q(t)e^{i\lambda t^2}\,dt + O(\lambda^{-\infty}).$$

The function q is smooth and even, so that $q(t) \sim \sum_{k=0}^\infty C_k t^{2k}$. Therefore

$$I(\lambda) \sim e^{i\lambda S(x_0)}\sum_{k=0}^\infty C_k \int_0^\infty t^{2k}e^{i\lambda t^2}\,dt \sim e^{i\lambda S(x_0)}\sum_{k=0}^\infty C_k\lambda^{-k-1/2}\int_0^\infty s^{2k}e^{is^2}\,ds.$$

Here $C_0 = q(0) = 2f(x_0)\sqrt{2/S''(x_0)}$ and $\int_0^\infty e^{is^2}\,ds = (\sqrt{\pi}/2)e^{i\pi/4}$. Hence the result follows.

Since

$$I^{(j)}(\lambda) = e^{i\lambda S(x_0)}\int_0^\infty q(t)[it^2 + iS(x_0)]^j e^{i\lambda t^2}\,dt + O(\lambda^{-\infty}),$$

the asymptotics for $I^{(j)}(\lambda)$ can be obtained in the same way.

If $S''(x_0) < 0$, the result follows from the preceding one after consideration of the integral $\overline{I(\lambda)}$. □

Lemma 15. *Let $f(x,\alpha)$ and $S(x,\alpha)$ be smooth functions of $x \in (a,b)$ and $\alpha \in \overline{Q}$, where Q is a bounded domain in R^m and for each α there is a unique nondegenerate stationary point $x_0(\alpha) \in (a,b)$. Then*

$$I(\lambda,\alpha) = \int_a^b f(x,\alpha)e^{iS(x,\alpha)}\,dx \sim e^{i\lambda S(x_0,\alpha)} \sum_{k=0}^{\infty} a_k(\alpha)\lambda^{-1/2-k}$$

and $a_0(\alpha) = f(x_0(\alpha),\alpha)\sqrt{2\pi/|S''_{xx}(x_0(\alpha),\alpha)|}e^{(i\pi/4)\,\mathrm{sign}\,S''_{xx}(x_0(\alpha),\alpha)}$.

This asymptotic expansion is uniform in α and can be differentiated with respect to λ and α any time.

Proof. The uniformity of the expansion is proved easily after the estimates of the remainder. The smoothness of the function $x_0(\alpha)$ follows from the implicit function theorem, applied to the equation $S'_x(x_0(\alpha),\alpha) = 0$, defining $x_0(\alpha)$. □

Now we consider the asymptotics of a multi-dimensional integral

$$I(\lambda) = \int f(x)e^{i\lambda S(x)}\,dx,$$

where $x = (x_1,\dots,x_n) \in \Omega,\ f \in C_0^\infty(\Omega),\ S(x) \in C^\infty(\Omega)$.
A point $x_0 \in \Omega$ is called *stationary*, if $\nabla S(x_0) = 0$. A stationary point x_0 is nondegenerate, if $\det\ S''_{x_ix_j}(x_0) \neq 0$.

Lemma 16. *Let x_0 be a nondegenerate stationary point of $S(x)$ and $S(x) \in C^\infty(\Omega)$, where Ω is a neighborhood of x_0. Then there are a neighborhood U of x_0, a neighborhood V of the origin in R^n_y, and a diffeomorphism φ of V onto U such that*

$$S(\varphi(y)) = S(x_0) + \frac{1}{2}\sum_{j=1}^{n} \mu_j y_j^2,$$

where μ_j are the eigenvalues of the matrix $S_{xx}(x_0)$. The Jacobian of the diffeomorphism does not vanish and $\det[\partial\varphi(0)/\partial x)] = 1$.

Proof. Using a linear change of variables one can reduce S to the form

$$S(x) = S(x_0) + \frac{1}{2}\sum_{j=1}^{n} \mu_j (x_j - x_j^0)^2 + r(x),$$

where $r(x) = \sum_{j,k=1}^n r_{jk}(x)(x_j - x_j^0)(x_k - x_k^0)$ and $r_{jk}(x) = o(|x - x_0|)$. One can apply the Jacobi method and represent S in the form

$$S(x) = S(x_0) + \frac{1}{2}S''_{11}(x)(x_1 - x_1^0 + \sum_{j=1}^{n} \frac{S''_{1j}(x)}{S''_{11}(x)}(x_j - x_j^0))^2$$

$$+ \frac{1}{2}\sum_{j,k=2}^{n} r_{jk}^{(1)}(x)(x_j - x_j^0)(x_k - x_k^0).$$

Continuing, we obtain that

$$S(x) = S(x_0) + \frac{1}{2}\sum_{j=1}^{n} \mu_j (x_j - x_j^0 + \sum_{k=1}^{n-j+1} \rho_{jk}(x)(x_k - x_k^0))^2,$$

where $\rho_{jk}(x) = O(|x - x^0|)$. Setting $y_j = x_j - x_j^0 + \sum_{k=1}^{n-j+1} \rho_{jk}(x)(x_k - x_k^0)$, $j = 1, \dots, n$, we see that $\det(\partial y(x^0)/\partial x) = 1$. The inverse function $x = \varphi(y)$ satisfies the requirements of Lemma 16. □

Lemma 17. *If there are no stationary points in* $\overline{\Omega}$, *then* $I(\lambda) = O(\lambda^{-N})$ *as* $\lambda \to +\infty$ *for any* $N > 0$.

Proof. Observe that $e^{i\lambda S(x)} = \frac{1}{i\lambda} L(e^{i\lambda S(x)})$, where $L = \sum_{j=1}^{n} \frac{\partial S/\partial x_j}{|\nabla S|^2}\frac{\partial}{\partial x_j}$. Therefore, $I(\lambda) = \frac{1}{i\lambda}\int e^{i\lambda S} L^* f(x)\, dx$, where

$$L^* f = -\sum \frac{\partial}{\partial x_j}[\frac{\partial S/\partial x_j}{|\nabla S|^2} f].$$

Repeating this argument N times, we get the formula

$$I(\lambda) = \frac{1}{(i\lambda)^N}\int e^{i\lambda S} L^{*N}(f(x))\, dx,$$

implying the statement. □

Theorem 18. *In* $\overline{\Omega}$ *let there be only one stationary point* x_0, *which is nondegenerate. Then* $I(\lambda) \sim e^{i\lambda S(x_0)} \sum_{k=0}^{\infty} a_k \lambda^{-n/2-k}$, *where*

$$a_0 = f(x_0)(2\pi)^{n/2} |\det\ S''(x_0)|^{-1/2} e^{(i\pi/4)\,\mathrm{sign}\, S''_{xx}(x_0)}.$$

This asymptotic expansion can be differentiated with respect to λ.

Proof. Let $h \in C_0^\infty(\Omega)$, $h(x) = 1$ for $|x - x_0| < \delta/2$, $h(x) = 0$ for $|x - x_0| \geq \delta$. Then by Lemma 17

$$\int (1 - h(x)) f(x) e^{i\lambda S(x)}\, dx = O(\lambda^{-\infty}).$$

By Lemma 16 we can take the number δ so small that in the neighborhood $|x - x_0| \leq \delta$ one can define a smooth nondegenerate transformation $y = y(x)$, transferring the point x_0 in $y = 0$ and the function $S(x)$ into the function $S(x_0) + \frac{1}{2}\sum_{j=1}^{n} \mu_j y_j^2$, where μ_j are the eigenvalues of the matrix $S_{xx}(x_0)$. We have $J(x_0) = 1$, if $J(x) = \det(\partial y/\partial x)$. Therefore

$$\begin{aligned} I(\lambda) &\sim \int h(x) f(x) e^{i\lambda S(x)}\, dx \\ &= e^{i\lambda S(x_0)} \int h(x(y)) f(x(y)) e^{(i/2)\lambda \sum_{j=1}^{n} \mu_j y_j^2} J(x(y))\, dy. \end{aligned}$$

Applying Lemma 15 to the integrals with respect to $y_1, y_2, \dots, y_n$ we obtain our statement. □

Remark 19. As in Lemma 15, one can assume that the functions f and S depend on a parameter $\alpha = (\alpha_1, \dots, \alpha_n) \in Q$. The obtained asymptotic expansion is then uniform with respect to α and can be differentiated with respect to λ and α.

Let M be a smooth closed $(n-1)$-dimensional surface in $\mathbf{R}^n$. M is called a *closed surface* if it is connected, is a finite sum of surfaces as defined in Section 2.1, and it is such a surface in a neighbourhood of each of its points. Assume that M is bounded. Consider the function

$$F(x) = \int_M f(\xi) e^{i(x,\xi)}\, d\xi.$$

Let $x = \omega r$, $\omega \in R^n$, $|\omega| = 1$. We are interested in the asymptotics of the function $F(r\omega)$ as $r \to \infty$.

Theorem 20. *Let $\omega_0 \in \mathbf{R}^n, |\omega| = 1$. On the surface M let there be m points $\sigma_j(\omega_0)$, at which the normal vector to M is parallel to the vector ω_0 and the complete curvature $k_j(\omega_0) \neq 0$ for $j = 1, \dots, m$. Let γ_j be the difference between the numbers of positive and negative main curvatures at the point $\sigma_j(\omega)$. Then there is $\varepsilon > 0$ such that for $|\omega - \omega_0| < \varepsilon$ and $r \to +\infty$ we have*

$$F(x) \sim \sum_{j=1}^{m} e^{-i\mu_j(\omega) r} \sum_{k=0}^{\infty} a_{jk}(\omega) r^{(1-n)/2-k},$$

where $$a_{j0} = (2\pi)^{(n-1)/2} |k_j(\omega)|^{-1/2} f(\sigma^j(\omega)) e^{-i\gamma_j \pi/4}.$$

This expansion can be differentiated with respect to x.

Proof. Each point $\sigma_j(\omega_0)$ with $1 \le j \le m$ has a neighborhood U_j, such that U_j are mutually disjoint and U_j is smoothly projected on the plane, tangent to M at any point $\sigma \in \overline{U}_j$.

We can assume that ε is so small that for $|\omega - \omega_0| \le \varepsilon$ the numbers m and γ_j do not depend on ω, the points $\sigma_j(\omega)$ and the functions $k_j(\omega)$ smoothly depend on ω, $|k_j(\omega)| \ge \delta > 0$, and the set $\{\sigma^j(\omega),\ |\omega - \omega_0| \le \varepsilon\}$ is lying in U_j.

Complement the set of domains $U_1, \dots, U_m$ until the covering $U_1, \dots, U_N$ of the surface M by coordinate neighborhoods and construct a subordinated partition of unity $\sum_{j=1}^{N} \varphi_j = 1$. Then

$$F(x) = \sum_{j=1}^{N} \int_M \varphi_j(\xi) f(\xi) e^{i(x,\xi)}\, dS_\xi = \sum_{j=1}^{N} \int_{U_j} \varphi_j(\eta) f(\eta) e^{i(\omega,\sigma(\eta))r} J(\eta)\, d\eta,$$

where $\eta = (\eta_1, \dots, \eta_{n-1})$ are local coordinates in U_j, and $J(\eta)$ is the Jacobian $|d\xi/d\eta|$, $x = r\omega$. Suppose that local coordinates $\xi_1, \dots, \xi_n$ are chosen in the neighborhood U_j for $j = 1, \dots, m$ in such a way that the axis ξ_n is directed along the vector ω, and $\xi_1, \dots, \xi_{n-1}$ are local coordinates in M. Then we can assume that $\eta_1 = \xi_1, \dots, \eta_{n-1} = \xi_{n-1}$, and V_j is a domain in R^{n-1}, serving as a local card for M.

Now we apply the method of stationary phase. According to this method, the main contribution (up to terms of order r^{-N_1} with arbitrary N_1) is given by the points η at which $D_{\eta_k}(\omega, \sigma(\eta)) = 0$ for $k = 1, \dots, n-1$. However, this equality means that the vectors $D_{\eta_k}\sigma(\eta)$ are orthogonal to ω. The vectors $D_{\eta_k}\sigma(\eta)$, $k = 1, \dots, n-1$, form a basis in the tangent plane so that this condition means the orthogonality of ω to the surface M. If $j > m$, then there are no stationary points and the corresponding integrals are equal to $O(r^{-N_1})$ for any $N_1 \geq 1$.

Let us consider an integral

$$F_j = \int_{U_j} \varphi_j(\eta) f(\eta) e^{i(\omega,\sigma(\eta))r} J(\eta)\, d\eta,$$

with $j \leq m$. In coordinates η the point $\sigma_j(\omega)$ corresponds to $\eta = \eta^j$, $\xi_n^j = \mu_j(\omega)$, and the vector x has the form $(0, \dots, 0, r)$. Therefore

$$F_j(x) = \int_{V_j} \varphi_j(\eta) f(\eta) e^{-ir\xi_n^j(\eta,\omega)} J(\eta)\, d\eta,\ j \leq m.$$

At the stationary point $\eta = \eta^j$ we have $\xi_n^j = \mu_j(\omega)$. It is clear that if $(\xi_n^j)'' = \partial^2 \xi_n^j / \partial\eta_k \partial\eta_l$, $1 \leq k, l \leq n-1$, then at $\eta = \eta^j$ we have

$$|\det(\xi_n^j)''| = |k_j(\omega)|,\ \operatorname{sign}(-\xi_n^j)'' = -\gamma_j.$$

Besides, since the point $\sigma_j(\omega)$ is lying in U_j and does not belong to U_k with $k \neq j$, we have $\varphi_j(\sigma^j(\omega)) = 1$. It is obvious that $J(\eta) = 1$ at $\eta = \eta^j$. Applying Theorem 18, we obtain our statement. □

7.6 Asymptotics of a fundamental solution of an elliptic operator with constant coefficients

Let $P(D) = \sum_{|\alpha| \leq m} a_\alpha D^\alpha$ be an elliptic operator with constant coefficients, i.e.

$$|P(\xi)| \geq C_0 |\xi|^m - C_1,\ \xi \in \mathbf{R}^n,$$

where $C_0 > 0$.

Suppose that the operator $P(D)$ satisfies the following three conditions:

1. If $P(\xi) = 0$, $\xi \in \mathbf{R}^n \setminus 0$, then $\nabla P(\xi) \neq 0$;
2. The set $\{\xi \in \mathbf{R}^n,\ P(\xi) = 0\}$ consists of a finite number k of smooth connected closed surfaces $S_1, \dots, S_k$ of dimension $n-1$;
3. At any point on the surfaces S_j, $j = 1, \dots, k$, the full curvature does not vanish.

Theorem 21. *If an elliptic operator $P(D)$ satisfies conditions 1–3, then there exists a fundamental solution $E(x)$ (i.e. a solution of the equation $P(D)E(x) = \delta(x)$) such that*

$$E(x) \sim \sum_{j=1}^{2k} e^{-i\mu_j(\omega)r} \sum_{l=0}^{\infty} a_{jl}(\omega) r^{(1-n)/2-l}$$

as $r \to \infty$, *where* $a_{jl} \in C^\infty(S^{m-1})$,

$$a_{j0} = \frac{(2\pi)^{(1-n)/2} e^{i\delta_j (n-1)\pi/4}}{i\sqrt{|k_j(\omega)|}\frac{\partial P}{\partial \nu}(\sigma^j(\omega))},$$

$\sigma_j(\omega)$ *are the points on* S_j *for* $j = 1, \dots, k$, *at which the outer normal vector* ν *is parallel to the vector* ω *and has the same direction, and* $\sigma_{j+k}(\omega)$ *are the points at which the normal vector has the opposite direction,* $k_j(\omega)$ *is the full curvature of the surface* S_j *at the point* $\sigma_j(\omega)$, *the value of* δ_j *is* 1 *for* $j = 1, \dots, k$ *and* -1 *for* $j = k+1, \dots, 2k$, *and* $\mu_j(\omega) = (\omega, \sigma_j(\omega))$. *This asymptotic expansion is uniform in* ω *and can be differentiated with respect to* ω.

The proof of Theorem 21 is based on the following two auxiliary lemmas.

Lemma 22. *If* $f \in C^1(\mathbf{R})$, *then* $|v.p. \int_{|t|<\delta} \frac{f(t)}{t}\, dt| \le 2\delta \sup_{|t|<\delta} |f'(t)|$.

Proof. It is evident that

$$v.p. \int_{|t|<\delta} \frac{f(t)}{t}\, dt = \int_{|t|<\delta} \frac{f(t) - f(0)}{t}\, dt = \int_{|t|<\delta} f'(\theta t)\, dt,$$

where $\theta = \theta(t)$, $0 \le \theta \le 1$. Hence the statement follows readily. □

Lemma 23. *Let* $f(t,\omega)$, $S(t,\omega) \in C^\infty(\Omega \times \mathbf{R})$, $f = 0$, *for* $|t| \ge \delta > 0$ *and* $S_t'(0,t) \ne 0$ *for* $|t| \le \delta$, $\operatorname{Im} S \equiv 0$. *Then*

$$v.p. \int_{-\infty}^{\infty} \frac{1}{t} f(t,\omega) e^{i\lambda S(t,\omega)}\, dt = \alpha\pi i f(0,\omega) e^{iS(0,\omega)\lambda} + F(\omega, \lambda),$$

where $\alpha = \operatorname{sign} S_t'(0,\omega)$, *and the function* F *and its derivatives with respect to* ω *and* λ *of any order tend to* 0 *as* $\lambda \to \infty$ *faster than* λ^{-N} *for any* N.

Proof. Set $\tau = S(t,\omega) - S(0,\omega)$. The considered integral has the form

$$I = \alpha e^{i\lambda S(0,\omega)} v.p. \int_{-\infty}^{\infty} \frac{1}{\tau} g(\tau,\omega) e^{i\tau\lambda}\, d\tau, \ g \in C^\infty,$$

$$g(\tau,\omega) = f(t,\omega) \frac{S(t,\omega) - S(0,\omega)}{t|S_t'(t,\omega)|},$$

so that $g(0,\omega) = f(0,\omega)$. Let $h = h(\tau) \in C_0^\infty$, $h(\tau) = 1$ for $|\tau| \le \delta/2$. By Lemma 13 the function

$$\int_{-\infty}^{\infty} \frac{1}{\tau} [g(\tau,\omega) - g(0,\omega)] e^{i\tau\lambda}\, d\tau$$

tends to zero as $\lambda \to \infty$ faster than λ^{-N} for any N as well as all its derivatives. Therefore

$$I \sim \alpha f(0,\omega) e^{i\lambda S(0,\omega)} \cdot v.p. \int_{-\infty}^{\infty} \frac{1}{\tau} h(\tau) e^{i\lambda\tau}\, d\tau.$$

It remains to verify that

$$v.p. \int_{-\infty}^{\infty} \frac{1}{\tau} h(\tau) e^{i\tau\lambda}\, d\tau \sim \pi i.$$

Remark that if l_ε is the demi-circle $\{z:\ |z|=\varepsilon,\ \operatorname{Im} z>0\}$ in the complex plane, then

$$\lim_{\varepsilon\to 0} \int_{l_\varepsilon} (e^{i\lambda z}/z)\, dz = \lim_{\varepsilon\to 0} \int_0^{\pi} e^{i\lambda\varepsilon e^{i\phi}}\, i\, d\phi = \pi i.$$

Therefore

$$\begin{aligned}\lim_{\varepsilon\to 0} \int_{\varepsilon<|\tau|<\delta/2} \frac{1}{\tau} h(\tau) e^{i\tau\lambda}\, d\tau &= \lim_{\varepsilon\to 0}\Big(- \int_{l_{\delta/2}} \frac{1}{\tau} e^{i\tau\lambda}\, d\tau + \int_{l_\varepsilon} \frac{1}{\tau} e^{i\tau\lambda}\, d\tau\Big)\\ &= - \int_{l_{\delta/2}} \frac{1}{\tau} e^{i\tau\lambda}\, d\tau + \pi i\end{aligned}$$

and

$$\begin{aligned} v.p. \int_{-\infty}^{\infty} \frac{1}{\tau} h(\tau) e^{i\tau\lambda} d\tau &= \pi i - \int_{l_{\delta/2}} \frac{1}{\tau} e^{i\tau\lambda} d\tau + \int_{-\delta}^{-\delta/2} \frac{h(\tau)}{\tau} e^{i\tau\lambda} d\tau + \int_{\delta/2}^{\delta} \frac{h(\tau)}{\tau} e^{i\tau\lambda} d\tau\\ &= \pi i + \int_\gamma h(\tau) \frac{1}{\tau} e^{i\tau\lambda} d\tau,\end{aligned}$$

where γ is the contour consisting of the segments $(-\delta, -\delta/2)$, $(\delta/2, \delta)$ and the arc $|\tau| = \delta/2$, on which $h(t) \equiv 1$. Since $\tau \neq 0$ on this contour, the function $h(\tau)\tau^{-1} \in C^\infty$ and one can integrate by parts several times. It follows that $\int_\gamma h(\tau)\tau^{-1} e^{i\lambda\tau}\, d\tau = O(\lambda^{-\infty})$ and the proof is complete. □

Proof of Theorem 21. Let $E(\xi)$ be the distribution from the space $\mathcal{S}'(R^n)$ such that

$$E(\phi) = v.p. \int \frac{\phi(\xi)\, d\xi}{\overline{P}(\xi)},\ \phi \in \mathcal{S}.$$

Here

$$v.p. \int \frac{\phi(\xi)\, d\xi}{\overline{P}(\xi)} = \lim_{\varepsilon\to 0} \int_{|P(\xi)|>\varepsilon} \frac{\phi(\xi)\, d\xi}{\overline{P}(\xi)}.$$

Set $\mathcal{F}^{-1}_{\xi\to x} E(\xi) = \mathcal{E}(x)$. We have

$$P(D)\mathcal{E}(x)(\phi) = (2\pi)^{-n} P(\xi) E(\xi)(\tilde{\phi}) = (2\pi)^{-n} \int \tilde{\phi}(\xi)\, d\xi = \phi(0).$$

Thus $P(D)\mathcal{E}(x) = \delta$, i.e. $\mathcal{E}(x)$ is a fundamental solution.

Condition 1 guarantees the smoothness of the surface $P(\xi) = 0$. Moreover, there is a δ such that the surfaces $S_j(t)$ defined by the equality $P(\xi) = t$ for $|t| \le \delta$, $j = 1, \dots, k$, are smooth, too. Let U_j be the neighborhood of S_j formed by the surfaces $S_j(t)$ as $-\delta < t < \delta$. One can assume that the domains U_j are mutually nonintersecting. Let $h_j(\xi) \in C_0^\infty(R^n)$, $h_j(\xi) = 0$ for $\xi \notin U_j$, $h_j(\xi) = 1$ at the points $\xi \in U_j$ at which the inequality $|P(\xi)| < \delta/2$ holds. Then

$$\mathcal{E}(x) = (2\pi)^{-n} \sum_{j=1}^{k} v.p. \int_{U_j} \frac{h_j(\xi)}{P(\xi)} e^{ix\xi}\, d\xi + \mathcal{F}^{-1}_{\xi\to x}((1 - \sum h_j(\xi))E(\xi)).$$

Remark that in the integral over U_j we have $d\xi = ds\, dl$, where dl is the length element of the normal line to $S_j(t)$. Since $dP = |\nabla P| dl$, we have $d\xi = |\nabla P|^{-1}\, ds\, dt$. Set

$$\Phi_j(t, \omega, r) = (2\pi)^{-n} \pi i \int_{S_j(t)} \frac{e^{ix\xi} h_j(\xi)}{|\nabla P(\xi)|}\, ds.$$

Then

$$\mathcal{E}(x) = \sum_{j=1}^{k} \frac{1}{\pi i} v.p. \int_{-\delta}^{\delta} \frac{1}{t} \Phi_j(t, \omega, r)\, dt + \mathcal{F}^{-1}_{\xi\to x}((1 - \sum h_j)E(\xi)).$$

Conditions 1–3 imply that each surface S_j for $j = 1, \ldots, k$ is convex and contains exactly one point $\sigma_j(\omega)$ at which the normal vector is collinear to ω and one point $\sigma_{j+k}(\omega)$, at which these vectors have the opposite directions.

In virtue of Theorem 20 we have

$$\Phi_j(t, \omega, r) = \sum_{j=1}^{2k} e^{-i\mu_j(t,\omega) r} \sum_{l=0}^{N} a_{jl}(t, \omega) r^{(1-n)/2-l} + O(r^{(1-n)/2-N-1}),$$

where N is arbitrary, the functions $a_{jl}(t, \omega) \in C^\infty$, and the remainders as well as their derivatives with respect to ω, t, r are decreasing uniformly in ω and t as $r \to \infty$. Moreover,

$$a_{j0}(\omega) = i\pi (2\pi)^{(-1-n)/2} |k_j(\omega)|^{-1/2} \frac{1}{|\nabla P(\sigma_j(\omega))|} e^{-i(n-1)\pi/4}.$$

Substituting the obtained asymptotics of the functions Φ_j in the expression for $\mathcal{E}(x)$, we obtain, applying Lemma 23, that as $r \to \infty$

$$\mathcal{E}(x) = \sum_{j=1}^{2k} b_j e^{-i\mu_j(0,\omega) r} \sum_{l=0}^{N} a_{jl}(0, \omega) r^{(1-n)/2-l} + O(r^{(1-n)/2-N-1})$$
$$+ \mathcal{F}^{-1}_{\xi\to x}[(1 - \sum h_j)E(\xi)].$$

To estimate the last term $v(x)$ remark that $E(\xi) = 1/P(\xi)$ outside of the support of the function $1 - \sum h_j$, so that

$$|D_\xi^\alpha E(\xi)| \le C_\alpha (1 + |\xi|)^{-m-|\alpha|}.$$

It is obvious that

$$\mathcal{F}^{-1}_{\xi\to x}[(1-\Delta_\xi)^N[(1-\sum h_j)E(\xi)]] = (1+|x|^2)^N v(x).$$

If $N \geq N_0$, then the function $(1-\Delta_\xi)^N[(1-\sum h_j)E(\xi)]$ is summable so that $|(1+|x|^2)^N v(x)| \leq C_N$. This means that $v(x)$ decreases as $r\to\infty$ faster than any power of r^{-1}. In addition, the same is true for any derivatives $D^\beta_x v(x)$.

The proof is complete. □

7.7 Estimates of the eigenfunctions of an elliptic operator with constant coefficients

Let $L(D) = \sum_{|\alpha|\leq 2m} a_\alpha D^\alpha$ be an elliptic symmetric differential operator with constant complex coefficients a_α. Let Ω be a bounded domain in $\mathbf{R}^n$ with smooth boundary $\partial\Omega$. We will study the eigenfunctions of the operator L, i.e. the nontrivial solutions of the problem

$$L(D)u(x) = \lambda^{2m}u(x),\ x\in\Omega;\quad u(x) = \frac{\partial u(x)}{\partial\nu} = \cdots = \frac{\partial^{m-1}u(x)}{\partial\nu^{m-1}} = 0,\ x\in\partial\Omega, \tag{$\mathcal{P}$}$$

such that

$$\int_\Omega |u(x)|^2\,dx = 1.$$

Our aim is to estimate $\max_{\overline{\Omega}}|u(x)|$. Such estimates for the Laplace operator $L=\Delta$ were obtained by Eidus. He proved that $|u(x)|^2 \leq C\lambda^{n-1}\ln\lambda$ for $\lambda\geq 2$ with a constant C depending only on Ω.

We are proving a similar estimate for all operators L of order $2m$ satisfying the following condition:

A. *Let $L_0(\xi)$ be the characteristic form of the operator L, i.e.*

$$L_0(\xi) = \sum_{|\alpha|=2m} a_\alpha\xi^\alpha.$$

The set $S=\{\xi\in R^n;\ L_0(\xi)=1\}$ is a finite number of smooth connected closed surfaces of dimension $n-1$ and the full curvature of each surface does not vanish.

Theorem 24. *If an elliptic operator satisfies condition A, then for all normed eigenfunctions of the problem ($\mathcal{P}$) the inequality*

$$|u(x)|^2 \leq C\lambda^{n-1}|\ln\lambda|$$

is fulfilled for $\lambda\geq 2$ with a constant C independent of λ.

The proof of Theorem 24 is based on some lemmas.

Lemma 25. *If $0\leq|\alpha|\leq 2m,\ \lambda\geq\lambda_0\geq 1$, then*

$$\int_\Omega |D^\alpha u(x)|^2\,dx \leq C\lambda^{2|\alpha|}.$$

Proof. Let $Lv = f$ in Ω and v vanish on $\partial\Omega$ with its derivatives up to order $m-1$. Then by Theorem 12 of Chapter 3 we have

$$\sum_{|\alpha|\le 2m} \int_\Omega |D^\alpha v(x)|^2\, dx \le C_1\Big(\int_\Omega |f(x)|^2\, dx + \int_\Omega |v(x)|^2\, dx\Big).$$

Therefore

$$\sum_{|\alpha|\le 2m} \int_\Omega |D^\alpha u(x)|^2\, dx \le C_1(\lambda^{4m}+1) \int_\Omega |u(x)|^2\, dx \le C_2\lambda^{4m}.$$

The interpolational inequality

$$\sum_{|\alpha|=j} \int_Q |D^\alpha u(y)|^2\, dy \le C_3\Big(\sum_{|\alpha|=2m} \int_Q |D^\alpha u(y)|^2\, dy + \int_Q |u(y)|^2\, dy\Big),$$

where $1 \le j < 2m$, $Q \subset \mathbf{R}^n$, and the constant C_3 is independent of Q, implies after the change of variables $y = \lambda x$ that

$$\sum_{|\alpha|=j} \int_\Omega |D^\alpha u(x)|^2\, dx\cdot\lambda^{-2j} \le C_4\Big(\sum_{|\alpha|=2m} \int_\Omega |D^\alpha u(x)|^2\, dx\cdot\lambda^{-4m} + \int |u(x)|^2\, dx\Big). \ \square$$

Lemma 26. *The normed eigenfunctions of the problem* ($\mathcal{P}$) *satisfy for* $\lambda > 1$ *the inequality*

$$\int_{\partial\Omega} |D_\nu^m u(x)|^2\, dS_x \le C\lambda^{2m}.$$

Proof. Let x_0 be an arbitrary point of the boundary $\partial\Omega$. There exists a neighborhood U_{x_0} of this point diffeomorphic to a cube Q in $\mathbf{R}^n$, such that the part of the surface $\partial\Omega$ lying in U_{x_0} passes by this diffeomorphism in a part Γ of the plane $y_n = 0$, and the image of $\Omega \cap U_{x_0}$ is lying in the domain $y_n > 0$.

The covering of the surface $\partial\Omega$ by such neighborhoods contains a finite subcovering by domains $U_1, \dots, U_N$. Let $\sum_{j=1}^{N+1} h_j^2(x) = 1$ be a subordinate partition of the unity such that $\operatorname{supp} h_j \subset U_j$ for $j = 1, \dots, N$, $h_j \ge 0$ are smooth functions. Let $v_j(y)$ be the image of the function $h_j(x)u$ under this diffeomorphism. It is obvious that

$$\int_{\partial\Omega} |D_\nu^m u(x)|^2\, dS_x \le C \sum_{j=1}^N \int_{Q_j\cap\{y_n=0\}} |D_n^m v_j(y)|^2\, dS_y,$$

where $D_n = \partial/\partial y_n$, and Q_j is the image of U_j.

Applying Lemma 25 it is easy to check that $\int_{Q_j} |D^\alpha v_j(y)|^2\, dy \le C_j\lambda^{2|\alpha|}$ for $0 \le |\alpha| \le 2m$, where the constant C_j depends on j but not on λ.

Let v be one of the functions v_j and $Q = Q_j$. Then $L_1(x, D)v = \lambda^{2m}v + g$ and $\int_Q |g|^2\,dy \le C\lambda^{4m-2}$. The operator L_1 is elliptic and its coefficient $b(y)$ before the derivative $D_n^{2m}v$ does not vanish. Without loss of generality we can assume that $b(y) = 1$, since otherwise we can replace the operator L by L/b.

Let $w \in C^{2m}(Q)$. We have

$$\operatorname{Re}\int_Q L_1 v\cdot\overline{w}\,dy = \operatorname{Re}\sum_{\substack{|\alpha|\le m\\|\beta|\le m}}\int_Q b_{\alpha\beta}(y)D^\alpha v(y)\overline{D^\beta w(y)}\,dy \\ + \operatorname{Re}\sum_{\substack{|\alpha+\beta|\le 2m-1\\|\beta|<m-1}}\int_\Gamma C_{\alpha\beta}(y)D^\alpha v(y)\cdot\overline{D^\beta w(y)}\,dS_y. \tag{6}$$

Let us take now $w(y) = \partial v/\partial y_n$. The boundary conditions imply that the latter sum is equal to $-\frac{1}{2}\operatorname{Re}\int_\Gamma |D_n^m v(y)|^2\,dS_y$. The first sum on the right-hand side in (6) is equal to

$$\operatorname{Re}\sum_{\substack{|\alpha|\le m\\|\beta|\le m}}\int_Q b_{\alpha\beta}(y)D^\alpha v(y)\overline{D^\beta w(y)}\,dy = \operatorname{Re}\sum_{\substack{|\alpha|=m\\|\beta|=m}}\frac{1}{2}\int_Q\frac{\partial}{\partial y_n}(b_{\alpha\beta}(y)D^\alpha v(y)\overline{D^\beta v(y)})\,dy \\ + \operatorname{Re}\sum_{\substack{|\alpha|\le m\\|\beta|\le m}}\int_Q b'_{\alpha\beta}(y)D^\alpha v(y)\overline{D^\beta v(y)}\,dy.$$

In virtue of Lemma 25 this implies that

$$\operatorname{Re}\int_Q L_1 v\overline{D_n v}\,dy = -\frac{1}{2}\int_\Gamma |D_n^m v(y)|^2\,dS_y + O(\lambda^{2m}).$$

On the other hand, we have

$$\operatorname{Re}\int_Q L_1 v\cdot\overline{D_n v}\,dy = \operatorname{Re}\lambda^{2m}\int_Q v(y)\cdot\overline{D_n v(y)}\,dy + \operatorname{Re}\int_Q g(y)\cdot\overline{D_n v(y)}\,dy = O(\lambda^{2m}).$$

Therefore

$$\operatorname{Re}\int_\Gamma |D_n^m v(y)|^2\,dS_y = O(\lambda^{2m}).$$

Summing over all j we get the result. □

Lemma 27. *The normed eigenfunctions of the problem* $(\mathcal{P})$ *satisfy the inequality*

$$\int_{\partial\Omega}|D^\alpha u(x)|^2\,dS \le C\lambda^{2|\alpha|},$$

where $|\alpha| \le 2m-1$ *and the constant* C *is independent of* λ.

Proof. Let us fix a point $x_0 \in \partial\Omega$. In a neighborhood U_{x_0} of this point one can define a diffeomorphic map in $\mathbf{R}^n$, under which the surface $\partial\Omega \cap U_{x_0}$ passes in the plane $y_n = 0$ and the image Q of $\Omega \cap U_{x_0}$ is lying in the domain $y_n > 0$. Let $h \in C_0^\infty(R^n)$, $\operatorname{supp} h \subset U_{x_0}$, and $h(x) \geq 0$ and $h(x) = 1$ in a neighborhood of the point x_0. Let $v(y)$ be the function corresponding to hu by this map.

It is obvious that $L_1 v = \lambda^{2m} v + g$ and $\int_\Omega |g|^2\,dx = O(\lambda^{4m-2})$. Multiplying the equation by $\overline{D_n^{2m-1} v}$ and using Lemma 25, we see that for $\lambda > 1$

$$\int_\Gamma |D_n^{2m-1} v|^2\,dS_y \leq C\lambda^{4m-2} + C \sum_{\substack{|\alpha|\leq 2m-1\\ m\leq\alpha_n\leq 2m-2}} \int_\Gamma |D^\alpha v|^2\,dS_y.$$

Analogously, multiplying the equation $L_1 v = \lambda^{2m} v + g$ by $\overline{D_j^2 D_n^{2m-3} v}$ with $j = 1, \ldots, n-1$, we obtain that

$$\sum_{j=1}^{n-1} \int_\Gamma |D_n^{2m-2} D_j v|^2\,dS_y \leq C\lambda^{4m-2}$$

$$+ C \sum_{\substack{|\alpha|\leq 2m-1\\ m\leq\alpha_n\leq 2m-3}} \int_\Gamma |D^\alpha v|^2\,dS_y + C\sum_{j=1}^{n-1} \int_\Gamma |D_n^{2m-3} D_j^2 v| \cdot |\overline{D_n^{2m-1} v}|\,dS_y.$$

The last sum is estimated in the following way:

$$C\sum_{j=1}^{n-1} \operatorname{Re} \int_\Gamma D_n^{2m-3} D_j^2 v \cdot \overline{D_n^{2m-1} v}\,dS_y$$

$$\leq \delta \int_\Gamma |D_n^{2m-1} v|^2\,dS_y + \frac{1}{4\delta} \sum_{j=1}^{n-1} \int_\Gamma |D_n^{2m-3} D_j^2 v|^2\,dx,$$

where δ is a positive number. Using the above estimate and choosing δ sufficiently small, we see that the terms with $D_n^{2m-2} D_j v$ and $D_n^{2m-2} v$ on the right-hand side can be omitted, so that

$$\sum_{j=1}^{n-1} \int_\Gamma |D_n^{2m-2} D_j v|^2\,dS_y \leq C\lambda^{4m-2} + C \sum_{\substack{|\alpha|\leq 2m-1\\ \alpha_n\leq 2m-3}} \int_\Gamma |D^\alpha v|^2\,dS_y.$$

Now we use induction. Let k be such a number that $2 \leq k \leq m-2$ and let the estimate

$$\sum_\beta \int_\Gamma |D_n^{2m-1-|\beta|} D^\beta v|^2\,dS_y \leq C\lambda^{4m-2} + C \sum_{\substack{|\alpha|\leq 2m-1\\ \alpha_n\leq 2m-2-|\beta|}} \int_\Gamma |D^\alpha v|^2\,dS_y$$

be valid with $|\beta| \le k-1$. Let us prove that it is true for $|\beta| = k$, too. For this we multiply the equation $L_1 v = \lambda^{2m} v + g$ by $\overline{D_n^{2m-1-2|\beta|} D^{2\beta} v}$ and integrate over the domain $y_n > 0$. Integrating by parts with respect to y_n, we get that

$$\begin{aligned}\operatorname{Re}\int_Q D_n^{2m} v \cdot \overline{D_n^{2m-1-2|\beta|} D^{2\beta} v}\, dy \\ = O(\lambda^{4m-2}) - \operatorname{Re}\int_Q D_n^{2m-1} v \cdot \overline{D_n^{2m-2|\beta|} D^{2\beta} v}\, dy \\ - \operatorname{Re}\int_\Gamma D_n^{2m-1} v \cdot \overline{D_n^{2m-1-2|\beta|} D^{2\beta} v}\, dS_y.\end{aligned}$$

Now integrate by parts with respect to y' and y_n in such a way that the orders of the multiplied derivatives in the resulting integral are not more than $2m$ and $2m-1$, respectively. Finally we have

$$\begin{aligned}\operatorname{Re}\int_Q D_n^{2m} v \cdot \overline{D_n^{2m-1-2|\beta|} D^{2\beta} v}\, dy = \operatorname{Re}\int_Q D_n^{2m-|\beta|} D^\beta v \cdot \overline{D_n^{2m-|\beta|-1} D^\beta v}\, dy \\ + O(\lambda^{4m-2}) - \sum_{j=0}^{|\beta|-1} \sum_{|\gamma|=j} \operatorname{Re}\int_\Gamma D_n^{2m-1-j} D^\gamma v \cdot \overline{D_n^{2m-1+\gamma-2|\beta|} D^{2\beta-\gamma} v}\, dS_y.\end{aligned}$$

The first integral on the right-hand side is equal to $-\frac{1}{2}\int_\Gamma |D_n^{2m-|\beta|-1} D^\beta v|^2\, dS_y$. Considering the other terms of the operator L_1 in the same way, we obtain that

$$\begin{aligned}\int_\Gamma |D_n^{2m-|\beta|-1} D^\beta v|^2\, dS_y \\ \le C[\delta \sum_{\substack{|\alpha|=2m-1 \\ 2m-|\beta|\le\alpha_n\le 2m-2}} \int_\Gamma |D^\alpha v|^2\, dS_y + \lambda^{4m-2} + \frac{1}{\delta} \sum_{\substack{|\alpha|\le 2m-1 \\ m\le\alpha_n\le 2m-2-|\beta|}} \int_\Gamma |D^\alpha v|^2\, dS_y].\end{aligned}$$

Using induction and choosing δ sufficiently small we obtain the result. Therefore,

$$\sum_{|\alpha|=2m-1} \int_\Gamma |D^\alpha v|^2\, dS_y \le C\lambda^{4m-2}.$$

Using induction once again we can see that for $1 \le j \le m$ the estimate

$$\sum_{|\alpha|=2m-j} \int_\Gamma |D^\alpha v|^2\, dS_y \cdot \lambda^{-2|\alpha|} \le C + \sum_{|\beta|<2m-j} \int_\Gamma |D^\beta v|^2\, dS_y \cdot \lambda^{-2|\beta|}$$

is valid. For $j = 1$ this inequality has been proved above. Assume that it is true for $j = k-1$, where $2 \le k \le m$ and prove it for $j = k$. Let us multiply the equation $L_1 v = \lambda^{2m} v + g$ by $\overline{D_n^{2m-2k+1} v}$ and integrate the result over the domain $y_n > 0$.

Consider the principal term:

$$\operatorname{Re}\int_Q D_n^{2m}v\cdot\overline{D_n^{2m-2k+1}v}\,dy = O(\lambda^{4m-2k}) + \operatorname{Re}\int_Q D_n^{2m-k}v\cdot\overline{D_n^{2m-k+1}v}\,dy$$
$$-\sum_{j=1}^{k-1}\operatorname{Re}\int_\Gamma D_n^{2m-j}v\cdot\overline{D_n^{2m-2k+j}v}\,dS_y = O(\lambda^{4m-2k})$$
$$-\frac{1}{2}\int_\Gamma |D_n^{2m-k}v|^2\,dS_y - \sum_{j=0}^{k-1}\operatorname{Re}\int_\Gamma D_n^{2m-j}v\cdot\overline{D_n^{2m-2k+j}v}\,dS_y.$$

Considering the other terms in the same way, we see that

$$\int_\Gamma |D_n^{2m-k}v|^2\,dS_y \le C\lambda^{4m-2k} + \sum_{|\alpha|<2m-k}\int_\Gamma |D^\alpha v(y)|^2\,dy\lambda^{4m-2k-2|\alpha|}.$$

The further estimates are also based on induction. Assume that

$$\int_\Gamma |D_n^{2m-k-|\beta|}D^\beta v|^2\,dS_y \le C[\lambda^{4m-2k} + \sum_{\substack{|\alpha|=2m-k\\ m\le\alpha_n\le 2m-k-|\beta|-1}}\int_\Gamma |D^\alpha v|^2\,dS_y]$$

for $|\beta|\le l-1$, where $1\le l\le m-k$ and prove this estimate for $|\beta|=l$. For this multiply the equality $L_1v=\lambda^{2m}v+g$ by $D_n^{4m-2k-2|\beta|+1}D^{2\beta}v$ and integrate the result over the domain $y_n>0$. Integrating by parts with respect to the variables x_n, x' in such a way that the orders of the obtained derivatives are not higher than $2m$ for each factor, we obtain that

$$\begin{aligned}\operatorname{Re}&\int_Q D_n^{2m}v\cdot\overline{D_n^{2m-2k-2|\beta|+1}v(y)}\,dy\\ &= O(\lambda^{4m-2k}) - \sum_{j=1}^{k+|\beta|}\int_\Gamma D_n^{2m-j}v\cdot\overline{D_n^{2m-2k-2|\beta|+j}v(y)}\,dS_y\\ &\qquad + \operatorname{Re}\int_Q D_n^{2m-k-|\beta|}v\cdot\overline{D_n^{2m-k-|\beta|+1}D^{2\beta}v}\,dy.\end{aligned}$$

Integrating by parts in the latter integral $|\beta|$ times, we see that

$$\operatorname{Re}\int_Q D_n^{2m-k-|\beta|}D^\beta v\cdot\overline{D_n^{2m-k-|\beta|+1}D^\beta v}\,dy = \int_\Gamma |D_n^{2m-k-|\beta|}D^\beta v|^2\,dS_y.$$

The other terms are studied in the same way. In the integrals over the boundary, obtained after integrating by parts, the order of derivatives with respect to y_n will be less than in the considered case. As result we get the estimate

$$\sum_{|\beta|=l} \int_\Gamma |D_n^{2m-k-|\beta|} D^\beta v(y)|^2 \, dS_y$$

$$\leq C(\lambda^{4m-2k} + \frac{1}{\delta} \sum_{\substack{|\alpha|=2m-k \\ 2m-k-|\beta|+1\leq\alpha_n\leq 2m-k}} \int_\Gamma |D^\alpha v(y)|^2 \, dS_y + \delta \sum_{\substack{|\alpha|=2m-k \\ m\leq\alpha_n\leq 2m-k-|\beta|}} \int_\Gamma |D^\alpha v(y)|^2 \, dS_y).$$

The sum on the right-hand side having the factor $1/\delta$ is estimated by the inductive hypothesis. Taking δ sufficiently small, we obtain the result. □

Lemma 28. *If u is a nontrivial solution of the problem $(\mathcal{P})$, then the maximal value of $|u(x)|$ is attained at a point $x_0 \in \Omega$, whose distance to the boundary of the domain Ω is bigger than $C\lambda^{-1}$, where the constant C is positive and independent of λ.*

Proof. Let $y = \lambda x$. The equation $L(D)u = \lambda^{2m}u$ takes the form $L_0(D_y)v + L_1(\lambda, D_y)v = v$, where L_0 is the principal part of the operator L, and L_1 is an operator of order $2m-1$, whose coefficients are bounded uniformly in λ. If a point x_0 is at distance $d_1 \leq \lambda^{-1}$ from $\partial\Omega$, then in the y-coordinates this distance is less than 1. Let y_1 be the nearest to $y_0 = \lambda x_0$ point of $\partial\Omega$ and K be the ball of radius 2 with its center at a point y_1. Norming the function v so that $\max_y |v(y)| = |v(y_0)| = 1$, and using Theorem 11 of Chapter 3, we see that $\max_y |\operatorname{grad} v(y)| \leq C_1$, where the constant C_1 is independent of λ. By the Lagrange theorem,

$$1 = |v(y_0)| = |v(y_0) - v(y_1)| \leq |\operatorname{grad} v(\tilde{y})||y_0 - y_1|.$$

Therefore, $|y_0 - y_1| \geq C_1^{-1}$, so that $\operatorname{dist}(x_0, \partial\Omega) \geq C_1^{-1}\lambda^{-1}$. □

Lemma 29. *If $L_1(D)$ is an elliptic operator of order $2m$ with constant coefficients satisfying condition A, then there is a fundamental solution $F(x)$ of the operator $L_1(D) - I$ such that*

$$|D^\alpha F(x)| \leq C_\alpha |x|^{-(n-1)/2}, \ |x| \geq 1$$

for all α.

Proof. The proof follows from Theorem 21. □

Proof of Theorem 24. Let $\max |u(x)|$ be attained at a point $x_0 \in \Omega$. Assume at first that $\operatorname{dist}(x_0, \partial\Omega) \geq C_0$, where $C_0 = \text{const} > 0$. Let $y = \lambda x, \ v(y) = u(x)\lambda^{-n/2}$. Then $L_1 v = v$, where L_1 is an elliptic operator, whose principal part is independent of λ and the coefficients of the operator $\lambda(L_1 - L_0)$ are uniformly bounded in λ. Let $h \in C_0^\infty(R^n)$ and $h(y) = 0$ for $|y - y_0| \geq C_0\lambda$, where $y_0 = \lambda x_0, \ h(y) = 1$ for $|y - y_0| \leq C_0\lambda/2$, and $|D^\alpha h(y)| \leq C\lambda^{-|\alpha|}$. Then

$$L_1(hv) - hv = \sum k_{\alpha\beta} D^\alpha h(y) D^\beta v(y)$$

and in the latter sum we have $|\alpha+\beta| \leq 2m,\ |\alpha| \geq 1$. Let $F(y)$ be the fundamental solution of the operator $L_1 - I$ indicated in Lemma 29. Then

$$hv(y) = \int F(y-z) \sum k_{\alpha\beta} D^\alpha h(z) D^\beta v(z)\, dz.$$

If the point y_0 coincides with the origin, then the integral here is taken over the domain $C_0\lambda/2 \leq |z| \leq C_0\lambda$. Integrating by parts we obtain that

$$hv(0) = \int v(z) \sum k'_{\alpha\beta} D^\alpha h(z) D^\beta F(-z)\, dz.$$

The estimates $|D^\beta F(-z)| \leq C_\beta |z|^{-(n-1)/2}$ imply that

$$\begin{aligned}|v(0)|^2 = |hv(0)|^2 &\leq C \int |v(z)|^2 dz \int \sum |D^\alpha h(z) \cdot D^\beta F(-z)|^2\, dz \\ &\leq C_1 \lambda^{-2} \lambda^{-(n-1)} \int_{C_0\lambda/2 \leq |z| \leq C_0\lambda} dz \leq C_2 \lambda^{-1},\end{aligned}$$

where the constant C_2 is independent of λ. Therefore, max $|u(x)|^2 \leq C_2 \lambda^{n-1}$ and the proof is complete.

In the general case, by Lemma 28 there is a constant C_0 such that if $|u(x)| =$ max $|u(x)|$, then $\operatorname{dist}(x_0, \partial\Omega) \geq C_0/\lambda$. Let $y = \lambda x,\ v(y) = u(x)\lambda^{-n/2}$. Then $L_1 v = v$. Let F be the fundamental solution of the operator $L_1 - I$ from Lemma 29. Then by the Green formula we have

$$v(y_0) = \sum_{j=m}^{2m-1} \int_{\partial\Omega} \alpha_j D_\nu^j v \cdot D_\nu^{2m-1-j} F(y_0 - z)\, dS_z,$$

where α_j are constants. The function $F(y-z)$ and all its derivatives are continuous for $y \neq z$, and by Lemma 29 we have $|D_\nu^{2m-1-j} F(z)|^2 \leq C|z|^{1-n}$ and by Lemma 27 $\int_{\partial\Omega} |\partial_\nu^j v(z)|^2\, dS_z \leq C\lambda^{-1}$.

Therefore

$$\begin{aligned}|v(y_0)|^2 &\leq C \sum_{j=m}^{2m-1} \int_{\partial\Omega} |D_\nu^j v(z)|^2\, dS_z \cdot \int_{\partial\Omega} |D_\nu^{2m-1-j} F(y_0 - z)|^2\, dS_z \\ &\leq C_1 \lambda^{-1} \Big(\int_{|y_0 - z| \leq \rho} dS_z + \int_{\rho \leq |y_0 - z| \leq C\lambda} |y_0 - z|^{1-n} dS_z \Big).\end{aligned}$$

Thus if $C_0 < \rho < \lambda$, then $|v(y_0)|^2 \leq C_2 \lambda^{-1}(\rho^{n-1} + \ln \frac{C\lambda}{\rho})$. Setting, for example, $\rho = 2C$, we obtain the estimate $|v(y_0)|^2 \leq C_3 \lambda^{-1} \ln\ \lambda$. Hence $|u(x)|^2 \leq C_3 \lambda^{n-1} \ln\ \lambda$ for all $x \in \overline{\Omega}$. □

Remark 30. The estimates $\|u\|_{L_2(\Omega)} = 1$ and $\|u\|^2_{L_\infty(\Omega)} \le C\lambda^{n-1} \ln\, \lambda$ imply for $2 \le p \le \infty$ the inequality

$$\|u\|^2_{L_p(\Omega)} \le C_p \lambda^{(n-1)(1-2/p)} |\ln\, \lambda|^{1-2/p}.$$

7.8 Estimates of the first eigenvalue of an elliptic operator in a multi-connected domain

Let

$$Lu = \sum_{|\alpha| \le 2m} a_\alpha(x) D^\alpha u$$

be an elliptic operator of order $2m$ in a domain Ω of the space $\mathbf{R}^n$. Assume that

$$\operatorname{Re}(Lu, u) \ge c_0 |u|^2_m \tag{7}$$

for all functions u from $C_0^\infty(\Omega)$, where $c_0 = \text{const} > 0$, $(,)$ is a scalar product in $L_2(\Omega)$, and

$$|u|^2_m = \int \sum_{|\alpha|=m} |D^\alpha u|^2\, dx.$$

As we know the estimate (7) is valid for the Laplacian, for its powers, and for any strongly elliptic operator after adding to it a big enough constant C.

As usual we consider a generalized solution of the equation $Lu = \lambda u$ from the space $H_0^m(\Omega)$.

We are estimating the minimal eigenvalue λ of this operator in terms of the inner radius ρ of Ω and the number of connected components of the supplement of Ω.

Such estimates for the Laplace operator with $n = 2$ were studied by Osserman [Os] and Taylor [Ta]. Taylor has proved that $\lambda \ge C/\rho^2 k$ with some constant C. Osserman has constructed an example of the domain for which $\lambda \le C_1/\rho^2 k$, i.e. proved that the estimate is sharp.

For $2m \ge n$ such estimates for simply connected domains were obtained by Kondratiev [Ko2] and Hayman [Hay].

As we have seen above, for $2m < n$ such estimates are in principle impossible since there are simply connected domains with arbitrary small inner radii, for which $\lambda = 1$. (See Theorem 4.)

Let k_1 be the number of connected components of $\mathbf{R}^n \setminus \Omega$, whose diameter is less than $3\rho\sqrt{n}$. If this number is equal to zero, we set $k_1 = 1$.

Theorem 31. *If $n = 2m$, then $\operatorname{Re}\lambda \ge C/\rho^n k_1$, where $C > 0$ is a constant depending only on c_0 and n.*

Proof. Let u be a nontrivial generalized solution of the equation $Lu = \lambda u$. By condition, $\operatorname{Re}\lambda(u, u) = \operatorname{Re}(Lu, u) \ge c_0|u|^2_m$, so that $c_0|u|^2_m \le \operatorname{Re}\lambda\|u\|^2_0$. Change the coordinates, setting $x = x'\rho k_0^{1/n}$, where k_0 is such a number that $k_0 = 2^{n[\log_2 3k_1^{1/n}]}$ and $[a]$ is the integer part of a.

In new coordinates $c_0|u|_m^2 \le \operatorname{Re}\lambda(\rho k_0^{1/n})^{2m}\|u\|_0^2$. Now we are going to prove the estimate

$$\|u\|_0^2 \le A|u|_m^2 \tag{8}$$

for any function u from $C_0^\infty(\Omega')$, where Ω' is the image of the domain Ω in the space of variables x'. Combining the above estimates, we obtain that $c_0 \le A(\operatorname{Re}\lambda)(\rho k_0^{1/n})^{2m}$ and therefore

$$\operatorname{Re}\lambda \ge \frac{c_0}{A(\rho k_0^{1/n})^{2m}} \ge \frac{c_0}{A(3\rho k_1^{1/n})^{2m}} = \frac{C^{2m}}{k_1\rho^{2m}},$$

where $C = \text{const} > 0$, i.e. our statement is valid.

To prove (8), we will represent the space $\mathbf{R}^n$ as a sum of closed cubes Π with nonintersecting interiors, having their edges parallel to the coordinate axes and their sizes equal to 4.

Let Π be one of these cubes. Bisect it on k_0 equal cubes with their edges parallel to those of Π and equal to $4k_0^{-1/n}$. Recall that $k_0^{1/n}$ is integer and has the form 2^{nN} with some natural N.

Let Π' be one of the smaller cubes and Π'' the cube, concentric with Π' and having the edge twice less. Since the size of Π'' is larger than the inner radius of Ω', it contains a point of $\mathbf{R}^n \setminus \overline{\Omega'}$. Since the domain $\mathbf{R}^n \setminus \overline{\Omega'}$ has not more than k_1 connected components of diameters $\le 3\sqrt{n}k_0^{-1/n}$, at least $k_0 - k_1$ cubes $\Pi'_1, \dots, \Pi'_{k_0-k_1}$ of type Π' and of size $4k_0^{-1/n}$ contain connected components of the set $\mathbf{R}^n \setminus \Omega'$, having limit points on the boundaries of Π'_j and of concentric cubes Π''_j with twice smaller edges. Let Γ be a part of such a component, lying in the cube Π'.

Its projection on one of the axes, say x_n, contains a segment γ of length $\ge k_0^{-1/n}$. Let us consider the hyperplane $x_n = t$, where $t \in \gamma$. Let us denote its intersection with Π' by Π'_t. This $(n-1)$-dimensional cube contains at least one point of Γ, at which $u = 0$. By the Sobolev inequality, valid for $2m > n - 1$, we have

$$|v(x)|^2 \le Ch^{1-n} \int_{\Pi'_t} \sum_{|\alpha|\le m} h^{2|\alpha|}|D^\alpha v|^2\, dx',\ dx' = dx_1 \dots dx_{n-1}$$

for any function v from $W_2^m(\Pi'_t)$ and for any $x \in \Pi'_t$, where we have set $h = 4k_0^{-1/n}$.

Let $\overline{u}$ be the mean value of u on Π'_t and $\overline{u} \ne 0$. Setting $v = u - \overline{u}$ in the above inequality and taking as x the point at which $u(x) = 0$, we get the estimate

$$|\overline{u}|^2 \le Ch^{1-n} \int_{\Pi'_t} |u - \overline{u}|^2\, dx' + C \int_{\Pi'_t} \sum_{|\alpha|=1}^{m} h^{2|\alpha|-n+1}|D^\alpha u|^2\, dx'.$$

By the Poincaré inequality we have

$$\int_{\Pi'_t} |u-\overline{u}|^2\,dx' \le C_1 h^2 \int_{\Pi'_t} |\operatorname{grad}\ u|^2\,dx'.$$

Hence

$$\begin{aligned}\int_{\Pi'_t} |u|^2\,dx' &\le 2\int_{\Pi'_t} |u-\overline{u}|^2\,dx' + 2\int_{\Pi'_t} |\overline{u}|^2\,dx' \\ &\le 2(C+1)\int_{\Pi'_t} |u-\overline{u}|^2\,dx' + 2C\int_{\Pi'_t}\sum_{|\alpha|=1}^{m} h^{2|\alpha|}|D^\alpha u|^2\,dx' \\ &\le 2C_1(C+1)h^2\int_{\Pi'_t} |\operatorname{grad}\ u|^2\,dx' + 2C\int_{\Pi'_t}\sum_{|\alpha|=1}^{m} h^{2|\alpha|}|D^\alpha u|^2\,dx' \\ &\le C_2\int_{\Pi'_t}\sum_{|\alpha|=1}^{m} h^{2|\alpha|}|D^\alpha u|^2\,dx'.\end{aligned}$$

Integrating this inequality over all $t\in\gamma$, we obtain

$$\int |u|^2\,dx \le C_2\int \sum_{|\alpha|=1}^{m} h^{2|\alpha|}|D^\alpha u|^2\,dx,$$

where the integral is taken over the part of Π' that consists of all planes $x_n = t$ intersecting Γ. Note that by construction the measure of this set is $> 1/(4k_0)$.

After summing the obtained inequalities over all k_0-k_1 cubes $\Pi'_1,\dots,\Pi'_{k_0-k_1}$, we obtain a similar inequality, where the integral is taken over a set E of measure $> (k_0-k_1)/(4k_0)$ times the volume of the cube Π. It is evident that

$$3k_1^{1/n}/2 \le k_0^{1/n} \le 3k_1^{1/n}$$

and therefore

$$\frac{k_0-k_1}{k_0} \ge \frac{(3k_1^{1/n}/2)^n - k_1}{3^n k_1} = \frac{(3/2)^n-1}{3^n}.$$

It is easy to see that in this case the inequality

$$\int_\Pi |u|^2\,dx \le C_3\Big(\int_E |u|^2\,dx + \int_\Pi |grad\ u|^2\,dx\Big)$$

is true with the constant C_3 depending on n only.

Using the obtained inequality and summing over all cubes Π, we see that

$$\int_{\mathbf{R}^n} |u|^2\,dx \le C_4\int_{\mathbf{R}^n}\sum_{|\alpha|=1}^{m} |D^\alpha u|^2\,dx.$$

We assume here that $u = 0$ outside of Ω.

Using for $1 \leq |\alpha| \leq m-1$ the interpolatory inequalities

$$\int |D^\alpha u|^2\,dx \leq \varepsilon \int |u|^2\,dx + C_\varepsilon \int \sum_{|\beta|=m} |D^\beta u|^2\,dx,$$

which are valid for any $\varepsilon > 0$, and choosing ε sufficiently small, we obtain finally that $\|u\|_0^2 \leq C_5 \|u\|_m^2$, i.e. inequality (8) is true. □

Theorem 32. *If $2m > n$, then $\operatorname{Re}\lambda \geq C/\rho^{2m}$, where $C > 0$ is a constant, depending on c_0 and n.*

Proof. Let u be a nontrivial generalized solution of the equation $Lu = \lambda u$. By the condition, $\operatorname{Re}\lambda(u,u) = \operatorname{Re}(Lu,u) \geq c_0|u|_m^2$, so that $c_0|u|_m^2 \leq \operatorname{Re}\lambda\|u\|_0^2$. Change the coordinates, setting $x = x'\rho$.

In new coordinates $c_0|u|_m^2 \leq \operatorname{Re}\lambda\rho^{2m}\|u\|_0^2$. Now we are going to prove the estimate

$$\|u\|_0^2 \leq A|u|_m^2 \tag{9}$$

for any function u from $C_0^\infty(\Omega')$, where Ω' is the image of the domain Ω in the space of variables x'. Combining the above estimates, we obtain that $c_0 \leq A\operatorname{Re}\lambda\rho^{2m}$ and therefore our statement is valid.

To prove (9), we represent the space $\mathbf{R}^n$ as a sum of cubes Π with nonintersecting interiors, with their edges parallel to the coordinate axes and their sizes equal to 1.

Let Π be one of these cubes and $x \in \Pi$.

By the Sobolev inequality

$$|v(x)|^2 \leq C \int \sum_{|\alpha|\leq m} |D^\alpha v|^2\,dx,\ v \in C^\infty(\Pi).$$

Substituting $v = u - \overline{u}$, where $u \in C^\infty(\Pi)$, $\overline{u}$ is the mean value of u in the cube Π and taking as x a point at which $u(x) = 0$, we obtain the inequality

$$|\overline{u}|^2 \leq C \int_\Pi |u-\overline{u}|^2\,dx + C \int_\Pi \sum_{|\alpha|=1}^{m} |D^\alpha u|^2\,dx.$$

By the Poincaré inequality we have

$$\int_\Pi |u-\overline{u}|^2\,dx \leq C_1 \int_\Pi |\operatorname{grad}\ u|^2\,dx.$$

Therefore

$$\begin{aligned}\int_\Pi |u|^2\,dx &\le 2\int_\Pi |u-\overline{u}|^2\,dx + 2\int_\Pi |\overline{u}|^2\,dx\\ &\le 2(C+1)\int_\Pi |u-\overline{u}|^2\,dx + 2C\int_\Pi \sum_{|\alpha|=1}^m |D^\alpha u|^2\,dx\\ &\le 2C_1(C+1)\int_\Pi |\operatorname{grad}\ u|^2\,dx + 2C\int_\Pi \sum_{|\alpha|=1}^m |D^\alpha u|^2\,dx\\ &\le C_2\int_\Pi \sum_{|\alpha|=1}^m |D^\alpha u|^2\,dx.\end{aligned}$$

After summing over all cubes, we obtain the inequality

$$\int_{R^n} |u(x)|^2\,dx \le C_2\int_{R^n} \sum_{|\alpha|=1}^m |D^\alpha u(x)|^2\,dx,$$

and the interpolatory inequalities imply that $\|u\|_0^2 \le C_3|u|_m^2$, i.e. the inequality (9) is true and the proof is complete. □

Now we consider a non-self-adjoint elliptic operator

$$L_1 = \sum_{i,j=1}^2 A_{ij}(x)\frac{\partial^2}{\partial x_i \partial x_j} + C(x),\quad \sum_{i,j=1}^2 A_{ij}(x)\xi_i\xi_j \ge c_0|\xi|^2$$

of second order with real coefficients in a plane domain Ω.

Theorem 33. *If $A_{ij}(x)$ and $C(x)$ are uniformly bounded, the operator L_1 is uniformly elliptic in $\overline{\Omega}$, and $C(x) \le 0$, then $|\lambda| \ge C_1\gamma^{k_1}/\rho^2$, where ρ is the inner radius of Ω, C_1 and γ are constants depending only on $\max(\sum |A_{ij}|+|C|)$ and the ellipticity constants of L_1, $\gamma < 1$, and k_1 is the number of connected components of $\mathbf{R}^n \setminus \Omega$ having their diameters smaller than $3\rho\sqrt{2}$.*

Our proof is based on the following lemmas.

Lemma 34. *Let Π be a unit square and Π' be the concentric square of size 1/2. Let a connected component of the complement to Ω have points on the boundaries of the squares Π and Π'. Let $L_1 u = 0$ in $\Omega \cap \Pi$, $u = 0$ on the part of the boundary $\partial\Omega$ lying in Π, and $C(x) \le 0$. Then $\max_{\Pi'}\ u \le \alpha \max_\Pi\ u$, where the constant $\alpha < 1$ depends on the maxima of coefficients and the ellipticity constant of L_1.*

Proof. Let $\max_\Pi\ u = 1$. The function $v = 1 - u$ is nonnegative in Π and $L_1 v \le 0$. Let us take an ellipse, having common points with the connected component of the complement to Ω, indicated in the conditions of the Lemma, and such that its part E_1 bounded by this component and the contour of the ellipse, is lying in the square Π and contains inside a disk K of radius $1/10$.

Let $x_1^2/a^2 + x_2^2/b^2 = 1$ be the equation of this ellipse in some coordinate system and $\sigma(x) = x_1^2/a^2 + x_2^2/b^2$. Then the function $w = (e^{-k\sigma} - e^{-k})/(1 - e^{-k})$ is such that $0 \le w \le 1$ in the ellipse, $w = 0$ on its boundary, and $L_1 w > 0$ for $x_1 \ge a^2/2$, if k is large enough. We can assume that $a = 2$ and the value k depends on the maxima of coefficients of the operator L_1 and its ellipticity constant.

Remark that $L_1(v - w) < 0$ in E_1 and $v - w \ge 0$ on the boundary of E_1, since $v = 1$ in the complement to Ω and $w = 0$ on the contour of the ellipse. By the maximum principle, $v - w \ge 0$ in E_1, so that $v \ge c_0 > 0$ in K. Therefore, $u \le 1 - c_0$ in the disk K.

Now let x be an arbitrary point of Π'. Let us consider such an ellipse that the radius of the disk K divides it on two parts and the lesser of them, E_2, contains a disk of radius $1/20$ with its center at the point x and is lying in the square Π. Let us construct a function w_1 for this new ellipse as was done above and apply the maximum principle to the function $v - c_0 w_1$ in the domain E_2. As before, we obtain that $v \ge c_1 > 0$ in K_1 and hence $u \le 1 - c_1$ at the point x. The constant c_1 depends only on the maxima of coefficients of the operator L_1 and its ellipticity constant. □

Lemma 35. *Let Π be an open unit square and Π' be a similar square, nonintersecting Π but having a common side with it. Let $v \ge 0$, $L_1 v \le 0$ in the $1/2$-neighborhood of the set $\Pi \cup \Pi'$ and $v \ge 1$ in the square Π. Then $v \ge \beta$ in Π', where $\beta > 0$ is a number depending only on the maxima of the coefficients and the ellipticity constant of L_1.*

Proof. The proof is similar to the proof of Lemma 34. Let x be a point of Π' and the ellipse $x_1^2/a^2 + x_2^2/b^2 = 1$, having its center in the square Π, contain the disk with its center at x and having the radius $1/10$. We assume that this ellipse is divided by the common side of squares Π and Π' on two parts, the lesser of which, E, is lying in the $1/2$-neighborhood of the square Π'. Set $\sigma = x_1^2/a^2 + x_2^2/b^2$ and $w = (e^{-k\sigma} - e^{-k})/(1 - e^{-k})$. Choosing k we can achieve that $Lw > 0$ in E. Since $L(v - w) \le 0$ in E and $v - w \ge 0$ on the boundary of E, we see that $v - w \ge 0$ everywhere in E and in particular at the point x. Since at this point $\sigma \le c_0 < 1$, we have $w(x) \ge \beta > 0$ and therefore, $v(x) \ge \beta$. □

Proof of Theorem 33. Let $L_1 u = -\lambda u$, $\lambda \in \mathbf{R}$, $u = 0$ on $\partial\Omega$, $u \not\equiv 0$. By the maximum principle, $\lambda > 0$. Without loss of generality, we can assume that $\max |u(x)| = 1$ and the maximal value is attained at a point $x_0 \in \Omega$. Assume that $u(x_0) = 1$.

Set $h = 2\rho$ and consider the strip, generated by two parallel lines, distant from x_0 on ρ. Then x_0 is the center of the square Π_0 of size h, lying in the strip. Consider the sequence of $2k_1$ nonintersecting squares $\Pi_0, \Pi_1, \ldots, \Pi_{2k_1-1}$, lying in the strip and such that the squares Π_j and Π_{j+1} have a common side for $j = 0, \ldots, 2k_1 - 2$. Let Π'_j be the square concentric to Π_j of size ρ.

Each square Π'_j contains points of the domain $\mathbf{R}^2 \setminus \Omega$ and not more than k_1 of them contain a connected component of the complement. Hence there is a square Π_j, for which a connected component of the complement to Ω has points lying on the boundaries of the squares Π_j and Π'_j.

Let the direction of the lines bounding the strip coincide with the direction of the axis x_1 and the point x_0 coincide with the origin. Consider the function u_1, such that $u = u_1(1 - c_0\lambda x_2^2)$, where c_0 is a constant such that $2A_{22}(x)c_0 > 1$. We can assume also that $1 - c_0\lambda x_2^2 > 1/2$ for $|x_2| \leq h$, since otherwise we get the estimate $2c_0\lambda h^2 \geq 1$ readily.

It is clear that u_1 satisfies the equation $L_2(u_1) = 0$, where L_2 is an elliptic operator and $L_2(1) < 0$. Set $M = (1 - c_0\lambda h^2)^{-1}$. Then the function $u_2 = u_1/M$ satisfies the equation $L_2(u_2) = 0$ and $|u_2| \leq 1$.

After the change of variables $x = yh$ the squares Π_j become unit squares and the squares Π'_j become squares of size $1/2$. The coefficients of the operator $h^2L_2(y, D_y)$ are bounded uniformly in h, since $c_0\lambda^2h^2 \leq 1/2$, and the ellipticity constant is bounded below by a constant independent of h. Lemma 34 implies that $|u_2| \leq \alpha$ in Π'_j.

Let us join Π'_j with Π'_0 by a chain of squares of size $1/2$, adding to $\Pi'_0, \Pi'_1, \ldots$, Π'_j intermediate squares $\Pi''_0, \Pi''_1, \ldots, \Pi''_{j-1}$. Apply Lemma 35 to the squares Π''_{j-1}, Π'_j for the function $v = (1 - u_2)/(1 - \alpha)$. As we have seen before, $v \geq 1$ in Π'_j and therefore, $v \geq \beta$ in Π''_{j-1}. Repeating this argument for the squares Π'_{j-1}, Π''_{j-1}, we obtain that $v \geq \beta^2$ in Π'_{j-1}, $v \geq \beta^3$ in Π''_{j-2} and so on, $v \geq \beta^{2k_1}$ in Π'_0. Then $u_2 \leq 1-(1-\alpha)\beta^{2k_1}$ in Π'_0, so that $u_1 \leq [1-(1-\alpha)\beta^{2k_1}]/[1-c_0\lambda h^2]$ in Π'_0. However, $u_1(x_0) = 1$ and hence $1 - c_0\lambda h^2 \leq 1 - (1-\alpha)\beta^{2k_1}$, i.e. $c_0\lambda h^2 \geq (1-\alpha)\beta^{2k_1}$. □

7.9 Estimates of the first eigenvalue of the Schrödinger operator in a bounded domain

Let Ω be a bounded domain in $\mathbf{R}^n$ with a regular boundary. Consider the boundary value problem:

$$(-\Delta)^m u = \lambda P(x)u, \; u \in W_{2,0}^m(\Omega),$$

where Δ is the Laplace operator, $m \geq 1$, and the function P is positive and belongs to R_α, i.e. $\int_\Omega P(x)^\alpha \, dx = 1$.

Our aim is to find

$$m_\alpha = \inf_{P \in R_\alpha} \lambda_1, \; M_\alpha = \sup_{P \in R_\alpha} \lambda_1,$$

where λ_1 is the first eigenvalue of the Dirichlet problem, i.e.

$$\lambda_1 = \inf_{u \in W_{2,0}^m(\Omega)} \frac{\int_\Omega \sum_{|\alpha|=m} |D^\alpha u|^2 \, dx}{\int_\Omega P(x)|u(x)|^2 \, dx}.$$

Here we have the following result.

Theorem 36. *If $n \geq 2m$, $\alpha > n/2m$, then $m_\alpha > 0$.*
If $n \geq 2m$, $\alpha \leq n/2m$, then $m_\alpha = 0$.
If $n < 2m$, $\alpha \geq 1$, then $m_\alpha > 0$.
If $n < 2m$, $\alpha < 1$, then $m_\alpha = 0$.
If $\alpha \geq 1/2m$, then $M_\alpha = \infty$.

The proof of this statement is very similar to that of Theorem 25 in Chapter 5 and we omit it.

The following theorem shows that the regularity of the boundary is essential in this problem.

Let us recall that the Minkovsky dimension of a set A is defined as

$$\mu = n - \overline{\lim}_{\rho\to 0}\frac{\ln \operatorname{meas} A_\rho}{\ln \rho},$$

where A_ρ is the ρ-neighborhood of A.

Theorem 37. *If the boundary $\partial\Omega$ is smooth enough (for example, satisfies the Lipschitz condition, if $m = 1$) and $\alpha < 1/2m$, then $M_\alpha < \infty$.*

If $m = 1$, $n \geq 2$, $\alpha > n/2$, then $m_\alpha \geq \lambda$, where λ is the least eigenvalue in the following problem:

$$\Delta u + \lambda u^{(\alpha+1)/(\alpha-1)} = 0,\ u|_{\partial K} = 0,\ \int_K |u(x)|^{2\alpha/(\alpha-1)}\, dx = 1$$

in the ball K whose measure is equal to the measure of Ω.

If $m = 1$, $n \geq 2$, the Minkovsky dimension of $\partial\Omega$ is μ, and $\mu < n - 2\alpha$, then $M_\alpha < \infty$. However, for any μ such that $\mu \geq n - 2\alpha$ there exists a domain Ω with the Minkovsky dimension of $\partial\Omega$ equal to μ and such that $M_\alpha = \infty$.

Proof. If $\alpha < 0$, then

$$\int_\Omega u(x)^p\, dx \leq (\int_\Omega P(x)|u(x)|^2\, dx)^{p/2}(\int_\Omega P(x)^\alpha\, dx)^{1/(1-\alpha)},$$

where $p = 2\alpha/(\alpha - 1) > 0$. Therefore

$$M_\alpha \leq \frac{\int_\Omega |D^m u_0|^2\, dx}{(\int_\Omega |u_0|^p\, dx)^{2/p}},$$

where u_0 is a function whose integrals on the right-hand size of the latter inequality are finite. For instance, if $u_0 = d(x)^\gamma$ in some neighborhood of the boundary, where $d(x)$ is the distance of x from the boundary, then it is true when $\gamma > m - 1/2$, $\gamma > -1/p$.

If $0 < \alpha < 1$, then $p < 0$ and

$$\int_\Omega P(x)^\alpha \, dx \le (\int_\Omega P(x)u(x)^2 \, dx)^\alpha (\int_\Omega u(x)^p \, dx)^{1-\alpha}.$$

In this case the function u_0 satisfies the conditions of convergence if $-1/p > \gamma > m - 1/2$. Therefore, such a γ exists, if $1 > 2m\alpha$.

Now let $m = 1$. If the boundary of Ω is irregular and its Minkovsky content is $\mu > n-1$, then the function u_0 satisfies the required conditions when $(\mu - n)/p > \gamma > 1 - (n - \mu)/2$, what is possible when $\mu < n - 2\alpha$.

On the other hand, if $\mu > n - 2\alpha$, let us take as Ω the domain in the plane of x, y, contained in the square $0 < x < 1$, $0 < y < 1$, and obtained from the square by removing the segments $x = A_n, 1/3 < y < 2/3$, where $A_n = k_0 \sum_{j=1}^n j^{-s}$, $s > 1$, $n = 1, 2, \dots$ and k_0 is such that $k_0 \sum_{j=1}^\infty j^{-s} = 1$. Let us set $a_n = k_0 n^{-s}$, $P(x, y) = h(y) b_n$, for $A_n < x < A_{n+1}$, where $h \in C_0^\infty(1/3, 2/3)$, $h(y) = 0$ for $y < 1/3$ and for $y > 2/3$, and the constants b_n are chosen so that $b_n a_n^2 \to 0$ as $n \to \infty$, $\sum_{n=1}^\infty b_n^\alpha a_n = \infty$. Given $\varepsilon > 0$ one can take $P_\varepsilon = 0$ for $x < A_m$ and $x > A_k$, where m is such that $b_m a_m^2 < \varepsilon$ and k is such that

$$\int_{A_m}^{A_k} \int_0^1 P_\varepsilon(x, y)^\alpha \, dx \, dy \sim 1.$$

Then

$$\int_0^1 \int_0^1 P_\varepsilon(x) u(x)^2 \, dx \, dy \le \varepsilon^2 \int_0^1 \int_0^1 |\nabla u(x)|^2 \, dx \, dy,$$

for all $u \in W_2^2(\Omega)$ and therefore $M_\alpha = \infty$. It is easy to see that the Minkovsky dimension μ of the boundary is equal to $n - 1 + 1/s$. So we can set $b_n = n^{(s-1)/\alpha}$ and all the conditions can be satisfied, if $\mu > n - 2\alpha$. If $\mu = n - 2\alpha$, one can take $b_n = n^{2/(1-2\alpha)} (\ln n)^{-1/\alpha}$. □

Chapter 8

Negative Spectra of Elliptic Operators

8.1 Introduction

We are going to discuss here estimates of the number of points in the negative spectrum of an elliptic operator. Such estimates are important for physical applications, especially in quantum mechanics. They are of interest for spectral theory; from the point of view of the theory of partial differential equations they provide an opportunity to define subspaces of finite codimension on which some embedding theorems are valid. This can be useful, for instance, in stating and studying boundary value problems. In particular, these estimates allow us to find stable subspaces of solutions to the corresponding parabolic equations.

Let Ω be a domain in $\mathbf{R}^n$ which can be bounded or unbounded or even coincide with the whole space $\mathbf{R}^n$. The completion of the space $C_0^\infty(\Omega)$ with respect to the norm

$$\|u\|_m \equiv \Big(\int (|u|^2 + \sum_{|\alpha|=m} |D^\alpha u(x)|^2)\, dx\Big)^{1/2}$$

is denoted by $H_0^m(\Omega)$.

Let measurable functions $a_{\alpha\beta}$ be defined in Ω for $|\alpha| \le m,\ |\beta| \le m$ and

$$L_0 u = \sum_{|\alpha|\le m,\ |\beta|\le m} D^\alpha (a_{\alpha\beta}(x) D^\beta u(x)).$$

Assume this operator to be symmetric, i.e. $a_{\alpha\beta} \equiv (-1)^{|\alpha+\beta|}\overline{a}_{\beta\alpha}$ for all α and β. Everywhere below $(L_0 u, v)$ for $u, v \in H^m(\Omega)$ means

$$\int_\Omega \sum_{|\alpha|\le m,\, |\beta|\le m} (-1)^{|\alpha|} a_{\alpha\beta}(x) D^\beta u(x) \overline{D^\alpha v(x)}\, dx.$$

We assume that the operator L_0 is elliptic and normed in such a way that

$$(L_0u, u) \geq a_0 \int_\Omega \sum_{|\alpha|=m} |D^\alpha u(x)|^2 \, dx \text{ for all } u \in H_m(\Omega), \; a_0 = \text{const} > 0.$$

For example, this condition is satisfied by the operator $L_0 = (-\Delta)^m$, where Δ is the Laplace operator, $m \in \mathbf{N}$. The main object of our study is the operator

$$L \equiv L_0 - V(x),$$

where the real-valued function $V(x)$ is called a *potential.* If $V(x) \equiv 0$, then the spectrum of the operator L is continuous and lies on the positive real semi-axis. For a nonzero potential the spectrum can contain points of the negative semi-axis. However, it is well known (and it will be proved in what follows) that if the potential has a compact support and is bounded or is sufficiently high degree-summable, then the operator L can have only a finite number of negative eigenvalues of finite multiplicity. Estimates of the number N of these eigenvalues is the object of our study.

We stress that no conditions of regularity are imposed on the coefficients $a_{\alpha\beta}$ or on the boundary of Ω. Therefore, the eigenfunctions of the operator L are generalized solutions to the equation $Lu = \lambda u$, i.e. functions of the space $H_0^m(\Omega)$ such that

$$(Lu, \varphi) \equiv (L_0u, \varphi) - (Vu, \varphi) = \lambda(u, \varphi)$$

for all functions φ of the space $C_0^\infty(\Omega)$. Here and in what follows $(\; , \;)$ means the scalar product in the space $L_2(\Omega)$. (The meaning of the expression (L_0u, φ) was explained earlier).

Using simple considerations based on the variational principle (see Section 4.1) it is easy to show that

1. If there exists a k-dimensional subspace F_k in C_0^∞ such that

$$(Lu, u) < 0 \text{ for } u \in F_k,$$

then $N \geq k$.

2. If there exists a k-dimensional subspace F_k in C_0^∞ such that

$$(Lu, u) \geq 0 \text{ for } u \in G_k,$$

where G_k is the orthogonal complement to F_k in $L_2(\Omega)$, then $N \leq k$.

Therefore, the estimation of the number of points in the negative spectrum is reduced to the problem of finding a corresponding subspace.

In the case $m \geq 1$, $n > 2m$, Rozenblyum [Ro1] obtained the following important estimate:

$$N \leq c_{n,m} \int V_+(x)^{n/2m} \, dx, \qquad V_+(x) = \max(0, V(x)).$$

Other proofs of this estimate for $m = 1$ were found by Lieb and Cwikel (see [ReSi]). There are generalizations of this estimate for the case $m = 1$ in [Fef], [KerSa], [EgKo5], [EgKo8].

We note that it is impossible to obtain such estimates for $n \leq 2m$. In fact, such an estimate would imply $N = 0$ if the integral of $V_+^{n/2m}$ is small enough. However, Example 33 of Chapter 4 shows that for the operator $(-\Delta)^m + V(x)$ the negative spectrum is always nonvoid if $V(x) \leq 0, V(x) \not\equiv 0,\ n \leq 2m$. Thus study of the negative spectrum in the spaces of small dimensions involves additional difficulties.

Sharp conditions for discreteness of the negative spectrum in terms of capacity (for $m = 1$) are stated in the book of Mazya [Maz1]. The Birman-Schwinger estimate (see [Bi], [ReSi]) plays an important role in spectral theory (it concerns the case $m = 1$, $n = 3$):

$$N \leq \frac{1}{16\pi^2} \int\int \frac{V_+(x)V_+(y)}{|x-y|^2} dxdy.$$

In the work of Birman-Solomyak [BiSo] these inequalities are improved with the help of interpolation techniques. In the papers of Lieb [L1-L3], Li-Yau [LiYau], and others sharp constants in estimates of the considered type are obtained for the Schrödinger operator. We also note the results of Ivrii [Iv] on the asymptotics of the negative spectrum for the Schrödinger operator.

Fefferman's work [Fef] contains the following more precise estimates of the number N (in the case $m = 1$). Let $Q_1, Q_2, \dots, Q_M$ be M nonintersecting cubes in $\mathbf{R}^n$ that remain disjoint when doubled and satisfy the inequality

$$\int_{Q_j} V_+(x)\, dx \geq c|Q_j|^{1-2/n},$$

where $|Q|$ is the volume of Q and c is a positive constant, depending on n only. It turns out that the number of negative eigenvalues is not less than the number M. Conversely, if $p > 1$, the corresponding upper bound is given by the number of minimal dyadic nonintersecting cubes Q_j satisfying the inequality

$$\int_{Q_j} V_+(x)^p dx \geq c'|Q_j|^{1-2p/n}.$$

Kerman-Sawyer [KerSa] showed that the same result is valid if the considered cubes satisfy the inequality

$$\int_{Q_j}\int_{Q_j} \frac{V_+(x)V_+(y)}{|x-y|^{n-2}}\, dx\, dy \geq c \int_{Q_j} V_+(x)\, dx.$$

This result is stronger since for $p > 1$

$$|Q|^{2/n-1}\int_Q V_+(x)\,dx \le c\int_Q\int_Q \frac{V_+(x)V_+(y)}{|x-y|^{n-2}}\,dx\,dy \Big/ \int_Q V_+(x)\,dx$$
$$\le \sup_{Q'\subset Q} |Q'|^{2/n-1/p}\Big(\int_{Q'} V_+(x)^p\,dx\Big)^{1/p}.$$

Schechter proposed an alternative proof in [Sche2].

8.2 One-dimensional case

Consider at first the case when $n = 1$ and $L = (-d^2/dx^2)^m - V(x)$.

Theorem 1. *If $n = 1$, then for any $q \ge 1$ the estimate*

$$N \le m\Big(1 + a_q\int_{-\infty}^{\infty} V_+(t)^q |t|^{2mq-1}\,dt\Big)$$

is true, where a_q is a constant, depending on q only.

For the proof we need some auxiliary results. These lemmas will also be used later, in the study of the higher dimensional case.

Lemma 2. *Let $y \in C_0^m(\mathbf{R}_+)$, $y(0) = y'(0) = \cdots = y^{(m-1)}(0) = 0$. Then*

$$|y(x)|^2 \le a|x|^{2m-1}\int_0^{\infty} |y^{(m)}(t)|^2\,dt,\ a = \frac{1}{(m-1)!^2(2m-1)}.$$

Proof. Let the maximum value of the function $|y(x)|^2/|x|^{2m-1}$ be attained at $x = R > 0$. Let $t = x/R$, $z(t) = y(x)$. Since $z(0) = z'(0) = \cdots = z^{(m-1)}(0) = 0$, we can use the Cauchy formula

$$z(t) = \int_0^t \frac{(t-s)^{m-1}}{(m-1)!} z^{(m)}(s)\,ds,$$

which implies for $|t| \le 1$ the inequality $|z(t)|^2 \le a\int_{-1}^{1} |z^{(m)}(t)|^2\,dt$. Therefore, for $|x| \le R$ the inequality $|y(x)|^2 \le aR^{2m-1}\int_{-R}^{R} |y^{(m)}(x)|^2\,dx$ is true and hence

$$|y(x)|^2|x|^{1-2m} \le |y(R)|^2R^{1-2m} \le a\int_{-\infty}^{\infty} |y^{(m)}(x)|^2\,dx. \qquad \square$$

Lemma 3. *$y \in C_0^{2m}(\mathbf{R}^n)$. Let (r,ω) be the spherical coordinates in $\mathbf{R}^n$, $|\omega| = 1$, $\{\varphi_j\}$ be a complete orthonormal system of spherical functions, $j = 0, 1, \ldots$ and $u(x) = \sum_{j=0}^{\infty} R_j(r)\varphi_j(\omega)$. Then*

$$\int_{\mathbf{R}^n} (-\Delta)^m u(x)\overline{u}(x)\,dx = \sum_{j=0}^{\infty}\int L_j(D_t)T_j(t)\overline{T_j(t)}\,dt,$$

where $r = e^t$, $D_t = d/dt$, $T_j(t) = R_j(r)r^{(n-2m)/2}$, and L_j is a differential operator of order $2m$ with real constant coefficients and the characteristic form

$$L_j(\lambda) = \prod_{i=0}^{m-1} [(k - m + n/2 + 2i)^2 - \lambda^2].$$

The value of k in this formula is found from the relation $\mu_j = -k(k+n-2)$, where μ_j is the jth eigenvalue of the Laplace-Beltrami operator on the unit sphere.

If $n = 1$, then

$$\int_0^\infty (-1)^m u^{(2m)}(x)\overline{u}(x)\,dx = \int_{-\infty}^{\infty} L(D_t)T(t)\overline{T(t)}\,dt,$$

where $x = e^t$, $D_t = d/dt$, $T(t) = u(x)x^{(1-2m)/2}$, and $L(D_t)$ is a differential operator of order $2m$ with real constant coefficients and the characteristic form

$$L(\lambda) = \prod_{j=0}^{m-1} [(1/2 + j)^2 - \lambda^2].$$

Proof. It is well known that

$$\Delta = \frac{d^2}{dr^2} + (n-1)\frac{d}{rdr} + \frac{L_\omega}{r^2},$$

where L_ω is the Laplace-Beltrami operator acting on the unit sphere. The functions φ_j are eigenfunctions of the operator $L_\omega :\ L_\omega\varphi_j = \mu_j\varphi_j$, and $\mu_j = -k(k+n-2)$ for some $k = k(j) \in \mathbf{N}$. Therefore

$$\Delta u = \sum_{j=0}^{\infty}(R_j'' + (n-1)R_j'/r + \mu_j R_j/r^2)\varphi_j(\omega)$$

$$\text{and}\quad \Delta^m u = \sum_{j=0}^{\infty}(\frac{d^2}{dr^2} + (n-1)\frac{d}{rdr} + \frac{\mu_j}{r^2})^m R_j(r)\varphi_j(\omega)$$

so that

$$\int_{\mathbf{R}^n} \Delta^m u(x)\overline{u(x)}\,dx = \int_0^\infty \sum_{j=0}^{\infty}(\frac{d^2}{dr^2} + (n-1)\frac{d}{rdr} + \frac{\mu_j}{r^2})^m R_j(r)\overline{R_j(r)}r^{n-1}\,dr.$$

If $n = 1$, one can set $\mu_j = 0$. Set $r = e^t$, $R_j(r) = T_j(t)e^{t(m-n/2)}$. Then

$$\int_{\mathbf{R}^n} (-\Delta)^m u(x) \cdot \overline{u(x)}\,dx = \sum_{j=0}^{\infty} \int_{\mathbf{R}} L_j(D_t)T_j(t) \cdot \overline{T_j(t)}\,dt,$$

where L_j is the differential operator indicated above.

Indeed, it is clear that the leading coefficient of the characteristic polynomial $L_j(\lambda)$ is equal to $(-1)^m$. Its roots can be found from the following considerations. To any root λ there corresponds a solution $u = e^{\lambda t}$ of the equation $L(D_t)u = 0$. Since the equation

$$\frac{d^2u}{dr^2} + (n-1)\frac{du}{rdr} + \mu_j \frac{u}{r^2} = 0, \quad \text{where} \quad \mu_j = -k(k+n-2),$$

has the solutions $u = r^k$ and $u = r^{2-k-n}$, the polynomial $L_j(\lambda)$ has the roots k and $2-k-n$. It follows that the equation

$$(\frac{d^2}{dr^2} + (n-1)\frac{d}{rdr} + \frac{\mu_j}{r^2})^2 u = 0$$

has the solutions r^k, r^{2-k-n}, r^{k+2}, and r^{4-k-n}. Moreover, all roots of the equation $L_j(\lambda) = 0$ are obtained from the numbers k and $2-k-n$ by adding to them the numbers $2r-m+n/2$ for $r = 0, 1, \dots, m-1$. Hence one can see that the polynomial $L_j(\lambda)$ has the indicated form.

If $n = 1$, then the roots of the characteristic equation $L(\lambda) = 0$ are obtained in the same way from $-m+1/2$ by adding the numbers $0, 1, \dots, 2m-1$. □

Lemma 4. *Let $p \geq 1$. Then the inequality*

$$(\int_{\mathbf{R}} |u(t)|^{2p}\, dt)^{1/p} \leq a_p \int_{\mathbf{R}} (|u'(t)|^2 + |u(t)|^2/4)\, dt$$

is true for $u \in C_0^\infty(\mathbf{R})$, where

$$a_p = \frac{2^{1+1/p}}{p} \left[\frac{(p-1)\Gamma(p/(p-1)+1/2)}{\sqrt{\pi}\Gamma(p/(p-1))}\right]^{(p-1)/p} \tag{1}$$

for $p > 1$ and $a_1 = 4$.

Proof. Let $p > 1$ and $m = \inf\limits_{u \in C_0^\infty(\mathbf{R})} \dfrac{\int(|u'(t)|^2 + |u(t)|^2/4)\, dt}{(\int |u(t)|^{2p}\, dt)^{1/p}}$.

If a sequence $\{u_k\}$ is such that $\int(|u_k'(t)|^2 + |u_k(t)|^2/4)\, dt \to m$ and $\int |u_k(t)|^{2p}\, dt = 1$, then there is a subsequence converging in L_2 and L_{2p} and weakly in H^1 to a function $u \in H^1$. Moreover, then $\int |u(t)|^{2p}\, dt = 1$ and $\int(|u'(t)|^2 + |u(t)|^2/4)\, dt \leq m$. Since m is the minimal value, we have the equality $\int(|u'(t)|^2 + |u(t)|^2/4)\, dt = m$. The function u satisfies the Euler-Lagrange equation

$$u'' - u/4 + mu^{2p-1} = 0.$$

Multiplying by u' and integrating we obtain that $u'^2 = u^2/4 - mu^{2p}/p + C$. Since $u \to 0$, $u' \to 0$ as $t \to \infty$, we see that $C = 0$. Then the equation is integrable and has the functions

$$u(x) = [\frac{2\sqrt{m}}{p} ch \frac{p-1}{2}(x + C_1)]^{1/(1-p)}$$

as its solutions. Integrating, we obtain the indicated value of a_p.

It is easy to see that $\lim_{p \to 1} = 4$ and, on the other hand, the inequality for $p = 1$ is valid with $a_1 = 4$. □

Lemma 5. *Let $p \geq 1$. Then the inequality*

$$[\int_{\mathbf{R}} \frac{|y^{2p}(t)|}{|t|^{p+1}}\, dt]^{1/p} \leq a_p \int_{\mathbf{R}} |y'(t)|^2\, dt$$

is valid for the functions y from $C_0^\infty(\mathbf{R})$, vanishing at $t = 0$, where a_p is defined by (1).

Proof. It is sufficient to prove the inequality for functions u vanishing for $t < 0$. Set $t = e^s$ and $y = ue^{s/2}$. Then the inequality will take the form

$$(\int_{\mathbf{R}} |u(s)|^{2p}\, ds)^{1/p} \leq a_p \int_{\mathbf{R}} (|u'(s)|^2 + |u(s)|^2/4)\, ds$$

and the result follows from Lemma 4. □

Lemma 6. *Let $p \geq 1$, L_j be the differential operator defined in Lemma 3, and $k(j)$ be such that $\mu_j = -k(k+n-2)$, where μ_j is the jth eigenvalue of the Laplace-Beltrami operator. If $k(j) \geq m+1-n/2$ or n is odd, then the inequality*

$$(\int_{\mathbf{R}} |u(t)|^{2p}\, dt)^{1/p} \leq b_p k(j)^{1-1/p-2m} \int_{\mathbf{R}} L_j(D_t)u(t) \cdot \overline{u(t)}\, dt$$

holds for all functions $u(t)$ from $C_0^\infty(\mathbf{R})$.

Proof. It is easy to see that

$$\int_{\mathbf{R}} L_j(D_t)u(t) \cdot \overline{u(t)}\, dt \geq \int_{\mathbf{R}} ((-1)^m D_t^{2m} + a_j^0)u(t)\overline{u(t)}\, dt,$$

where

$$a_j^0 = \prod_{i=0}^{m-1} (k(j) - m + \frac{n}{2} + 2i)^2. \tag{2}$$

If n is odd, then $a_j^0 \geq c_0 k(j)^{2m}$. If n is even, but $k(j) \geq m+1-n/2$, then we also have the inequality $a_j^0 \geq c_0 k(j)^{2m}$ for all j.

In the same way as in Lemma 4 we can see that

$$(\int_{\mathbf{R}} |u(t)|^{2p}\, dt)^{1/p} \leq b_p \int_{\mathbf{R}} (|u^{(m)}(t)|^2 + c_0|u(t)|^2)\, dt.$$

Replacing t by kt, we obtain the estimate

$$(\int_{\mathbf{R}} |u(t)|^{2p}\, dt)^{1/p} \leq b_p k^{1-2m-1/p} \int_{\mathbf{R}} (|u^{(m)}(t)|^2 + a_j^0|u(t)|^2)\, dt$$

for functions $u(t)$ from $C_0^\infty(\mathbf{R})$, which completes the proof. □

Lemma 7. *If $p \geq 1$, then for all $j \geq 1$*

$$(\int_{\mathbf{R}} |u(t)|^{2p}/|t|^{p+1}\,dt)^{1/p} \leq b_p k(j)^{-2m} \int_{\mathbf{R}} L_j(D_t)u(t)\overline{u(t)}\,dt$$

for $u \in C_0^\infty(\mathbf{R})$ such that $u(0) = 0$, where $L_j(D_t)$ is the differential operator indicated in Lemma 3.

Proof. It is easy to see that

$$\int_{\mathbf{R}} L_j(D_t)u(t)\overline{u(t)}\,dt \geq \int_{\mathbf{R}} ((-1)^m D_t^{2m} + b_j^0 D_t^2)u(t)\overline{u}(t)\,dt,$$

where

$$b_j^0 = \sum_{l=0}^{m-1} \prod_{i=0, i\neq l}^{m-1} (k(j) - m + 2i + n/2)^2 > 0.$$

It is evident that $b_j^0 \geq c_0 k^{2m-2}$, $c_0 > 0$. Using the dilatation $t' = kt$ one can obtain the result from the estimate

$$[\int_{\mathbf{R}} \frac{|y(t)|^{2p}}{|t|^{p+1}}\,dt]^{1/p} \leq a_p \int_{\mathbf{R}} |y'(t)|^2\,dt,$$

which by Lemma 5 is valid for functions y from $C_0^\infty(\mathbf{R})$, vanishing at $t = 0$. □

Lemma 8. *For any $p > 1$ the estimate*

$$\left[\int_0^\infty |y(x)|^{2p}|x|^{p-1-2mp}\,dx\right]^{1/p} \leq c_p \int_0^\infty |y^{(m)}(x)|^2\,dx \tag{3}$$

is true if $y \in C_0^\infty(\mathbf{R})$, $y(0) = y'(0) = \cdots = y^{(m-1)}(0) = 0$.

Proof. Let $x = e^t$, $y = ze^{(m-1/2)t}$. The inequality (3) is equivalent to the following one

$$\left[\int_{\mathbf{R}} |z(t)|^{2p}\,dt\right]^{1/p} \leq c_p \int_{\mathbf{R}} L(D_t)z(t)\cdot\overline{z(t)}\,dt,$$

where $L(D_t)$ is the differential operator found in Lemma 3, with constant coefficients and the characteristic form $L(\lambda) = \prod_{j=0}^{m-1}[(m - 1/2 - 2j)^2 - \lambda^2]$. As in Lemma 7, one can see that

$$\int_{\mathbf{R}} L(D_t)z(t)\overline{z}(t)\,dt \geq a_m \int_{\mathbf{R}} |z(t)|^2\,dt + b_m \int_{\mathbf{R}} |z'(t)|^2\,dt,$$

where

$$a_m = \prod_{j=0}^{m-1}(m - 1/2 - 2j)^2 > 0, \quad b_m = \sum_{i=0}^{m-1} \prod_{j=0, j\neq i}^{m-1} (m - 1/2 - 2j)^2 > 0.$$

Therefore the inequality (3) follows from Lemma 4. □

Proof of Theorem 1. It suffices to consider the case when $V \in C_0^\infty(\mathbf{R})$, $V(x) \geq 0$. Let at first $q > 1$. Set $I = \int_{\mathbf{R}} V(x)^q |x|^{2mq-1}\, dx$. Divide the line $\mathbf{R}$ on the $k+1$ segment by some points $x_1, \ldots, x_k$ so that

$$\int_{x_i}^{x_{i+1}} V(x)^q |x|^{2mq-1}\, dx \leq I/k,\ i = 0, 1, \ldots, k,$$

where $x_0 = -\infty$, $x_{k+1} = \infty$, and one of the points x_i is 0. Let $K = \{y \in C_0^\infty(\mathbf{R}),\ y(x_i) = y'(x_i) = \cdots = y^{(m-1)}(x_i) = 0 \text{ for } i = 1, \ldots, k\}$. Our aim is to prove the inequality

$$\int_{\mathbf{R}} V(x)|y(x)|^2\, dx \leq \int_{\mathbf{R}} |y^{(m)}(x)|^2\, dx \text{ for } y \in K.$$

By the Hölder inequality and Lemma 8

$$\begin{aligned}
&\int_{\mathbf{R}} V(x)|y(x)|^2\, dx \\
&\quad = \sum_{i=0}^{k} \int_{x_i}^{x_{i+1}} V(x)|y(x)|^2\, dx \\
&\quad \leq \sum_{i=0}^{k} [\int_{x_i}^{x_{i+1}} V(x)^q |x|^{2mq-1}\, dx]^{1/q} [\int_{x_i}^{x_{i+1}} |y(x)|^{2p}|x|^{p-1-2pm}\, dx]^{1/p} \\
&\quad \leq c_p (I/k)^{1/q} \sum_{i=0}^{k} \int_{x_i}^{x_{i+1}} |y^{(m)}(x)|^2\, dx = c_p (I/k)^{1/q} \int_{\mathbf{R}} |y^{(m)}(x)|^2\, dx.
\end{aligned}$$

Setting $k = [I/c_p^q] + 1$, we obtain the above inequality. Since $N = km$, the proof is complete.

If $q = 1$, then the estimate

$$\int_{\mathbf{R}} V(x)|u(x)|^2\, dx \leq \max(|u(x)|^2/|x|^{2m-1}) \int_{\mathbf{R}} V_+(t)|t|^{2m-1}\, dt$$

is true and the result follows from Lemma 2 as above. □

Theorem 9. *Let $n = 1$, $\ln_1 x = \ln|x|$, and $\ln_{j+1} x = \ln|\ln_j x|$ for $j \geq 1$. Let the positive numbers $b_j^{(i)}$ be such that*

$$\prod_{j=0}^{i-1} [\lambda^2 + (i - \frac{1}{2} - 2j)^2] \equiv \sum_{j=0}^{i} b_j^{(i)} \lambda^{2j}.$$

Set

$$S_0(x) = b_m^{(m)} x^{-2m},$$

$$S_1(x) = S_0(x) + x^{-2m} \sum_{j=1}^{m} b_j^{(m)} b_0^{(j)} (\ln_1 x)^{-2j},$$

$$\dots\dots\dots\dots$$

$$S_k(x) = S_{k-1}(x) + x^{-2m} \sum_{1 \le i_k \le \ldots \le i_1 \le m} b_{i_1}^{(m)} b_{i_2}^{(i_1)} \ldots b_0^{(i_k)} (\ln_1 x)^{-2i_1} (\ln_2 x)^{-2i_2} \ldots (\ln_k x)^{-2i_k}.$$

If $V(x) \le S_k(x)$, then $N \le 2^{k+1} - 1$. If $W(x) \ge S_k(x)$ and $W(x) \ge S_k(x) + \delta$ on some nonempty interval, $\delta > 0$, then there exists a function $V(x) \in C_0^\infty(\mathbf{R})$ such that $V(x) \le W(x)$ and for the operator

$$H = (-d^2/dx^2)^m - V(x)$$

the number of points in the negative spectrum is not less than 2^{k-1}.

Let us remark that for $m = 1$ the functions S_k have the form

$$S_0(x) = (2x)^{-2}; S_k(x) = S_{k-1}(x) + (2x \ln_1 x \ldots \ln_k x)^{-2}.$$

Proof. As before we have to show that

$$\int_{\mathbf{R}} V(x)|y(x)|^2 \, dx \le \int_{\mathbf{R}} |y^{(m)}(x)|^2 \, dx \text{ for } y \in K_N,$$

where K_N is a subspace in $C_0^\infty(\mathbf{R})$ of codimension N.

Let $V(x) \le S_k(x)$. Then

$$(Ly, y) = \int_{\mathbf{R}} [|y^{(m)}(x)|^2 - V(x)|y(x)|^2] \, dx \ge \int_{\mathbf{R}} [|y^{(m)}(x)|^2 - S_k(x)|y(x)|^2] \, dx.$$

For $N = m(2^{k+1} - 1)$ denote by K_N the linear subspace of functions in $C_0^\infty(\mathbf{R})$, vanishing together with their derivatives of order $\le m - 1$ at the points x such that $\ln_j x = 0$ for some $j \le k$.

It is evident that after the change $x = e^t$, $y = ze^{(m-1/2)t}$ the equality

$$\int_0^\infty [|y^{(m)}(x)|^2 - S_k(x)|y(x)|^2] dx = \int_{-\infty}^\infty [L(D_t)z(t)\overline{z}(t) - S_{k-1}(t)|z(t)|^2] dt,$$

is valid, where $S_{k-1}(t) = S_k(e^t)$. For $x < 0$ a similar result is obtained after the change $x = -e^t$, $y = ze^{(m-1/2)t}$. Since

$$\int_{\mathbf{R}} L(D_t) z\overline{z} \, dt = \sum_{j=0}^{m} b_j^{(m)} \int_{\mathbf{R}} |z^{(j)}(t)|^2 \, dt$$

and all $b_j^{(m)}$ are positive, one can repeat the argument for $1 \le j \le m$. After k repetitions one obtains that $(Ly, y) \ge 0$ for $y \in K_N$, where $N = (2^{k+1} - 1)m$.

If $W(x) \ge S_k(x)$, then the integral $\int_\Delta [|y^{(m)}(x)|^2 - V(x)|y(x)|^2]\,dx$ over a segment Δ, on which $W(x) \ge S_k(x) + \delta$, after the k changes of variables turns into an integral of the form

$$\int [P(D_t)z\bar{z} + (S_k(x) - V(x))|z|^2]\,dt,$$

where $P(D_t)$ is a differential operator with constant coefficients and $P(0) = 0$. One can find a function $V \in C_0^\infty(\mathbf{R})$ such that $S_k(x) < V(x)$ on an interval I. If $z \in C_0^\infty(\mathbf{R})$, $z = 1$ in I, and the derivatives of z are small enough, then this integral is < 0. Let y_1 be the function corresponding to z in the original coordinates. The functions $y_2, \dots, y_{2^{k+1}}$ from $C_0^\infty(\mathbf{R})$ are chosen in such a way that their supports are disjoint and lie in small neighborhoods of the zeroes of the functions $\ln_j x$ for $1 \le j \le k$. It is obvious that it is possible to find the functions y_i so that $(Ly_i, y_i) < 0$, $y_i \in C_0^\infty$, and S_k has large enough values on the supports of y_i.Therefore, we have the inequalities $(Ly_i, y_i) < 0$, $i = 1, \dots, 2^{k+1}$, for the corresponding potential $V(x)$, which proves our theorem. □

The following theorem gives sufficient conditions for N to be infinite.

Theorem 10. *Let $V(x) \ge 0$. Assume that there exists an infinite sequence of disjoint intervals $I_j = \{x : |x - M_j| < M_j/2\}$ and*

$$\int_{I'_j} V(x)|x|^{2m-1}\,dx \ge a_0,$$

where $I'_j = \{x : |x - M_j| < M_j/4\}$ and a_0 is an appropriate constant indicated below. Then the spectrum of the operator $L = L_0 - V(x) = (-1)^m d^{2m}/dx^{2m} - V(x)$ contains an infinite set of points on $(-\infty, 0)$.

Proof. We can assume that $M_j \to \infty$ or $M_j \to 0$.

Let $h_0 \in C_0^\infty(-1, +1)$ and $h_0(x) = 1$ for $|x| < 1/2$. Set $h_j(x) = h_0(2(x - M_j)/M_j)$. Observe that

$$|D^\alpha h_j(x)| \le C M_j^{-\alpha} \text{ for } |\alpha| \le m.$$

Set now $\varphi_j(x) = h_j(x)|x|^{m-1/2}$. Then

$$(L_0\varphi_j,\ \varphi_j) \le C_1 \int_{I_j} |D^m \varphi_j|^2\,dy \le c,$$

where the constant c does not depend on j.

On the other hand,

$$\int_{\mathbf{R}} V(y)\varphi_j^2(y)\,dy \ge \int_{I'_j} V(x)|x|^{2m-1}\,dx \ge a_0 > 0.$$

The relation $(L\varphi_j, \varphi_j) < 0$ is true if $a_0 > c$. Since the set of φ_j is infinite, the proof is complete. □

Theorem 11. *Let $V(x) \geq 0$ and $0 < 2\gamma m < 1$. There exists a positive constant b_0 such that if*

$$\int_0^\infty V(x)^\gamma |x|^{2m\gamma-1}(1+|\ln|x||)^{-1-\varepsilon} dx \geq b_0 k$$

for some positive ε and a natural number k, then $N \geq k$.

Proof. Let us denote by $Q(R,d)$ the segment $\{x: \ x \in \mathbf{R}^1, \ R < x < R+d\}$. Set $u(x) = ((x-R)(R+d-x))^{m-1/2+\varepsilon}$ for $R < x < R+d$, where $d > 0$ and $u(x) = 0$ for $x < R$ and for $x > R+d$. It is clear that $u(x) \in H_0^m(Q(R,d))$ and

$$\int_R^{R+d} |D^m u(x)|^2 \, dx \leq C d^{2m-1+4\varepsilon}.$$

On the other hand, by the Hölder inequality

$$\int_R^{R+d} V(x)^\gamma \, dx \leq (\int_R^{R+d} V(x) u(x)^2 \, dx)^\gamma (\int_R^{R+d} u(x)^{-2\gamma/(1-\gamma)} \, dx)^{1-\gamma},$$

so that

$$\int_R^{R+d} V(x) u(x)^2 \, dx \geq (\int_R^{R+d} V(x)^\gamma \, dx)^{1/\gamma} (\int_R^{R+d} u(x)^{-2\gamma/(1-\gamma)} \, dx)^{1-1/\gamma}.$$

Note that

$$\int_R^{R+d} u(x)^{-2\gamma/(1-\gamma)} \, dx \leq C_1 d^{1-2\gamma(2m-1+2\varepsilon)/(1-\gamma)}.$$

Therefore

$$\int_R^{R+d} V(x) u(x)^2 \, dx \geq C_2 (\int_R^{R+d} V(x)^\gamma \, dx)^{1/\gamma} d^{1-1/\gamma+2(2m-1+2\varepsilon)},$$

and $(Lu,u) < 0$ if

$$C_2 (\int_R^{R+d} V(x)^\gamma \, dx)^{1/\gamma} d^{1-1/\gamma+2(2m-1+\varepsilon)} > c_0 C d^{2m-1+4\varepsilon},$$

i.e. if

$$\int_{Q(R,d)} V(x)^\gamma \, dx > b_1 d^{1-2\gamma m}. \tag{4}$$

Thus (4) implies that $N \geq 1$.

Let us show now that if

$$\int_{Q(1,1)} V(x)^\gamma \, dx > b_1 k, \tag{5}$$

then there exist k functions $u_1, \dots, u_k$ from $W^m_{2,0}(Q(1,1))$ with nonintersecting supports for which $(Lu_j, u_j) < 0$. For this we will find on the segment $(1,2)$, $k-1$ points $r_1, \dots, r_{k-1}$ such that $1 = r_0 < r_1 < \dots < r_{k-1} < r_k = 2$ and

$$\int_{Q(r_j, r_{j+1}-r_j)} V(x)^\gamma \, dx > b_1 \text{ for } j = 0, 1, \dots, k-1.$$

Since on any segment $Q(r_j, r_{j+1} - r_j)$ the condition (4) is fulfilled, the desired function u_j does exist.

A similar statement applies for the segment $Q(2^j, 2^j)$, where $-\infty < j < +\infty$ is true if

$$\int_{Q(2^j,2^j)} V(x)^\gamma \, dx > b_1 k 2^{j(1-2m\gamma)}. \tag{6}$$

We have

$$\begin{aligned} b_0 k &\le \int_{\mathbf{R}} V(x)^\gamma |x|^{2m\gamma-1}(1 + |\ln|x||)^{-1-\varepsilon} \, dx \\ &= \sum_{j=-\infty}^{\infty} \int_{Q(2^j,2^j)} V(x)^\gamma |x|^{2m\gamma-1}(1 + |\ln|x||)^{-1-\varepsilon} \, dx \\ &\le \sum_{j=-\infty}^{\infty} 2^{j(2m\gamma-1)}(1 + |j|\ln 2)^{-1-\varepsilon)} \int_{Q(2^j,2^j)} V(x)^\gamma \, dx. \end{aligned}$$

Hence it follows that there exists a j_0 such that

$$2^{j_0(1-2m\gamma)} \int_{Q(2^{j_0},2^{j_0})} V(x)^\gamma \, dx > b_1 k,$$

where $b_1 = b_0 / \sum (1 + |j|\ln 2)^{-1-\varepsilon}$. Therefore, for the segment $Q(2^{j_0}, 2^{j_0})$ the condition (6) is fulfilled and the theorem is proved. □

8.3 Some inequalities and embedding theorems

In this section we state a number of inequalities that play an essential role in the proofs of the main results below. These inequalities quite naturally imply embedding theorems, which are of independent interest.

Lemma 12. *Let $n > 2m$ and $1 \le p \le n/(n-2m)$. Then there exists a constant $a = a(n,m,p)$ such that*

$$\left[\int_{\mathbf{R}^n} |u(x)|^{2p} |x|^{-n+p(n-2m)} \, dx\right]^{1/p} \le a \int_{\mathbf{R}^n} \sum_{|\alpha|=m} |D^\alpha u(x)|^2 \, dx \tag{7}$$

for all functions $u \in C_0^\infty(\mathbf{R}^n)$.

Proof. By the Hölder inequality

$$\int_{\mathbf{R}^n} |u(x)|^{2p}|x|^{-n+p(n-2m)}\,dx$$
$$\leq \left[\int_{\mathbf{R}^n} |u(x)|^2|x|^{-2m}\,dx\right]^{(n-pn)/2m+p}\left[\int_{\mathbf{R}^n} |u(x)|^{2n/(n-2m)}\,dx\right]^{(n-2m)(p-1)/2m}.$$

Therefore, it is sufficient to verify that

$$[\int_{\mathbf{R}^n} |u(x)|^{2n/(n-2m)}\,dx]^{(n-2m)/n} \leq a_1 \int_{\mathbf{R}^n} \sum_{|\alpha|=m} |D^\alpha u(x)|^2\,dx, \tag{8}$$

$$\int_{\mathbf{R}^n} |u(x)|^2|x|^{-2m}\,dx \leq a_2 \int_{\mathbf{R}^n} \sum_{|\alpha|=m} |D^\alpha u(x)|^2\,dx. \tag{9}$$

Inequality (8) follows from Theorem 57 of Chapter 2. In order to prove the second estimate we will use the spherical coordinates r, ω. Let $u(x) = \sum R_j(r)\varphi_j(\omega)$ be the series in spherical functions. Substitute $r = e^t$, $T_j(t) = R_j(r)r^{n/2-m}$. Inequality (9) will take the form

$$\sum_{j=0}^{\infty} \int |T_j(t)|^2\,dt \leq a_2 \sum_{j=0}^{\infty} \int L_j(D_t)T_j(t)\overline{T_j(t)}\,dt.$$

Inequality (7) follows now from Lemma 6, applied to the case $p = 1, n > 2m$. □

Lemma 13. *Let $n < 2m$ and n be odd. If $p \geq 1$, then there exists a constant $C = C(m,n,p)$ such that*

$$(\int_{\mathbf{R}^n} |u(x)|^{2p}|x|^{-n+p(n-2m)}\,dx)^{1/p} \leq C \int_{\mathbf{R}^n} \sum_{|\alpha|=m} |D^\alpha u(x)|^2\,dx \tag{10}$$

for all functions $u \in C_0^\infty(\mathbf{R}^n)$ such that $D^\alpha u(0) = 0$ if $|\alpha| < m - n/2$.

Proof. We may assume that u has real values. Let us pass to spherical coordinates (r, ω), where $0 < r < \infty$ and $|\omega| = 1$. Let $r = e^t$ and $u(t,x) = v(t,\omega)e^{(m-n/2)t}$. The desired inequality takes the form

$$(\int_M |v(t,\omega)|^{2p}\,dt\,d\omega)^{1/p} \leq C_p \int_M L(D_t, D_\omega)v(t,\omega)\overline{v(t,\omega)}\,dt\,d\omega,$$

where L is a differential operator of order $2m$ with constant coefficients and $M = \mathbf{R} \times \{\omega : |\omega| = 1\}$.

Let $v(t,\omega) = \sum T_k(t)\varphi_k(\omega)$, where $T_k(t)\varphi_k(\omega)$ is the orthogonal projection in $L_2(\mathbf{R}^n)$ of the function v on the eigenspace Π_k of the Beltrami-Laplace operator $L_\omega(D_\omega)$ on the unit sphere, corresponding to the eigenvalue $-\mu_k$,

$$\int_{|\omega|=1} \varphi_k^2(\omega)\,d\omega = 1.$$

It is well known (see [Mir]) that if the eigenvalues are enumerated so that $0 = \mu_0 < \mu_1 < \ldots$, then $\mu_k = k(k+n-2) \geq k^2$ and $\dim \Pi_k \leq Ck^{n-2}$. Moreover, $|\varphi_k(\omega)| \leq Ck^{(n-2)/2}$, $k = 1, 2, \ldots$. Therefore for all $k \geq 1$

$$(\int_{|\omega|=1} |\varphi_k(\omega)|^{2p} \, d\omega)^{1/p} \leq [\max |\varphi_k|^{2p-2} \int_{|\omega|=1} |\varphi_k^2(\omega)| d\omega]^{1/p} \leq C_1 k^{(n-2)/q}.$$

Note that

$$(\int_M |v(t,\omega)|^{2p} \, dt \, d\omega)^{1/p} = [\int_M |\sum_{k=0}^{\infty} T_k(t)\varphi_k(\omega)|^{2p} dt d\omega]^{1/p}$$
$$\leq [\sum_{k=0}^{\infty} (\int_M |T_k(t)\varphi_k(\omega)|^{2p} \, dt \, d\omega)^{1/(2p)}]^2.$$

By Lemma 6, for $k \geq 0$

$$(\int_{\mathbf{R}} |T_k(t)|^{2p} \, dt)^{1/p} \leq a_p k^{1/q-2m} (-1)^m \int_{\mathbf{R}} L_k(D_t) T_k(t) \overline{T_k(t)} \, dt,$$

where q is defined by the equality $p^{-1} + q^{-1} = 1$. Hence

$$\begin{aligned}
&(\int_M |v(t,\omega)|^{2p} \, dt \, d\omega)^{1/p} \\
&\leq [\sum_{k=0}^{\infty} (\int_M |T_k(t)\varphi_k(\omega)|^{2p} \, dt \, d\omega)^{1/2p}]^2 \\
&\leq C[\sum_{k=0}^{\infty} (1+k)^{(n-2)/2q} (\int_{\mathbf{R}} |T_k(t)|^{2p} \, dt)^{1/2p}]^2 \\
&\leq Ca_p[\sum_{k=0}^{\infty} (1+k)^{(n-2)/2q+1/2q-m} ((-1)^m \int_{\mathbf{R}} L_k(D_t) T_k(t) \overline{T_k(t)} \, dt)^{1/2p}]^2 \\
&\leq Ca_p \sum_{k=0}^{\infty} (1+k)^{(n-1)/q-2m} \sum_{k=0}^{\infty} (-1)^m \int_{\mathbf{R}} L_k(D_t) T_k(t) \overline{T_k(t)} \, dt \\
&\leq C_1 \int_{\mathbf{R}^n} \sum_{|\alpha| \leq m} |D^\alpha u(x)|^2 \, dx.
\end{aligned}$$

Here we have used the fact that $(n-1)/q - 2m < -1$, so that the series $\sum_{k=0}^{\infty} (1+k)^{(n-1)/q-2m}$ converges. The proof is complete. □

Lemma 14. *Let $n \leq 2m$, and n be even. If $p \geq 1$, then there exists a constant $C = C(m,n,p)$ such that*

$$(\int_{\mathbf{R}^n} |u(x)|^{2p} |x|^{-n+p(n-2m)} (1 + |\ln|x||)^{p-1-2mp} \, dx)^{1/p} \leq C \int_{\mathbf{R}^n} \sum_{|\alpha|=m} |D^\alpha u|^2 \, dx \tag{11}$$

for all functions $u \in C_0^\infty(\mathbf{R}^n)$ *such that* $D^\alpha u(0) = 0$ *for* $|\alpha| \le m - n/2$ *and*

$$\int_{|\omega|=1} \frac{\partial^j u(1,\omega)}{\partial r^j} \varphi_k(\omega)\, d\omega = 0$$

for $j = 0, 1, \dots, m-1$ *and all* k *such that* $|\mu_k| \le (m - n/2)(m + n/2 - 2)$.

Proof. The main idea of the proof is the same as for Lemma 13. After the change of variables $r = e^t$, $v = ur^{n/2-m}$, one can see that it is sufficient to verify the inequality:

$$(\int_M |v(t,\omega)|^{2p}(1+|t|)^{p-1-2mp}\, dt\, d\omega)^{1/p} \le C_p \int_M L(D_t, D_\omega)v(t,\omega)\overline{v(t,\omega)}\, dt\, d\omega,$$

where $M = \mathbf{R} \times \{\omega :\ |\omega| = 1\}$. We can represent v as the sum of the series

$$v(t,\omega) = \sum_{k=0}^\infty T_k(t)\varphi_k(\omega),$$

where $T_k(t)\varphi_k(\omega)$ is the orthogonal projection in $L_2(\mathbf{R}^n)$ of the function v on the eigenspace Π_k of the Beltrami-Laplace operator $L_\omega(D_\omega)$, corresponding to the eigenvalue $-\mu_k$, and $\int \varphi_k^2(\omega)\, d\omega = 1$.

Observe that

$$[\int_{\mathbf{R}^n} |u(x)|^{2p}|x|^{-n+p(n-2m)}(1+|\ln|x||)^{p-1-2mp}\, dx]^{1/p}$$

$$= [(\int_M |v(t,\omega)|^{2p}(1+|t|)^{p-1-2mp}\, dt\, d\omega)]^{1/p}$$

$$= [\int_M |\sum_{k=0}^\infty T_k(t)\varphi_k(\omega)|^{2p}(1+|t|)^{p-1-2mp}\, dt\, d\omega]^{1/p}$$

$$\le [\sum_{k=0}^\infty (\int_M |T_k(t)\varphi_k(\omega)|^{2p}(1+|t|)^{p-1-2mp}\, dt\, d\omega)^{1/2p}]^2$$

$$\le 2[\sum_{k=0}^{m-n/2} (\int_M |T_k(t)\varphi_k(\omega)|^{2p}(1+|t|)^{p-1-2mp}\, dt\, d\omega)^{1/2p}]^2$$

$$+ 2[\sum_{k=m-n/2+1}^\infty (\int_M |T_k(t)\varphi_k(\omega)|^{2p}(1+|t|)^{p-1-2mp}\, dt\, d\omega)^{1/(2p)}]^2.$$

By Lemma 8, if $T_k(0) = T_k'(0) = \cdots = T_k^{(m-1)}(0) = 0$, then

$$\begin{aligned}
&[(\int_M |T_k(t)\varphi_k(\omega)|^{2p}(1+|t|)^{p-1-2mp}\,dt\,d\omega)^{1/2p}]^2 \\
&\le C[\int_{\mathbf{R}} |T_k(t)|^{2p}(1+|t|)^{p-1-2mp}\,dt]^{1/p} \\
&\le C_1 \int_{\mathbf{R}} L_k(D_t)T_k(t)\overline{T_k(t)}\,dt \\
&\le C_2 \int_{\mathbf{R}^n} \sum_{|\alpha|=m} |D^\alpha u(x)|^2\,dx.
\end{aligned}$$

We use this estimate for all k such that $|\mu_k| \le (m-n/2)(m+n/2-2)$.

On the other hand, by Lemma 6,

$$\begin{aligned}
&2[\sum_{k\ge 1+m-n/2} (\int_M |T_k(t)\varphi_k(\omega)|^{2p}(1+|t|)^{p-1-2mp}\,dt\,d\omega)^{1/2p}]^2 \\
&\le 2[\sum_{k\ge 1+m-n/2} (\int_M |T_k(t)\varphi_k(\omega)|^{2p}\,dt\,d\omega)^{1/2p}]^2 \\
&\le C[\sum_{k\ge 1+m-n/2} k^{(n-2)/2q}(\int_{\mathbf{R}} |T_k(t)|^{2p}\,dt)^{1/2p}]^2 \\
&\le Ca_p[\sum_{k\ge 1+m-n/2} k^{(n-2)/q+1/2q-m}((-1)^m\int_{\mathbf{R}} L_k(D_t)T_k(t)\overline{T_k(t)}\,dt)^{1/2}]^2 \\
&\le Ca_p \sum_{k\ge 1+m-n/2} k^{(n-1)/q-2m} \sum_{k\ge 1+m-n/2} (-1)^m \int_{\mathbf{R}} L_k(D_t)T_k(t)\overline{T_k(t)}\,dt \\
&\le C_1 \int_{\mathbf{R}^n} \sum_{|\alpha|\le m} |D^\alpha u(x)|^2\,dx.
\end{aligned}$$

We have used here the fact that $(n-1)/q - 2m < -1$, so that the series $\sum k^{(n-1)/q-2m}$ converges. □

Lemma 15. *Let Q be a cube in $\mathbf{R}^n$ whose faces are parallel to the coordinates' planes and whose edges are equal to l. There exists a constant a depending on p, n, m only, such that for all functions $u \in C^\infty(Q)$, satisfying the relation*

$$\int_Q D^\alpha u(x)\,dx = 0 \quad \text{for} \quad |\alpha| < m, \tag{12}$$

the following is true:

1. *If $n > 2m$ and $p \leq n/(n-2m)$, then the inequality*

$$[\int_Q |u(x)|^{2p}|x|^{-n+p(n-2m)}\,dx]^{1/p} \leq a \int_Q \sum_{|\alpha|=m} |D^\alpha u(x)|^2\,dx \tag{13}$$

is valid.

2. *Let $n < 2m$, n be odd, and $p \geq 1$. If the point $x = 0$ is in Q, then let $D^\alpha u(0) = 0$ for $|\alpha| < m - n/2$. Then the inequality*

$$[\int_Q |u(x)|^{2p}|x|^{-n+p(n-2m)}\,dx]^{1/p} \leq a \int_Q \sum_{|\alpha|=m} |D^\alpha u(x)|^2\,dx \tag{14}$$

is true.

3. *Let $n \leq 2m$, n be even, and $p \geq 1$. If the point $x = 0$ is in Q, then let $D^\alpha u(0) = 0$ for $|\alpha| \leq m - n/2$. Let*

$$\int \frac{\partial^j (hu)(1,\omega)}{\partial r^j} \varphi_k(\omega)\,dS_\omega = 0$$

for $j = 0, 1, \ldots, m-1$; $k = 0, 1, \ldots, m - n/2$, where the function h will be defined below.

Then the inequality

$$[\int_Q |u(x)|^{2p}|x|^{-n+p(n-2m)}(1 + |\ln|x||)^{p-1-2mp}\,dx]^{1/p} \leq a \int_Q \sum_{|\alpha|=m} |D^\alpha u(x)|^2\,dx \tag{15}$$

is true.

Proof. Let $I(u, Q)$ be an integral on the left-hand side of one of the inequalities (13), (14), or (15), taken over the domain Q. Let us continue the function u on the cube Q' homothetic to the cube Q and having edge length $2l$. Here we can assume that

$$\int_{Q'} \sum_{|\alpha|\leq m} l^{2|\alpha|-2m}|D^\alpha u(x)|^2\,dx \leq c_0 \int_Q \sum_{|\alpha|\leq m} l^{2|\alpha|-2m}|D^\alpha u(x)|^2\,dx,$$

where the constant c_0 depends on m and n only. One can obtain that using the standard continuation operator for the unit cube and subsequently applying the dilatation $x' = lx$.

Let $h \in C_0^\infty(Q')$ be a nonnegative function equal to 1 in a neighborhood of the cube Q and such that

$$|D^\alpha h(x)| \leq Cl^{-|\alpha|} \text{ for } |\alpha| \leq m.$$

We have

$$\int_{Q'} \sum_{|\alpha|\le m} l^{2|\alpha|-2m} |D^\alpha(h(x)u(x))|^2\,dx \le c' \int_Q \sum_{|\alpha|\le m} l^{2|\alpha|-2m}|D^\alpha u(x)|^2\,dx.$$

Since by the Poincaré inequality (here we use the condition (12)!) the estimate

$$\int_Q \sum_{|\alpha|\le m} l^{2|\alpha|-2m}|D^\alpha u(x)|^2\,dx \le c'' \int_Q \sum_{|\alpha|=m} |D^\alpha u(x)|^2\,dx$$

is true, it is sufficient to prove that

$$I(hu, Q') \le a_3 \int_{Q'} \sum_{|\alpha|\le m} l^{2|\alpha|-2m}|D^\alpha(hu(x))|^2\,dx$$

(so that $a = a_1 c' c''$). The last inequality follows from the previous Lemmas 12, 13, and 14. □

Lemma 16. *Let $0 \le \gamma < 1$, $n > 2m$, $1 \le p \le n/(2h - n - n\gamma)$, and Q be a cube in $\mathbf{R}^n$. Then there exists a constant $C = C(\gamma, n, m, p)$ such that*

$$\int_Q |u(x)|^{2p}|x|^{-n+p(n-2m+n\gamma)}\,dx \le C\Big[\int_Q \Big(\sum_{|\alpha|=m} |D^\alpha u(x)|^2\Big)^{1/(1+\gamma)}\,dx\Big]^{p(1+\gamma)}$$

for all functions $u \in C^\infty(Q)$ such that

$$\int_Q D^\alpha u(x)\,dx = 0 \quad \text{for} \quad 0 \le |\alpha| \le m-1.$$

Proof. We have mainly to repeat the proof of Lemma 12. At first the desired inequality is proved for u from $C_0^\infty(\mathbf{R}^n)$; after this the same construction as in Lemma 15 is applied.

By the Sobolev theorem the inequality

$$\int_{\mathbf{R}^n} |u(x)|^{2n/(n+n\gamma-2m)}\,dx \le C\Big[\int_{\mathbf{R}^n} \Big(\sum_{|\alpha|=m} |D^\alpha u(x)|^2\Big)^{1/(1+\gamma)}\,dx\Big]^{(n+n\gamma)/(n+n\gamma-2m)} \tag{16}$$

is valid, if $u \in C_0^\infty(\mathbf{R}^n)$. On the other hand, we have the inequality

$$\int_{\mathbf{R}^n} |u(x)|^2|x|^{-2m}\,dx \le C \int_{\mathbf{R}^n} \sum_{|\alpha|=m} |D^\alpha u(x)|^2\,dx,$$

proved above, in Lemma 12.

Applying the interpolation one can deduce from the indicated inequalities the desired one.

It is also possible to obtain the result avoiding the interpolation. One can simply use the Hölder inequality (16) and the inequality

$$\int_{\mathbf{R}^n} |u(x)|^2 |x|^{n\gamma-2m}\, dx \le C\Big[\int_{\mathbf{R}^n} \Big(\sum_{|\alpha|=m} |D^\alpha u(x)|^2\Big)^{1/(1+\gamma)}\, dx\Big]^{1+\gamma}.$$

The proof is complete. □

8.4 Estimates of the number of points of the negative spectrum

The following lemma is a simplified variant of a theorem of Rozenblyum (cf. [Ro2]).

Lemma 17. *Let Q be any cube in $\mathbf{R}^n$ and $f(x)$ be a nonnegative measurable function in Q such that $\int_Q f(x)\, dx = 1$. Then for any $\varepsilon \in (0, 1/2)$ the cube Q can be covered by the union of closed cubes Q_j, $1 \le j \le M$, such that*

$$\int_{Q_j} f(x)\, dx \le (2^{n+1} - 1)\varepsilon, \quad M \le a_n(2^n - 1)\varepsilon^{-1},$$

where $a_n = \max(2^n - 2, 1)$ and each point of Q is covered by not more than a_n cubes.

Proof. Let at first $n = 1$. Moving from the left end a_0 of the segment Q we find a point a_1 such that the integral from f over $[a_0, a_1]$ is equal to 2ε. Again moving to the right we find a point a_2 such that the integral of f over $[a_1, a_2]$ is equal to 2ε and so on. It is evident that the number of intervals does not exceed $(2\varepsilon)^{-1} + 1$.

Let $n > 1$. Let us divide Q on 2^n cubes by planes through its center and parallel to its sides. Roughly speaking, we will take the smaller obtained cubes, the integrals of the function f over which are greater than ε, and divide them in the same way. However, we have to control the number of the cubes and for that we need a slightly more complicated procedure.

We shall denote by n_0 the number of cubes, the integrals over which are $\le \varepsilon$, and by n_1 the number of others. We will show that

$$n_0 \le n_1(2^n - 2).$$

This relation is true at the beginning, since then $n_0 = 0$ and $n_1 = 1$.

Let Q' be one of the cubes, for which $\int_{Q'} f(x)dx \ge \varepsilon$. If after dividing it in 2^n smaller cubes the integrals over each of them are $\le \varepsilon$, we annihilate the last partition. Then the numbers n_0 and n_1 will not be changed and $\int_{Q'} f(x)\, dx \le 2^n\varepsilon$.

If the division of the cube Q' produces at least two cubes with the integrals over them $\ge \varepsilon$, then we apply the same procedure on those cubes. Let then the number n_0 increase to n_0', then $n_0' \le n_0 + 2^n - 2$, while n_1 increases at least by 1: $n_1' \ge n_1 + 1$.

Therefore, we have

$$n_0' \le n_0 + 2^n - 2 \le (n_1 + 1)(2^n - 2) \le n_1'(2^n - 2),$$

and the relation between n_0 and n_1 is true again.

If after the partition there appears exactly one cube Q'', the integral over which is $\ge \varepsilon$, then we slightly change the procedure. If $\int_{Q''} f(x)\,dx < 2^n\varepsilon$, then we annihilate the last partition and have that

$$\int_{Q'} f(x)\,dx < (2^n + 2^n - 1)\varepsilon \le (2^{n+1} - 1)\varepsilon.$$

If $\int_{Q''} f(x)\,dx = A \ge 2^n\varepsilon$, then we will diminish it, leaving fixed its vertex common with the cube Q', until the integral over it becomes equal to $A - (2^n - 2)\varepsilon$, which is still $> \varepsilon$. At the same time we shall diminish the opposite cube to the same size and enlarge the other $2^n - 2$ cubes to the size equal to the difference of the sizes of Q' and Q'', keeping fixed their vertices common with those of Q'. Then for at least one of these bigger cubes the value of the integral over it $\ge \varepsilon$. Finally we obtain the system of overlapping cubes $Q_1'', \dots, Q_{2^n}''$, covering Q' and the values of at least two integrals over them are $\ge \varepsilon$. Therefore, the relation between n_0 and n_1 will be valid. Moreover, the integrals over each of these cubes except one are $\le (2^{n+1} - 4)\varepsilon$ and each point of Q' is covered by at most $2^n - 2$ smaller cubes. In what follows we will subdivide at most one of them, which does not intersect others, so the latter number will not increase.

We continue the process until the moment when the values of each integral of the covering are $\le (2^{n+1} - 1)\varepsilon$. Then the number of cubes in the final bisecting $M \le n_0 + n_1 \le n_1(2^n - 1)$. However, $n_1 \le (2^n - 2)\varepsilon^{-1}$. Therefore, $M \le (2^n - 2)(2^n - 1)\varepsilon^{-1}$. □

Theorem 18. *If $n > 2m$, then for $q \ge n/2m$ the following estimate:*

$$N \le C_{n,m,q} \int_{\mathbf{R}^n} V_+(x)^q |x|^{2mq-n}\,dx \tag{17}$$

is true.

Proof. We can assume without loss of generality that $V(x) \ge 0$, $V \in C_0^\infty(\mathbf{R}^n)$. Let the support of the function V be in a cube Q. Let us apply Lemma 17 with $f(x) = I^{-1}V(x)^q|x|^{2mq-n}$, where

$$I = \int_Q V(x)^q |x|^{2mq-n}\,dx$$

and the value of ε will be indicated later. We have

$$\int_Q V(x)|u(x)|^2 dx \le \sum_{i=1}^{M} \int_{Q_i} V(x)|u(x)|^2\,dx$$

$$\le (\varepsilon I)^{1/q} \sum_{i=1}^{M} [\int_{Q_i} |u(x)|^{2p}|x|^{-n+p(n-2m)}\,dx]^{1/p}.$$

By Lemma 15 there exists a constant $a = a(m, n, p)$ such that

$$[\int_{Q_i} |u(x)|^{2p}|x|^{-n+p(n-2m)}\,dx]^{1/p} \le a\int_{Q_i}\sum_{|\alpha|=m}|D^\alpha u(x)|^2\,dx,$$

if $\int_{Q_i} D^\alpha u(x)\,dx = 0$ for $|\alpha| \le m-1$, $i = 1,\dots,M$. Therefore

$$\int_{\mathbf{R}^n} V(x)|u(x)|^2\,dx \le a(\varepsilon 2^n I)^{1/q}\int_{Q_i}\sum_{|\alpha|=m}|D^\alpha u(x)|^2\,dx.$$

Therefore, our statement is true if $a(\varepsilon 2^n I)^{1/q} \le 1$. Set $\varepsilon = (2^n I a^q)^{-1}$. The common number of conditions for u is equal to $c_n\varepsilon^{-1} = c'_n I$. This argument shows also that $N = 0$, if

$$\int_{\mathbf{R}^n} V(x)^q|x|^{2mq-n}\,dx \le a_0^q a^{-q}. \qquad \square$$

Theorem 19. *If $n < 2m$ and n is odd, then for $q > 1$*

$$N \le C_{n,m,q}\int_{\mathbf{R}^n} V_+(x)^q|x|^{2mq-n}\,dx + K_{m,n}, \tag{18}$$

where $K_{m,n}$ is the number of derivatives of order α such that $0 \le |\alpha| < m - n/2$.

Proof. As usual, we assume that $V \in C_0^\infty(\mathbf{R}^n)$, $V(x) \ge 0$, and show that

$$\int_{\mathbf{R}^n} V(x)|u(x)|^2 dx \le a_0\int_{\mathbf{R}^n}\sum_{|\alpha|=m}|D^\alpha u(x)|^2\,dx \tag{19}$$

if a real-valued function u belongs to a subspace K_N in $C_0^\infty(\mathbf{R}^n)$ of codimension N.

Let the support of V be contained in a cube Q. As in the proof of Theorem 18, we apply Lemma 17 with

$$f(x) = I^{-1}V(x)^q|x|^{2mq-n}, \quad \text{where} \quad I = \int_{\mathbf{R}^n} V(x)^q|x|^{2mq-n}\,dx;$$

the value of ε will be indicated later. We have

$$\int V|u|^2 dx \le \sum_i\int_{Q_i} V|u|^2\,dx \le (\varepsilon 2^n I)^{1/q}\sum_i(\int_{Q_i}|u(x)|^{2p}|x|^{-n+p(n-2m)}\,dx)^{1/p}.$$

By Lemma 15, there exists a constant $a = a(n, m, p)$ such that

$$(\int_{Q_i}|u(x)|^{2p}|x|^{-n+p(n-2m)}\,dx)^{1/p} \le a\int_{Q_i}\sum_{|\alpha|=m}|D^\alpha u(x)|^2 dx$$

if (12) is valid and $D^\alpha u(0) = 0$ for $|\alpha| < m - n/2$. The inequality (19) follows from here, if ε is such that $a(\varepsilon 2^n I)^{1/q} = a_0$, i.e. if we take $\varepsilon = a_0^q a^{-q}2^{-n}I^{-1}$. Therefore the inequality (19) is proved for a subspace of functions u of the codimension $c_{m,n}\varepsilon^{-1} + K_{m,n} \le c'_{m,n}I + K_{m,n}$, where the constants $c_{m,n}$ and $c'_{m,n}$ depend on m and n only. Hence $N \le c'_{m,n}I + K_{m,n}$. $\square$

Theorem 20. *If $n \le 2m$ and n is even, then for $q > 1$*

$$N \le c_{n,m,q} \int_{\mathbf{R}^n} V_+(x)^q |x|^{2mq-n}(1 + |\ln|x||)^{2q-1}\, dx + K_{m,n},$$

where $K_{m,n}$ is the number of derivatives of order α such that $0 \le |\alpha| \le m - n/2$.

Proof. The proof is mainly a repetition of the proof of Theorem 19. The only difference is that instead of the inequality (14) the inequality (15) is used. □

Let $k(j)$ be defined by the relation $\mu_j = -k(k+n-2)$, where μ_j is the jth eigenvalue of the Laplace-Beltrami operator on the unit sphere. Let M_k be the set of $j \in \mathbf{N}$ such that $|\mu_j| \le k(k+n-2)$ and $M'_k = \mathbf{N} \setminus M_k$. Define integers $c_l^{(j)}$ by the formula

$$L_j(i\lambda) = \prod_{l=0}^{m-1} [\lambda^2 + (k - m + n/2 + 2l)^2] = \sum_{l=0}^{m-1} c_l^{(j)} \lambda^{2l}$$

and set

$$a_k = \min_{j \in M_k} c_0^{(j)}, \; b_k = \min_{j \in M_k} c_1^{(j)}, \; c_k = \min_{j \in M'_k} c_0^{(j)}.$$

Theorem 21. *If*

$$V(x) \le \frac{1}{r^{2m}} \min(c_k, a_k + \frac{b_k}{\ln^2 r}), \; r = |x|,$$

then $N \le \nu(k)$, where $\nu(k)$ is the number of elements of the set M_k.

Proof. Let us prove that

$$\int_{\mathbf{R}^n} V|u|^2\, dx \le a_0 \int_{\mathbf{R}^n} \sum_{|\alpha|=m} |D^\alpha u(x)|^2\, dx$$

for functions u from $F_{\nu(k)}$, where $F_{\nu(k)}$ is the subspace of the functions from $C_0^\infty(\mathbf{R}^n)$, satisfying the conditions

$$\int_{|x|=1} u(x)\varphi_j(x)\, dS_x = 0, \; j = 1, \dots, \nu(k), \tag{20}$$

and φ_j are eigenfunctions of the Beltrami-Laplace operator L_ω with the eigenvalues μ_j. In order to prove this we substitute $r = e^t$, $u = ve^{(m-n/2)t}$, so that the desired inequality will take the form

$$\int_{\mathbf{R}^n} V|v|^2 e^{2mt}\, dt\, d\omega \le \int_{\mathbf{R}^n} \sum_{|\alpha|=m} |P_\alpha(D_t, D_\omega)v|^2\, dt\, d\omega.$$

Let $v(t,\omega) = \sum \varphi_j(\omega)T_j(t)$. Replacing $V(t,\omega)$ by a majorant $w(t)$, depending on t only, we pass by Lemma 5, to the inequality

$$\int_{\mathbf{R}} w(t)|T_j(t)|^2 e^{2mt}\, dt \le \int_{\mathbf{R}} L_j(D_t)T_j(t)\overline{T_j(t)}\, dt.$$

This inequality results from the following one:

$$\int_{\mathbf{R}} w(t)|T_j(t)|^2 e^{2mt}\,dt \le \int_{\mathbf{R}} (c_0^{(j)}|T_j(t)|^2 + c_1^{(j)}|T_j'(t)|^2)\,dt.$$

If $j \in M_k$, then using the Hardy inequality we can take as w the function $e^{-2mt}(a_k+b_kt^{-2})$, assuming that $T_j(0)=0$, i.e. the conditions (20) hold. If $j \in M_k'$, then the inequality is true for $w(t)=e^{-2mt}c_k$. Therefore, the Theorem is proved if

$$V(x) \le |x|^{-2m}\min(c_k, a_k + b_k \ln^{-2}|x|). \qquad \square$$

8.5 Some generalizations

We consider here some generalizations of the preceding results: estimates for degenerated operators, estimates in unbounded domains, depending on their structure at infinity, and estimates of the number of eigenvalues that are not greater than a positive number.

Theorem 22. *Let*

$$\operatorname{Re}\int_{\mathbf{R}^n} L_0u\cdot\overline{u}\,dx \ge c_0\int_{\mathbf{R}^n}\mu(x)\sum_{|\alpha|=m}|D^\alpha u|^2\,dx \tag{21}$$

for $u \in H_0^m(\Omega)$. Let $0<\gamma<1$, $2m<n$. Then for any $q \ge n/(2m-n\gamma)$

$$N \le C_{n,m,q}\Big(\int_{\mathbf{R}^n}\mu(x)^{-1/\gamma}\,dx\Big)^{\gamma}\int_{\mathbf{R}^n} V_+(x)^q|x|^{q(2m-n\gamma)-n}\,dx.$$

Proof. The proof is a repetition of the proof of Theorem 18. However, here we use Lemma 16. Since by the Hölder inequality

$$\int_{\mathbf{R}^n}\Big(\sum_{|\alpha|=m}|D^\alpha u(x)|^2\Big)^{1/(1+\gamma)}\,dx$$

$$\le \Big(\int_{\mathbf{R}^n}\mu(x)^{-1/\gamma}\,dx\Big)^{\gamma/(1+\gamma)}\Big(\int_{\mathbf{R}^n}\mu(x)\sum_{|\alpha|=m}|D^\alpha u(x)|^2\,dx\Big)^{1/(1+\gamma)},$$

we can see that

$$(L_0u,u) \ge b_0\Big(\int_{\mathbf{R}^n}\mu^{-1/\gamma}\,dx\Big)^{-\gamma}\Big(\int_{\mathbf{R}^n}\Big(\sum_{|\alpha|=m}|D^\alpha u(x)|^2\Big)^{1/(1+\gamma)}\,dx\Big)^{1+\gamma},$$

where $b_0>0$. Now it remains to apply Lemma 16 in the same way as Lemma 15 has been applied in the proof of Theorem 18. $\square$

The following theorem improves the estimate in the case when the structure of Ω at infinity is known.

Theorem 23. *Assume that the operator L is elliptic and the domain Ω is such that the diameter of the intersection of Ω with the sphere $|x| = R$ is less than AR^α. If $n > 2m$, then for $q \geq n/2m$*

$$N \leq C_{n,m,q} \int_{\mathbf{R}^n} V_+(x)^q |x|^{(2mq-n)\alpha}\, dx.$$

Proof. We again repeat the proof of Theorem 18. However, now we will consider separately the cubes Q_j with diameter $\leq 2AR^\alpha$ and the others. For the cubes of the second group $u = 0$ on a set of measure bigger than half of the cube's volume. Therefore, the inequality (13) from Lemma 15 is valid for all these cubes without conditions (12) and it is sufficient to consider only cubes whose points x satisfy the inequality $|x - x_0^{(j)}| \leq 2A|x_0^{(j)}|^\alpha$, where $x_0^{(j)}$ is the center of the cube Q_j.

Let Q be such a cube, d be its diameter, and x_0 its center. We have

$$\int_Q V(x)|u(x)|^2\, dx \leq \Big(\int_Q |u^{2p}(x)||x - x_0|^{-\lambda p}\, dx\Big)^{1/p} \Big(\int_Q V^q(x)|x - x_0|^{\lambda q}\, dx\Big)^{1/q},$$

where $\lambda \geq 0$ and $p^{-1} + q^{-1} = 1$.

Considering the dimensions it is easy to see that under condition (12) and for $p < n/(n - 2m)$ the inequality

$$\Big(\int_Q |u(x)|^{2p}\, dx\Big)^{1/p} \leq C d^{2m-n+n/p} \int_Q \sum_{|\alpha|=m} |D^\alpha u(x)|^2\, dx$$

holds, where the constant C does not depend on d and u. If $\lambda = 2m - n/q$, then

$$\begin{aligned}\Big(\int_Q |u(x)|^{2p}|x - x_0|^{-\lambda p}\, dx\Big)^{1/p} &= \Big(\int_Q |u(x)|^{2p}|x - x_0|^{p(n-2m)-n}\, dx\Big)^{1/p} \\ &\leq d^{n-2m-n/p}\Big(\int_Q |u(x)|^{2p}\, dx\Big)^{1/p},\end{aligned}$$

and hence

$$\Big(\int_Q |u(x)|^{2p}|x - x_0|^{-\lambda p}\, dx\Big)^{1/p} \leq C \int_Q \sum_{|\alpha|=m} |D^\alpha u(x)|^2\, dx.$$

Therefore, if the condition (12) is fulfilled, then

$$\int_Q V(x)|u(x)|^2\, dx \leq a_0 \int_Q \sum_{|\alpha|=m} |D^\alpha u(x)|^2\, dx,$$

as long as

$$\left(\int_Q V^q(x)|x-x_0|^{\lambda q}\,dx\right)^{1/q} \le a_0 C^{-1}.$$

The total number of conditions imposed on u is equal to $N = Mc_m$, where M is the number of considered cubes and the inequality

$$\int_\Omega V(x)|u(x)|^2\,dx \le c_0 \int_\Omega \sum_{|\alpha|=m} |D^\alpha u(x)|^2\,dx$$

holds if

$$\int_\Omega V(x)^q (2A|x|^\alpha)^{\lambda q}\,dx = MC^{-q}a_0^q.$$

Therefore

$$N \le c_m C^q a_0^{-q}(2A)^{2mq-n} \int_\Omega V(x)^q |x|^{\alpha(2mq-n)}\,dx. \qquad \square$$

Theorem 24. *Let $x = (x', x'')$, where $x' = (x_1, \dots, x_k)$, $1 \le k \le n$, and $x'' = (x_{k+1}, \dots, x_n)$. Suppose that $\Omega \subset Q \times \mathbf{R}^{n-k}$, where Q is a bounded domain in $\mathbf{R}^k$. Suppose that an operator L be such that $L = L_1 + L_2$, where*

$$L_1 u = \sum_{\substack{|\alpha|\le m, |\beta|\le m\\ \alpha''=0,\beta''=0}} D^\alpha(b_{\alpha\beta}(x')D^\beta u),\; L_2 u = \sum_{\substack{|\alpha|\le m, |\beta|\le m\\ \alpha'=0,\beta'=0}} D^\alpha(c_{\alpha\beta}(x'')D^\beta u)$$

are elliptic operators in Q and $\mathbf{R}^{n-k}$, respectively. Let μ be the least eigenvalue of the operator L_1 in Q with zero Dirichlet conditions. Let $H = L - V(x)$ and N_μ be the number of the eigenvalues of the operator H that are smaller than μ.

If $n > 2m + k$, then N_μ is estimated in the following way:

$$N_\mu \le C_{n,m,q} \int_{\mathbf{R}^n} V_+(x)^q [|x|^{2mq-n} + |x''|^{2mq-n+k}]\,dx,$$

where q is any number, $q \ge n/(2m)$.

If $2m + k \ge n > 2m$ and $n - k$ is odd, then for $q \ge n/2m$

$$N_\mu \le C_{n,m,q}(1 + \int_{\mathbf{R}^n} V_+(x)^q [|x|^{2mq-n} + |x''|^{2mq-n+k}]\,dx).$$

If $2m + k \ge n > 2m$ and $n - k$ is even, then for $q \ge n/2m$

$$N_\mu \le C_{n,m,q} \int_{\mathbf{R}^n} V_+(x)^q [|x|^{2mq-n} + |x''|^{2mq-n+k}(1 + |\ln|x''||)^{2q-1}]\,dx.$$

If $n \le 2m$, n is odd, and k is even, then for $q > 1$

$$N_\mu \le C_{n,m,q}(1 + \int_{\mathbf{R}^n} V_+(x)^q [|x|^{2mq-n} + |x''|^{2mq-n+k}]\,dx).$$

If $n \le 2m$, n is odd, and k is odd, then for $q > 1$

$$N_\mu \le C_{n,m,q}\Big(1 + \int_{\mathbf{R}^n} V_+(x)^q \big[|x|^{2mq-n} + |x''|^{2mq-n+k}(1 + |\ln|x''||)^{2q-1}\big]\,dx\Big).$$

If $n \le 2m$, n is even, and k is odd, then for $q > 1$

$$N_\mu \le C_{n,m,q}\Big(1 + \int_{\mathbf{R}^n} V_+(x)^q \big[|x|^{2mq-n}(1 + |\ln|x''||)^{2q-1} + |x''|^{2mq-n+k}\big]\,dx\Big).$$

If $n \le 2m$, and n and k are even, then for $q > 1$

$$\begin{aligned} N_\mu \le C_{n,m,q}\Big(1 + \int_{\mathbf{R}^n} V_+(x)^q \big[&|x|^{2mq-n}(1 + |\ln|x||)^{2q-1} \\ &+ |x''|^{2mq-n+k}(1 + |\ln|x''||)^{2q-1}\big]\,dx\Big). \end{aligned}$$

Proof. Our aim is to find a subspace in $C_0^\infty(\Omega)$ of codimension N_μ of functions u such that the inequality

$$\mu \int_{\mathbf{R}^n} |u(x)|^2\,dx + \int_{\mathbf{R}^n} V(x)|u(x)|^2\,dx \le \int_{\mathbf{R}^n} \sum a_{\alpha\beta}(x) D^\alpha u(x) \overline{D^\beta u(x)}\,dx \tag{22}$$

is true. Represent u in the form

$$u = \varphi(x')u_1(x'') + u_2(x),$$

where $x' = (x_1, \ldots, x_k)$ and $x'' = (x_{k+1}, \ldots, x_n)$, φ is the first eigenfunction of the operator L_1, $\varphi \in H_0^m(Q)$, $\int |\varphi(x')|^2\,dx' = 1$, and

$$\int_Q \overline{u_2(x)}\varphi(x')\,dx' = 0 \qquad \forall x'' \in \mathbf{R}^{n-1}. \tag{23}$$

In order to prove the inequality (22) it is sufficient to show that

$$\begin{aligned} \mu \int_{\mathbf{R}^{n-k}} |u_1(x'')|^2\,dx'' + \mu \int_{\mathbf{R}^n} |u_2(x)|^2\,dx + \int_{\mathbf{R}^n} V_+(x)|u(x)|^2\,dx \\ \le \mu \int_{\mathbf{R}^{n-k}} |u_1(x'')|^2\,dx'' + \int_{\mathbf{R}^{n-k}} L_2 u_1 \overline{u}_1\,dx'' + \int_{\mathbf{R}^n} L u_2 \overline{u}_2\,dx, \end{aligned}$$

since $L(\varphi u_1) = \mu\varphi u_1 + \varphi L_2 u_1$ and it follows from (23) that

$$\begin{aligned} \int_{\mathbf{R}^n} L(\varphi u_1)\overline{u}_2\,dx = \mu \int_{\mathbf{R}^{n-k}} u_1(x'')\,dx'' \int_{\mathbf{R}^k} \varphi\overline{u_2(x)}\,dx' \\ + \int_{\mathbf{R}^{n-k}} L_2 u_1(x'')\,dx'' \int_{\mathbf{R}^k} \varphi\overline{u_2(x)}\,dx' = 0. \end{aligned}$$

Since

$$\int_{\mathbf{R}^n} V_+(x)|u(x')|^2\,dx' \le 2\int_{\mathbf{R}^n} V_+(x)|\varphi(x')|^2|u_1|^2\,dx + 2\int_{\mathbf{R}^n} V_+(x)|u_2|^2\,dx,$$

and

$$\int_{\mathbf{R}^k} L_1 u_2\overline{u}_2\,dx' \ge \mu_1 \int_{\mathbf{R}^k} |u_2|^2\,dx', \text{ where } \mu_1 > \mu,$$

we have to verify two inequalities:

$$2\int_{\mathbf{R}^{n-k}} V''(x'')|u_1(x'')|^2\,dx'' \le \int_{\mathbf{R}^{n-k}} L_2 u_1(x'')\overline{u_1(x'')}\,dx'', \tag{24}$$

$$2\int_{\mathbf{R}^n} V_+(x)|u_2(x)|^2\,dx \le (\mu_1-\mu)/\mu_1 \int_{\mathbf{R}^n} Lu_2\overline{u_2}\,dx. \tag{25}$$

Here $V''(x'') = \int_Q V_+(x)|\varphi(x')|^2\,dx'$. By Theorem 18 the inequality (24) is fulfilled if $q \ge (n-k)/2m$, $n > k+2m$, and u_1 is in the appropriate subspace in $C_0^\infty(\mathbf{R}^{n-k})$ of codimension

$$N_1 \le C_{n,q}\int_{\mathbf{R}^{n-k}} \Big(\int_Q V_+(x)|\varphi(x')|^2\,dx'\Big)^q |x''|^{2mq-n+k}\,dx''.$$

The inequality (25) is true if u_2 belongs to the appropriate subspace in $C_0^\infty(\Omega)$ of codimension

$$N_2 \le C_{n,q}\int_{\mathbf{R}^n} V_+(x)^q|x|^{2mq-n}\,dx.$$

Since $N_\mu \le N_1 + N_2$, Theorem 24 is proved for $n > k+2m$.

In order to prove it in the other cases we need only test the inequalities (24) and (25) using Theorems 19 and 20 instead of Theorem 18. □

The following is a variant of Theorem 22 for the Schrödinger operator in a strip.

Theorem 25. *Let $Q = (0,1)\times\mathbf{R}^{n-1}$ and $\Omega \subseteq Q$. Let $m = 1$ and $n \ge 4$. Let $L = -\Delta - V(x)$, and let N_1 be the number of the eigenvalues of L, not exceeding π^2. Then for all $q \ge (n-1)/2$ the inequalities*

$$N_1 \le C_{q,n}\int_Q V_+(x)^q\big(|x|^{2q-n} + |x'|^{2q-n+1}\big)\,dx,$$

are fulfilled, where $x' = (x_2,\dots,x_n)$.

Remark 26. One can prove theorems similar to Theorems 22 and 23 for $n \le 2m$, as was done for Theorem 24. Depending on the parity of n one obtains statements similar to Theorem 19 or 20.

8.6 Lower estimates for the number N

This section contains two theorems similar to Theorems 10 and 11: one gives the conditions that are sufficient for the negative spectrum to contain infinitely many points, the second gives lower estimates for the number of points of the negative spectrum. In Theorems 27 and 28 we assume that $|a_{\alpha\beta}(x)| \le C$.

The following theorem gives conditions sufficient for $N = \infty$.

Theorem 27. *Let $V(x) \ge 0$. Assume that there exists an infinite sequence of disjoint cubes $Q_j = \{x : \ |x_k - M_k^{(j)}| < R_j, \ k = 1, \dots, n\}$ such that $2R_j \le |M^{(j)}| \le 4R_j$ and*

$$\int_{Q'_j} V(x)|x|^{2m-n}\, dx \ge a_0,$$

where $Q'_j = \{x : \ |x_k - M_k^{(j)}| < R_j/2, \ k = 1, \dots, n\}$ and a_0 is the appropriate constant indicated below. Let the operator L_0 have the form

$$L_0 u = \sum_{|\alpha|=\beta|=m} D^\alpha (a_{\alpha\beta}(x) D^\beta u(x)).$$

Then the spectrum of the operator $L = L_0 - V(x)$ contains an infinite set of points on $(-\infty, 0)$.

Proof. We can assume that the centers of cubes Q_j and their sizes tend to ∞ or to 0 as $j \to \infty$. Let $h_0 \in C_0^\infty(Q_0)$, where $Q_0 = \{x : \ |x_k| < 1, \ k = 1, \dots, n\}$ $h_0(x) = 1$ for x from the cube $Q'_0 = \{x : \ |x_k| < 1/2, \ k = 1, \dots, n\}$. Set $h_j(x) = h_0((x - M^{(j)})/R_j)$. Then $|D^\alpha h_j(x)| \le C_\alpha R_j^{-|\alpha|}$ for $|\alpha| \le m$. Let $\varphi_j(x) = h_j(x)|x|^{m-n/2}$. We have

$$(L_0\varphi_j, \varphi_j) \le C_1 \int_{Q_j} \sum_{|\alpha|\le m} |D^\alpha \varphi_j|^2\, dy \le c,$$

where the constant c does not depend on j and on the size of Q_j. On the other hand,

$$\int V(x)\varphi_j^2(x)\, dx \ge \int V(x) h_j^2(x) |x|^{2m-n}\, dx \ge a_0 > 0.$$

The relation $(L\varphi_j, \varphi_j) < 0$ is fulfilled if $a_0 > c$. Thus if a_0 is such that this inequality is true, we have found a sequence of functions φ_j with disjoint supports such that $(L\varphi_j, \varphi_j) < 0$. □

Theorem 28. *Let $V(x) \ge 0$ and $2\gamma m < n$. There exists a positive constant b_0 such that if*

$$\int_{\mathbf{R}^n} V(x)^\gamma |x|^{2m\gamma - n} (1 + |\ln|x||)^{-1-\varepsilon}\, dx \ge b_0 k$$

for some positive $\varepsilon > 0$ and natural number k, then $N \ge k$.

Proof. Let $Q(R,d)$ be the domain in $\mathbf{R}^n$, consisting of the points x, for which $R < |x| < R+d$. Let us prove first that if

$$\int_{Q(R,d)} V(x)^\gamma \, dx \geq b_1 d^{1-2m\gamma}(R+d)^{n-1},$$

then there exists a function $u(x) \in H_0^m(Q(R,d))$, for which $(Lu,u) < 0$, so that $N \geq 1$. Set

$$u(x) = ((r-R)(R+d-r))^{m-1/2+\varepsilon} \quad \text{for} \quad R < r < R+d,$$

where $r = |x|$, and $u(x) = 0$ for $r < R$ and for $r > R+d$. It is obvious that

$$\int \sum_{|\alpha|=m} |D^\alpha u(x)|^2 \, dx \leq C(R+d)^{n-1} d^{2m-1+4\varepsilon}.$$

On the other hand, from the Hölder inequality it follows that

$$\int V(x)^\gamma \, dx \leq \Big(\int V(x)u(x)^2 \, dx\Big)^\gamma \Big(\int u(x)^{-2\gamma/(1-\gamma)} \, dx\Big)^{1-\gamma},$$

so that

$$\int V(x)u(x)^2 \, dx \geq \Big(\int V(x)^\gamma \, dx\Big)^{1/\gamma} \Big(\int u(x)^{-2\gamma/(1-\gamma)} \, dx\Big)^{1-1/\gamma}.$$

Note that

$$\int u(x)^{-2\gamma/(1-\gamma)} \, dx \leq C_1(R+d)^{n-1} d^{1-2\gamma(2m-1+2\varepsilon)/(1-\gamma)}.$$

Therefore

$$\int V(x)u(x)^2 \, dx \geq C_2\Big(\int V(x)^\gamma \, dx\Big)^{1/\gamma}(R+d)^{(n-1)(1-1/\gamma)} d^{1-1/\gamma+2(2m-1+2\varepsilon)},$$

and $(Lu,u) < 0$, if $C_2(\int V(x)^\gamma \, dx)^{1/\gamma}(R+d)^{(n-1)(1-1/\gamma)} d^{1-1/\gamma+2(2m-1+2\varepsilon)} > c_0 C(R+d)^{n-1} d^{2m-1+4\varepsilon}$, i.e. if

$$\int_{Q(R,d)} V(x)^\gamma \, dx > b_1(R+d)^{n-1} d^{1-2\gamma m}. \tag{26}$$

Let us show that if

$$\int_{Q(1,1)} V(x)^\gamma \, dx > b_1 k 2^{n-1},$$

then there exist k functions $u_1, \ldots, u_k$ from $W_{2,0}^m(Q(1,1))$ with nonintersecting supports, for which $(Lu_j, u_j) < 0$. For this we will find on the segment $(1,2)$ $k-1$ points $r_1, \ldots, r_{k-1}$ such that

$$1 = r_0 < r_1 < \cdots < r_{k-1} < r_k = 2$$

and

$$\int_{Q(r_j, r_{j+1}-r_j)} V(x)^\gamma \, dx > b_1 2^{n-1} \quad \text{for } j = 0, 1, \ldots, k-1.$$

Then in each layer $Q(r_j, r_{j+1} - r_j)$ the condition (26) is fulfilled and thus the desired function u_j does exist.

A similar statement for the layer $Q(2^j, 2^j)$, where $-\infty < j < +\infty$, is true if

$$\int_{Q(2^j,2^j)} V(x)^\gamma \, dx > b_1 k 2^{n-1+j(n-2m\gamma)}. \tag{27}$$

Now assume that the condition of the theorem is fulfilled. Then

$$\begin{aligned} b_0 k &\le \int_{\mathbf{R}^n} V(x)^\gamma |x|^{2m\gamma-n}(1+|\ln|x||)^{-1-\varepsilon} \, dx \\ &= \sum_{j=-\infty}^{\infty} \int_{Q(2^j,2^j)} V(x)^\gamma |x|^{2m\gamma-n}(1+|\ln|x||)^{-1-\varepsilon} \, dx \\ &\le \sum_{j=-\infty}^{\infty} 2^{j(2m\gamma-n)}(1+|j|\ln 2)^{-1-\varepsilon} \int_{Q(2^j,2^j)} V(x)^\gamma \, dx. \end{aligned}$$

It is easy to see that there exists a j_0 such that

$$2^{j_0(2m\gamma-n)} \int_{Q(2^{j_0},2^{j_0})} V(x)^\gamma \, dx > b_2 k,$$

where $b_2 = b_0 / \sum (1+|j|\ln 2)^{-1-\varepsilon}$.

Therefore, if

$$b_1 = b_0 2^{1-n} / \sum (1+|j|\ln 2)^{-1-\varepsilon},$$

then for the layer $Q(2^{j_0}, 2^{j_0})$ the condition (27) is fulfilled and the proof is complete. □

8.7 Other results

In Fefferman's work [Fef] the following estimate for the first negative eigenvalue of the operator $L = -\Delta - V(x)$ was obtained:

$$c \cdot \sup_{x,\delta} [Av_{B(x,\delta)} V - C\delta^{-2}] \le -\lambda_1 \le C \cdot \sup_{x,\delta} [(Av_{B(x,\delta)} V^p)^{1/p} - c\delta^{-2}],$$

where $B(x, \delta)$ is the ball of radius δ with its center at the point x and

$$Av_{B(x,\delta)} f(x) = \delta^{-n} \int_{B(x,\delta)} f(x) \, dx,$$

$1 < p < \infty$, and the constants c and C depend on n and p only.

This estimate was improved in Schechter [Sche2] to the form

$$\sup_{x,\delta}[c \cdot Av_{B(x,\delta)}V - \delta^{-2}] \le -\lambda_1 \le \sup_{x,\delta}[C \cdot (Av_{B(x,\delta)}V^p)^{1/p} - \delta^{-2}].$$

In Kerman-Sawyer's article [KerSa] the following estimate was obtained:

$$\sup_Q\{q^{-2} : F(V,Q) \ge C_1\} \le -\lambda_1 \le \sup_Q\{q^{-2} : F(V,Q) \ge C_2\},$$

where

$$F(V,Q) = \int_Q \int_Q V(x)V(y)|x-y|^{2-n}\,dx\,dy \Big/ \int_Q V(x)\,dx$$

and Q is a cube with side length q. (All this concerns only the case $m = 1$, $L = -\Delta$.)

Now we prove the following rather elementary theorem.

Theorem 29. *Let $m \ge 1$, $n > 2m$, and λ_1 be the first negative eigenvalue of the operator H. Then*

$$\begin{aligned}&\sup_Q\{\delta^{-2m} : \int_Q V(x)\,dx \ge C_1\delta^{n-2m}\} \le -\lambda_1\\ &\le sup_Q\{b_p\delta^{-2m} : \int_Q V(x)^p|x-x_0|^{2mp-n}\,dx \ge c_p\},\end{aligned} \tag{28}$$

where $p \ge n/2m$, and Q is a cube with its center at the point x_0 and side length δ.

Proof. It follows from the variational principle that

$$-\lambda_1 = \sup_u \frac{\int_{\mathbf{R}^n} V(x)|u(x)|^2\,dx - (L_0u,u)}{\int_{\mathbf{R}^n}|u(x)|^2\,dx}.$$

So we have $-\lambda_1 \ge C$, if C is a constant such that

$$\int_{\mathbf{R}^n} V(x)|u_0(x)|^2\,dx = (Lu_0,u_0) + C\int_{\mathbf{R}^n}|u_0(x)|^2\,dx$$

for some function $u_0 \in C_0^\infty(\mathbf{R}^n)$.

Let Q' be the cube concentrated with Q with its side length 2δ, and $u_0(x)$ a function of $C_0^\infty(Q')$, equal to 1 in Q. Then

$$\int_{\mathbf{R}^n} V(x)|u_0(x)|^2\,dx \ge C_1\delta^{n-2m},\ (Lu_0,u_0) \le C_2\delta^{n-2m},$$

$$\text{and} \quad \int_{\mathbf{R}^n}|u_0(x)|^2\,dx \ge C_3\delta^n.$$

The constants C_2 and C_3 do not depend on δ, and we can take C_1 to be equal to $(C_2+1)C_3$. Therefore, we get

$$-\lambda_1 \ge \delta^{-2m}(C_1 - C_2)/C_3 = \delta^{-2m}.$$

Thus the first inequality in (28) has been proved.

On the other hand, the variational principle gives the equality

$$-\lambda_1 = \inf\{\, \alpha > 0 : \int_{\mathbf{R}^n} V(x)|u(x)|^2\, dx \le (Lu, u) + \alpha \int_{\mathbf{R}^n} |u(x)|^2\, dx \,\}.$$

Therefore, $-\lambda_1 \le C_4$, if C_4 is a constant such that

$$\int_{\mathbf{R}^n} V(x)|u(x)|^2\, dx \le (Lu, u) + C_4 \int_{\mathbf{R}^n} |u(x)|^2\, dx, \quad u \in C_0^\infty(\mathbf{R}^n).$$

Let $q \ge n/2m$ and

$$\mu = \sup_Q\{\delta^{-2m} : \int_Q V(x)^p |x - x_0|^{2pm-n}\, dx \ge c_p\},$$

where Q is a cube with side length δ. It is clear that in fact the sign $\ge$ in the definition of μ can be substituted by the equals sign. Divide the space $\mathbf{R}^n$ on the cubes Q_j with nonintersecting inner parts and the size $\mu^{-1/2m}$. From the definition of the number μ it follows that

$$\int_{Q_j} V(x)^p |x - x_j|^{2pm-n}\, dx \le c_p,$$

where x_j is the center of the cube Q_j. We have

$$\begin{aligned}\int_{\mathbf{R}^n} V(x)|u(x)|^2\, dx &= \sum_j \int_{Q_j} V(x)|u(x)|^2\, dx\\ &\le \sum_j \Big(\int_{Q_j} V(x)^p|x - x_j|^{2pm-n}\, dx\Big)^{1/p}\Big(\int_{Q_j} |u|^{2q}|x - x_j|^{-n+q(n-2m)}\, dx\Big)^{1/q}\\ &\le c_p^{1/p} \sum_j \Big(\int_{Q_j} |u(x)|^{2q}|x - x_j|^{-n+q(n-2m)}\, dx\Big)^{1/q}.\end{aligned}$$

Here $q = p/(p-1)$. If Q_0 is a unit cube in $\mathbf{R}^n$ with its center at the origin, then there exists a constant a_p such that

$$\Big(\int_{Q_0} |u(y)|^{2q}|y|^{-n+q(n-2m)}\, dy\Big)^{1/q} \le a_p \int_{Q_0} \Big(\sum_{|\alpha|=m} |D^\alpha u(y)|^2 + |u(y)|^2\Big)\, dy.$$

After the substitution $x = x_j + \delta y$ we get the inequality

$$\Big(\int_{Q_j} |u(x)|^{2q}|x-x_j|^{-n+q(n-2m)}\, dx\Big)^{1/q} \le a_p \int_{Q_j} \Big(\sum_{|\alpha|=m} |D^\alpha u(x)|^2 + \delta^{-2m}|u(x)|^2\Big)\, dx,$$

valid for all $u \in H_m(Q_j)$. Therefore, if $c_p^{1/p} = a_p^{-1}a_0$, then

$$\int_{\mathbf{R}^n} V(x)|u(x)|^2\, dx \le (Lu, u) + a_p\delta^{-2m} \int_{\mathbf{R}^n} |u(x)|^2\, dx, \quad u \in C_0^\infty(\mathbf{R}^n).$$

Therefore, we have proved that $-\lambda_1 \le a_p\delta^{-2m}$. □

Similar arguments imply the following generalization of our results in [EgKo7]. Let $V(x)$ be a real potential in $\mathbf{R}^n$ such that $V(x) \geq 0$ and

$$\int_{\mathbf{R}^n} V(x)|x|^{2mp-n}\,dx < \infty, \; n > 2m, \; p \geq n/2m.$$

Let Q be a cube such that

$$\int_{\mathbf{R}^n \setminus Q} V(x)|x|^{2mp-n}\,dx < a_p.$$

Using plane sections parallel to the coordinate planes and containing the center of the cube Q, we can divide Q on 2^n smaller cubes $Q_1, \ldots, Q_{2^n}$ and choose those of them for which

$$\int_{Q_j} V(x)|x|^{2mp-n}\,dx \geq a_p.$$

Now we can divide these cubes in the same way, choose the smaller cubes with big values of integrals and stop the process only when the values of all integrals, over obtained cubes are $\leq a_p$. Let K_α be the number of those obtained cubes whose side length is $\leq \alpha^{-1/2m}$.

Theorem 30. *Let $n > 2m$, $p \geq n/2m$, $\int_{\mathbf{R}^n} V(x)\,dx < \infty$, and N_α be the number of the eigenvalues of the operator H that are not greater than $-\alpha$. Then*

$$N_\alpha \leq CK_\alpha \quad \text{and} \quad \sum |\lambda_j| \leq C \sum q_j^{-2m},$$

where q_j is the length of the side of the cube Q_j from the chosen set of K_α cubes, for which $q_j \leq \alpha^{-1/2m}$, and both the sums are taken over such indices j, and the constant C depends on n, m, p only.

Applying the results of Fefferman [Fef] and Kerman-Sawyer [KerSa] it is possible to prove the following.

Theorem 31. *Let $n > 2m$. There exist such positive constants c and C, depending on n only, that the following statements are valid:*

(a) Let $\alpha \geq 0$ and $Q_1, \ldots, Q_N$ be a set of cubes with the sides $\leq \alpha^{-1/2m}$, such that their doubles do not intersect and $F(V, Q_j) \geq C$. Then the operator H has at least N eigenvalues, not exceeding $-\alpha$.

(b) Inversely, let $\alpha \geq 0$ and H have at least CN eigenvalues that are $\leq -\alpha$. Then there are nonintersecting cubes $Q_1, \ldots, Q_N$ with its side lengths not exceeding $\alpha^{-1/2m}$, for which $F(V, Q_j) \geq c$, $j = 1, \ldots, N$.

Here

$$F(V, Q) = \frac{\int_Q \int_Q V(x)V(y)|x-y|^{2m-n}\,dx\,dy}{\int_Q V(x)\,dx}.$$

8.8 On moments of negative eigenvalues of an elliptic operator

Let Ω be a domain in $\mathbf{R}^n$, and let

$$L = \sum_{|\alpha|\le m, |\beta|\le m} D^\alpha a_{\alpha\beta}(x) D^\beta$$

be an elliptic symmetric positive operator of order $2m$ with measurable coefficients, considered with zero Dirichlet boundary conditions, such that

$$\int_\Omega \sum_{|\alpha|\le m, |\beta|\le m} a_{\alpha\beta}(x) D^\alpha u \overline{D^\beta u}\, dx \ge a_0 \int_\Omega \sum_{|\alpha|=m} |D^\alpha u(x)|^2\, dx,$$

where $a_0 = \text{const} > 0$ for all functions $u \in C_0^\infty(\Omega)$. Let $V(x)$ be a real-valued function defined in Ω.

We consider the problem: to estimate in terms of V the sum

$$S_\gamma = \sum_{\lambda_j<0} |\lambda_j|^\gamma,$$

where λ_j are negative eigenvalues of the operator $H = L - V$. We will assume for the operator H that the max-min principle is valid.

Such sums are very important for mathematical physics; they have been considered (for the Schrödinger operator) in the works of Ruelle [Ru], Lieb and Thirring [LTh], and others. In particular, Lieb and Thirring have obtained the estimate

$$S_\gamma \le C \int_{\mathbf{R}^n} V(x)^{\gamma+n/2}\, dx$$

in the case $m = 1$, $\gamma + n/2 > 1$. Note also Lieb's paper [L4], where sharp values of the constant C were indicated.

Lieb posed the question on the possibility of estimating the sum $S_{1/2}$ in terms of the integral $\int V(x)\, dx$ in the case $m = 1$, $n = 1$. For the operator

$$Ly = -y'' + V(x)y, \quad -\infty < x < \infty$$

with $V(x) \ge 0$ the estimate $S_\gamma \le C \int V(x)^{\gamma+1/2}\, dx$ is true for $\gamma > 1/2$. Theorem 32 below says that $S_{1/2}$ cannot be estimated by the integrals $\int V_+(x)^{1+r/2}|x - x_0|^r\, dx$ for any $r > 0$. Recently the estimate of the sum $S_{1/2}$ has been obtained by Weidl (see [W]. His result is stated below as Theorem 36. The proof is based on Lemmas 37 and 38, also from Weidl. We give slightly different proofs of his results.

The principal result of this section is the following theorem.

Theorem 32. *Let $r \ge 0$, $\gamma > 0$, $\gamma + (n + r)/2m > 1 + r/n$. Then*

$$S_\gamma \le C_{m,n,r,\gamma} \int_\Omega V_+(x)^{(n+r)/2m+\gamma} |x - x_0|^r\, dx, \tag{29}$$

where $V_+(x) = \max(V(x), 0)$, $x_0 \in \mathbf{R}^n$. *If* $1 + r/n \geq \gamma + (n+r)/2m > 1$ *and* $\int_\Omega V_+(x)^{(n+r)/2m+\gamma}|x - x_0|^r\,dx < \infty$, *then* $S_\gamma < \infty$. *However, in this case for any* $\varepsilon > 0$, $N > 0$ *there is a potential* $V(x)$ *such that* $\int_\Omega V_+(x)^{(n+r)/2m+\gamma}|x - x_0|^r\,dx < \varepsilon$, *and* $S_\gamma > N$.

Observe that (29) with $\gamma = 0$ gives our Theorem 15 for the counting function. Our proof is based on the following lemma, which is itself of interest.

Lemma 33. *Let* $\alpha > 0$, *and* N_α *be the number of eigenvalues that are smaller than* $-\alpha$. *Then*

$$N_\alpha \leq C_{m,n,q,r}\alpha^{-q+n/2m+r/2m}\int_\Omega V_\alpha(x)^q|x - x_0|^r\,dx + K_{m,n},$$

where $V_\alpha(x) = V(x)$ *if* $V(x) > \alpha$, *and* $V_\alpha(x) = 0$ *if* $V(x) \leq \alpha$; $q > 1$, $0 \leq r < 2mq - n$, *and* $K_{m,n}$ *is the number of different derivatives* D^β *of orders* β *such that* $0 \leq |\beta| \leq m - n/2$. *In particular,* $K_{m,n} = 0$ *if* $2m \leq n$.

Proof. Remark that N_α coincides with the codimension of the subspace of functions $u \in C_0^\infty(\Omega)$, for which the estimate

$$\int_\Omega V(x)|u(x)|^2\,dx \leq (Lu, u) + \alpha\int_\Omega |u(x)|^2\,dx$$

is valid. It is clear that this estimate is equivalent to the inequality

$$\int_\Omega V_\alpha(x)|u(x)|^2\,dx \leq (Lu, u) + \alpha\int_\Omega |u(x)|^2\,dx. \tag{30}$$

Let at first $n > 2m$. For a unit cube the estimate

$$\Big(\int_{Q_0}\frac{|u(x)|^{2p}}{|x - x_0|^s}\,dx\Big)^{1/p} \leq c_0\int_{Q_0}\Big(\sum_{|\beta|=m}|D^\beta u|^2 + |u|^2\Big)\,dx \tag{31}$$

is valid for all $u \in C^\infty(Q_0)$, if

$$0 \leq s < n,\ 1 \leq p < (n - s)/(n - 2m). \tag{32}$$

Indeed, by the Hölder inequality we have

$$\int_{Q_0}\frac{|u(x)|^{2p}}{|x - x_0|^s}\,dx \leq \Big(\int_{Q_0}|u(x)|^{2pp_1}\,dx\Big)^{1/p_1}\Big(\int_{Q_0}|x - x_0|^{-sq_1}\,dx\Big)^{1/q_1},$$

where $p_1^{-1} + q_1^{-1} = 1$. The latter integral is finite if $sq_1 < n$. On the other hand, the Sobolev estimate

$$\Big(\int_{Q_0}|u(x)|^{2pp_1}\,dx\Big)^{1/pp_1} \leq c_0\int_{Q_0}\Big(\sum_{|\beta|=m}|D^\beta u(x)|^2 + |u(x)|^2\Big)\,dx$$

is true if $2pp_1 \leq 2n/(n-2m)$. Therefore, such p_1 and q_1 do exist if the conditions (32) are satisfied.

Moreover, the estimate for a cube Q' with an edge b

$$\begin{aligned}(\int_{Q'} |u(x)|^{2p}|x-x_0|^{-s}\,dx)^{1/p} \\ \leq c_0 b^{2m-n/q-s/p} \int_{Q'} (\sum_{|\beta|=m} |D^\beta u(x)|^2 + b^{-2m}|u(x)|^2)\,dx,\end{aligned} \tag{33}$$

where $p^{-1}+q^{-1}=1$, holds under the conditions (32).

On the other hand, consider the class of smooth functions in Q_0, satisfying the conditions

$$\int_{Q_0} D^\beta u(x)\,dx = 0,\ 0 \leq |\beta| \leq m-1.$$

Repeating the above estimations, one can see that the estimate

$$(\int_{Q_0} \frac{|u(x)|^{2p}}{|x-x_0|^s}\,dx)^{1/p} \leq c_0 \int_{Q_0} \sum_{|\beta|=m} |D^\beta u|^2\,dx,$$

is true for functions of this class under the conditions (32). Moreover, the estimate

$$(\int_{Q'} \frac{|u(x)|^{2p}}{|x-x_0|^s}\,dx)^{1/p} \leq c_0 b^{2m-n/q-s/p} \int_{Q'} \sum_{|\beta|=m} |D^\beta u(x)|^2\,dx, \tag{34}$$

where $p^{-1}+q^{-1}=1$, is valid for functions of this class in a cube Q' with its edge equal to b under the conditions (32).

Let $\varepsilon > 0$ be a positive number, which will be indicated later. By Lemma 17, there is a covering of the support K of V (which we can assume to be compact) by a finite set of cubes $Q_1, \dots, Q_N$ with their edges parallel to the coordinate axes such that each point of K is covered by at most c_n cubes,

$$\int_{Q_j} V_\alpha(x)^q |x-x_0|^{s(q-1)}\,dx \leq \varepsilon |\alpha|^{q-n/2m-s(q-1)/2m} I, \tag{35}$$

where

$$I = \alpha^{-q+n/2m+s(q-1)/2m} \int_\Omega V_\alpha(x)^q |x-x_0|^{s(q-1)}\,dx$$

and $N \leq c_n \varepsilon^{-1}$. Let b_j be the edge of the cube Q_j.

If $2|\alpha|^{-1/2m} > b_j \geq |\alpha|^{-1/2m}$, then $b_j^{-2m} \leq |\alpha|$ and we have

$$\begin{aligned}(\int_{Q_j} |u(x)|^{2p}|x-x_0|^{-s}\,dx)^{1/p} \\ \leq c_0 |\alpha|^{-1+n/2mq+s/2mp} \int_{Q_j} (\sum_{|\beta|=m} |D^\beta u(x)|^2 + \alpha|u(x)|^2)\,dx.\end{aligned} \tag{36}$$

If $b_j \geq 2|\alpha|^{-1/2m}$, we can bisect the cube Q_j several times and represent it as a sum of intersecting cubes Q'_j having size b'_j such that

$$2|\alpha|^{-1/2m} > b'_j \geq |\alpha|^{-1/2m}.$$

We can write the above estimates (36) for each cube Q'_j and sum them over all such cubes. It gives (36) for the cube Q_j.

If $b_j \leq |\alpha|^{-1/2m}$ and the conditions (32) are fulfilled, then the estimate

$$(\int_{Q_j} \frac{|u(x)|^{2p}}{|x-x_0|^s}\,dx)^{1/p} \leq c_0|\alpha|^{-1+n/2mq+s/2mp} \int_{Q_j} \sum_{|\beta|=m} |D^\beta u(x)|^2\,dx \tag{37}$$

is true under the supplementary conditions

$$\int_{Q_j} D^\beta u(x)\,dx = 0,\ 0 \leq |\beta| \leq m-1. \tag{38}$$

Now using the estimates (35), (36), (37), we have

$$\begin{aligned}
&\int_\Omega V_\alpha(x)|u(x)|^2\,dx \\
&\leq \sum_{j=1}^N \int_{Q_j} V_\alpha(x)|u(x)|^2\,dx \\
&\leq \sum_{j=1}^N (\int_{Q_j} \frac{|u(x)|^{2p}}{|x-x_0|^s}\,dx)^{1/p} (\int_{Q_j} V_\alpha(x)^q |x-x_0|^{s(q-1)}\,dx)^{1/q} \\
&\leq c_0(\varepsilon I)^{1/q} \sum_{j=1}^N \int_{Q_j} (\sum_{|\beta|=m} |D^\beta u(x)|^2 + \alpha|u(x)|^2)\,dx \\
&\leq (Lu,u) + \alpha \int_\Omega |u(x)|^2\,dx,
\end{aligned}$$

if ε is such that $c_0(\varepsilon I)^{1/q} = \min(a_0, 1)$. Thus the estimate (30) holds, if the conditions (38) are fulfilled. Therefore, the number N_α does not exceed the number of the conditions (38), i.e.

$$N_\alpha \leq C\varepsilon^{-1} \leq C'I = C'\alpha^{-q+n/2m+s(q-1)/2m} \int_\Omega V_\alpha(x)^q |x-x_0|^{s(q-1)}\,dx,$$

where the constants C, C' depend on m, n, s. Setting $r = s(q-1)$, we obtain the result. Observe that

$$r = s(q-1) < n(q-1) \tag{39}$$

and r can be an arbitrary number such that $0 \leq r < nq - n$. It is easy to see that the conditions (32) are equivalent to the condition $0 \leq r < 2mq - n$.

If $n \le 2m$, then the estimates (31) and (34) are also true for $0 \le s < n$. However, if we assume that

$$D^\beta u(x_0) = 0 \quad \text{for} \quad 0 \le |\beta| \le m - n/2, \tag{40}$$

then applying the Hardy inequality (see Section 2.30) we can obtain the inequalities (33) and (34) for $0 \le s < n + (2m-n)p$, i.e. if $0 \le r < 2mq - n$. The proof stated above remains true, but the number N_α must be enlarged by $K_{m,n}$, equal to the number of conditions (40). Observe that, in particular, $K_{m,n} = 0$ if $2m \le n$, and $K_{m,n} = 1$ if $2m = n + 1$. □

Proof of Theorem 32. Observe that

$$S_\gamma = \sum_{\lambda_j < 0} |\lambda_j|^\gamma = -\int_0^\infty \alpha^\gamma \, dN_\alpha = \gamma \int_0^\infty \alpha^{\gamma-1}(N_\alpha - K_{m,n})\, d\alpha.$$

If $\gamma - q + n/2m + r/2m > 0$, then applying Lemma 33 we have

$$S_\gamma \le C \int_0^\infty \alpha^{\gamma-1-q+n/2m+r/2m} \int_\Omega V_\alpha(x)^q |x - x_0|^r \, dx\, d\alpha.$$

Taking into account the condition of Lemma 32, we can state that for any $\gamma > 0$, satisfying the inequality $\gamma + (n+r)/2m > 1 + r/n$, there is a $q > 1$ such that $\gamma + (n+r)/2m > q > 1 + r/n$ and

$$\begin{aligned} S_\gamma &\le C \int_\Omega |x - x_0|^r \, dx \int_0^{V_+(x)} \alpha^{\gamma-1-q+n/2m+r/2m} V_\alpha(x)^q d\alpha \\ &\le C_1 \int_\Omega |x - x_0|^r V_+(x)^{\gamma+n/2m+r/2m} \, dx. \end{aligned}$$

If $\gamma + (n+r)/2m \le 1 + r/n$, but $\gamma + (n+r)/2m > 1$, we cannot estimate the finite sum $\sum |\lambda_j|^\gamma$ corresponding to the first $K_{m,n}$ negative eigenvalues, but the sum of $|\lambda_j|^\gamma$ for other negative eigenvalues is estimated as above.

Let $(n+r)/2m + \gamma < 1 + r/n$. Set $V(x) = |x - x_0|^{-n-\kappa}$ if $\delta^2 < |x - x_0| < \delta$, $V(x) = \delta^{-2n-2\kappa}$, if $\delta^2 > |x - x_0|$, and $V(x) = 0$ if $\delta < |x - x_0|$. Here $\kappa > 0$ is such that $(1 + \kappa/n)((n+r)/2m + \gamma) < 1 + r/n$. Then $\int V(x)\, dx \to \infty$ as $\delta \to 0$, but

$$\int_{\mathbf{R}^n} |x - x_0|^r V_+(x)^{\gamma+n/2m+r/2m} \, dx < \varepsilon,$$

if δ is sufficiently small. Since the first negative eigenvalue λ_1 (with the maximal value of its module) can be defined as

$$\lambda_1 = \inf_u \int_\Omega \sum_{|\alpha| \le m, |\beta| \le m} (a_{\alpha\beta}(x) D^\alpha u \overline{D^\beta u} - V(x) u^2)\, dx \Big(\int u^2 \, dx\Big)^{-1},$$

we can see that λ_1 is less than

$$\int_\Omega \sum_{|\alpha|\le m, |\beta|\le m} (a_{\alpha\beta}(x) D^\alpha u_0 \overline{D^\beta u_0} - V u_0(x)^2)\, dx (\int_\Omega u_0^2\, dx)^{-1}$$
$$\le \int_\Omega \sum_{|\alpha|\le m, |\beta|\le m} a_{\alpha\beta}(x) D^\alpha u_0 \overline{D^\beta u_0}\, dx - \int_\Omega V(x)\, dx,$$

if u_0 is a function from C_0^∞ equal to 1 for $|x-x_0| < 1/4$ and such that $\int u_0^2\, dx = 1$. Therefore, $\lambda_1 \to -\infty$ and $S_\gamma \to +\infty$ as $\delta \to 0$.

If $(n+r)/2m + \gamma = 1 + r/n$, then we can get the same effect, setting $V(x) = |x-x_0|^{-n}/ln(1/|x-x_0|)$ for $\delta < |x-x_0| < 1/\ln(1/\delta)$.

The proof is complete. □

Theorem 34. *Let Ω be a domain in $\mathbf{R}^n$ such that its intersection with the sphere $|x-x_0| = R$ for $R > 1$ has a diameter not exceeding AR^σ with $\sigma < 1$. Let $r \ge 0,\ \gamma > 0,\ \gamma + (n+r)/2m > 1 + r/n$. Then*

$$S_\gamma \le C_{m,n,s,\gamma} \int_\Omega V_+(x)^{(n+r)/2m+\gamma} |x-x_0|^{\sigma r}\, dx,$$

where $V_+(x) = \max(V(x), 0),\ x_0 \in \mathbf{R}^n$.

Proof. As in the proof of Theorem 32 we show at first that

$$N_\alpha \le C_{m,n,r} I, \text{ where } I = \alpha^{-q+n/2m+r/2m} \int_\Omega V_\alpha(x)^q |x-x_0|^{\sigma r}\, dx. \tag{41}$$

After this the proof of Theorem 32 can be repeated.

To prove the inequality (41) we follow the same schema as above. However, we choose the cubes Q_j in such a way that

$$(2A)^r \int_{Q_j} V_\alpha(x)^q |x-x_0|^{\sigma s(q-1)}\, dx \le \varepsilon \alpha^{q-n/2m-s(q-1)/2m} I.$$

Let x_j be the center of a cube Q_j and b_j its size. We need no conditions (38) for cubes of size $b_j > 2A|x_j - x_0|^\sigma$, since for these cubes the estimate (34) is true, because $u(x) = 0$ on a subset of the cube of measure more than $c_0 b_j^n, c_0 > 0$.

Let Q_j be one of the other cubes. Then

$$\int_{Q_j} V_\alpha(x)|u(x)|^2\, dx \le (\int_{Q_j} \frac{u(x)^{2p}}{|x-x_j|^s}\, dx)^{1/p} (\int_{Q_j} V_\alpha(x)^q |x-x_j|^{s(q-1)}\, dx)^{1/q}.$$

If $n \le 2m$, then we assume that

$$u(x_j) = 0,\ D^\beta u(x_j) = 0 \ \text{ for } \ |\beta| < m - n/2.$$

If $s < n + p(2m - n)$, then the estimate

$$\left(\int_{Q_j} \frac{u(x)^{2p}}{|x - x_j|^s}\, dx\right)^{1/p} \le c_0 b_j^{2m-n/q-s/p} \int_{Q_j} \sum_{|\beta|=m} |D^\beta u(x)|^2\, dx$$

is true under the conditions (32) and since

$$\int_{Q_j} V_\alpha(x)^q |x - x_j|^r\, dx \le (2A)^r \int_{Q_j} V_\alpha(x)^q |x - x_0|^{\sigma r}\, dx \le \varepsilon \alpha^{q-n/2m-r/2m} I,$$

we can repeat the proof of Theorem 32. □

Remark 35. If the point x_0 is outside of the closure of the domain Ω, then the estimates in Theorems 32 and 34 are valid for $\gamma + (n + r)/2m > 1$.

Theorem 36. *Let $n = 1$, $V(x) \ge 0$, $V \in L_1(\mathbf{R})$, and*

$$S_{1/2} = \sum_{\lambda_j < 0} |\lambda_j|^{1/2},$$

where λ_j are negative eigenvalues of the Schrödinger operator

$$H = -d^2/\,dx^2 - V(x).$$

Then

$$S_{1/2} \le C \int_R V(x)\, dx.$$

Proof. As usual we can assume that the support of V is bounded.

Consider the Neumann problem on the interval $(0, l)$:

$$-y'' - V(x)y = \mu y, \quad y'(0) = y'(l) = 0.$$

If $\int_0^1 V(x)\, dx > 0$, then $(Hy, y) < 0$ for $y = 1$ and there exists at least one negative eigenvalue μ_1. On the other hand, by Lemma 37 below, for all functions y such that $y \in C^1(0, l)$, and $\int_0^l y(x)\, dx = 0$ we have

$$\max_{0 \le x \le l} y(x)^2 \le \frac{l}{3} \int_0^l y'(x)^2\, dx.$$

Therefore

$$\int_0^l V(x) y(x)^2\, dx \le \max_{0 \le x \le l} y(x)^2 \int_0^l V(x)\, dx,$$

so that $(Hy, y) \ge 0$, if $l \int_0^l V(x)\, dx \le 3$ and $\int_0^l y(x)\, dx = 0$.

Therefore, the Neumann problem has exactly one negative eigenvalue μ_1 if $0 < l \int_0^l V(x)\, dx \le 3$. Let

$$m = \inf\{\beta : a \max_{0 \le x \le l} y(x)^2 \le \int_0^l (y'(x)^2 + \beta y^2(x))\, dx\}.$$

The value $m(a) = -\rho^2$ is indicated in Lemma 38 below. Let

$$a = \int_0^l V(x)\,dx = 3/l$$

and ρ_0 be the minimal positive root of the equation $\rho_0 \cdot th\rho_0 = 3$. Then by Lemma 38 we have $l\rho \cdot th(\rho l) = al$ and

$$|\mu_1|^{1/2} \le \frac{\rho_0}{l} = \frac{\rho_0}{3}\int_0^l V(x)\,dx.$$

Let supp $V = \cup I_k$, where the intervals I_k are nonintersecting, and such that

$$|I_k| \int_{I_k} V(x)\,dx = 3.$$

Since $|I_k| \ge 3/\int_R V(x)\,dx$, these intervals can be found recurrently starting from an arbitrary point of $\mathbf{R}$. Let μ_k be the negative eigenvalue of the Neumann problem on I_k and the numeration be such that $|\mu_1| \ge |\mu_2| \ge \dots$. Let $\lambda_1 \le \lambda_2 \le \dots$ be the negative eigenvalues of H. According to the variational principle

$$|\lambda_1| = \{\inf \alpha : \int_R V(x)y(x)^2\,dx \le \int_R (y'(x)^2 + \alpha y^2(x))\,dx\} \le |\mu_1|.$$

Moreover,

$$\begin{aligned}|\lambda_2| \le &\{\inf \alpha : \int_R V(x)y(x)^2\,dx \le \int_R (y'(x)^2 + \alpha y^2(x))\,dx,\\ &y \in H^1(\mathbf{R}), \int_{I_1} y(x)\,dx = 0\} \le |\mu_2|\end{aligned}$$

and in general,

$$\begin{aligned}|\lambda_k| \le &\{\inf \alpha : \int_R V(x)y(x)^2\,dx \le \int_R (y'(x)^2 + \alpha y^2(x))\,dx,\\ &y \in H^1(\mathbf{R}), \int_{I_j} y(x)\,dx = 0, \quad j = 1, \dots, k-1\} \le |\mu_k|.\end{aligned}$$

Thus we obtain that

$$S_{1/2} \le \sum |\mu_k|^{1/2} \le \frac{\rho_0}{3}\sum \int_{I_k} V(x)\,dx \le \frac{\rho_0}{3}\int_R V(x)\,dx. \qquad \square$$

Lemma 37. *Let*

$$F[y] = \int_0^l y'(x)^2\,dx / \max_{0 \le x \le l} y(x)^2,$$

K be the set of functions from $H^1(0,l)$ such that $\int_0^l y(x)\,dx = 0$, and $m = \inf_K F[y]$. Then $m = 3/l$.

Proof. Let $F[y_k] \to m$, $\int_0^l y_k'(x)^2\,dx = 1$, and $\int_0^l y_k(x)\,dx = 0$. There exists a subsequence $\{y_{n_k}\}$ such that $y_{n_k} \to y_0(x)$ uniformly and $\int_0^l y_0(x)\,dx = 0$, $\max y_0(x)^2 = 1/m$ and $\int_0^l y_0'(x)^2\,dx \le 1$. Since $m = \inf_K F[y]$ we have $\int_0^l y_0'(x)^2\,dx = 1$. Without loss of generality we can assume that the functions y_{n_k} change sign in $[0,l]$ only once and are increasing, so that y_0 also has these properties and the maximal value is attained at $x = l$. The Euler-Lagrange equation has the form $y'' = C$ so that $y(x) = Cx^2/2 + ax + b$. Norming, one can put $C = 2$ or $C = 0$.

Let at first $C = 2$. Since $\int_0^l y_0(x)\,dx = 0$, we have $y_0(x) = x^2 - l^2/3 + a(x - l/2)$. Then

$$\max y_0(x)^2 = y(l)^2 = (2l^2/3 + al/2)^2$$

and $\int_0^l y_0'(x)^2\,dx = 4l^3/3 + 2al^2 + a^2 l$. Therefore

$$F[y_0] = \frac{1}{l}\frac{a^2 + 2al + 4l^2/3}{a^2/4 + 2al/3 + 4l^2/9} = \frac{1}{l}[3 + \frac{a^2/4}{(a/2 + 2l/3)^2}].$$

Therefore, $F[y_0] \ge 3/l$ and $F[y_0] = 3/l$, if $a = 0$.

If $C = 0$, then $y_0 = a(x - 1/2)$ and $F[y_0] = 4/l$, which is not minimal. □

Lemma 38. *Let $a > 0$,*

$$F[y] = \frac{\int_0^l y'(x)^2\,dx - a\max_{0\le x\le l} y(x)^2}{\int_0^l y(x)^2\,dx}$$

and $m = \inf_K F[y]$, where $K = H^1(0,l)$. Then $m = -\rho^2 < 0$ and ρ is the least positive root of the equation $\rho \cdot th(\rho l) = a$.

Proof. It is obvious that $m < 0$, since $F[1] < 0$. As in the proof of Lemma 37 it is easy to see that there exists a function $y_0 \in H^1(0,l)$ such that $F[y] = m$. Moreover, one can assume that $y(x) \ge 0$ and $y'(x) \ge 0$. The Euler-Lagrange equation has the form $y''(x) + my = 0$. Moreover, $y'(0) = 0$ and $y'(l) = ay(l)$. Therefore, $y = C \cdot ch(\rho x)$ and $\rho \cdot sh(\rho l) = a \cdot ch(\rho l)$. □

Bibliography

Adams, R. A.

[Ad] *Sobolev spaces*, Academic Press, New York, San Francisco, London, 1975.

Agmon, S.; Douglis, A.; and Nirenberg, L.

[ADN] *Estimates near the boundary for solutions of elliptic partial differential equations satisfying general boundary conditions.* I. Comm. Pure Appl. Math., **12** (1959), 623–727.

Agranovich, M. S.

[Ag1] *Elliptic singular integro-differential operators.* Uspehi Mat. Nauk **20**, 5 (1965), 3–120. (Russian Math. Surveys **20** (1965), 1–122).

[Ag2] *Boundary value problems for systems with a parameter.* Mat. Sb. **84** (1971), 27–65. (Math. USSR-Sb. **13** (1971), 25–64).

Agranovich, M. S., and Vishik, M. I.

[AgV] *Elliptic problems with a parameter and parabolic problems of general form.* Uspehi Mat. Nauk **19**, 3 (1964), 53–161. (Russian Math. Surveys **19**, 3 (1964), 53–157).

Bandle, C.

[Ba] *Isoperimetric inequalities and applications*, Pitman, Boston-London-Melbourne, 1980.

Berezin, F. A., and Shubin, M. A.

[BeSh] *Schrödinger operator.* Moscow University, Moscow, 1983.

Bers, L.; John, F.; and Schechter, M.

[BJS] *Partial Differential Equations.* American Mathematical Society, Providence, RI, 1964.

Birman, M. S.

[Bi] *On the spectrum of singular boundary-value problems.* Mat. Sb. **55(97)**, 2 (1961), 125–174 (Russian).

Birman, M. S., and Solomyak, M. Z.

[BiSo] *Application of interpolation to the estimates of the number of negative eigen-values of the Schrödinger operator.* Advances in Soviet Mathemat. **6** (1991) (AMS).

Blanchard, Ph.; Stubbe, Y.; and Rezende, Y.

[BlStRe] *New estimates on the number of bound states of Schrödinger operator.* Lett. Math. Phys. **14** (1987), 215–225.

Blaschke, W.

[Bl] *Kreis und Kugel.* Leipzig, 1916.

Bratus', A. S.

[Br] *Multiple eigenvalues in problems of optimizing. The spectral properties of systems with a finite number of degrees of freedom.* USSR Compt. Math. Math. Phys. **26** (1986), 1–7.

Bratus', A. S., and Seiranian, A. P.

[BrSe] *Bimodal solutions in eigenvalue optimization problems.* Prikl. Mat. Mech. USSR **47** (1983), 451–457.

Cox, S. J.

[Co] *The shape of the ideal column.* The Mathematical Intelligencer **14** (1992), 16–24.

Cox, S. J., and Overton, M. L.

[CoOv] *On the optimal design of columns against buckling.* SIAM J. Math. Anal. **23** (1992), 287–325.

Dyn'kin, E. M.

[Dy] EMS, vol. 33, 1990. (Springer Verlag, Berlin-Heidelberg 1991.)

Edmunds, D. E., and Evans, W. D.

[EdEv] *Spectral Theory and Differential Operators.* Clarendon Press, Oxford, 1987.

Egorov, Yu. V.

[Eg1] *Linear Differential Equations of Principal Type.* Nauka, Moscow, 1984 (Plenum Press, New York, 1985).

[Eg2] *Lectures on partial differential equations: Supplementary chapters.* Moscow University Press, 1985.

[Eg3] *Micro-local analysis.* EMS, vol. 33, 1990 (Springer Verlag, Berlin and Heidelberg, 1991).

Egorov, Yu. V., and Kondratiev, V. A.

[EgKo1] *On the oblique derivative problem.* Mat. Sb. **78** (1969), 148–176. (Math. USSR-Sb. **7** (1969), 139–169).

[EgKo2] *On an estimate of the first eigen-value for a self-adjoint elliptic operator,* Vestnik Mosk. un-ta, Mathem., Mechanics, **3** (1983), 46–52.

[EgKo3] *On estimates of the first eigen-value of the Sturm-Liouville problem.* Russian Math. Surveys **39** (2) (1984), 151–152.

[EgKo4] *On the estimate for the number of points of the negative spectrum of the Schrödinger operator.* Mat. Sb. **140** (1987), 556–570.

[EgKo5] *On an estimate for the first eigen-value of the Sturm-Liouville operator*, Vestnik Mosk. un-ta, Mathem., Mechanics **6** (1990), 75–78.

[EgKo6] *On an estimate for the principal eigen-value of the Sturm-Liouville operator.* Vestnik Mosk. un-ta, Mathem., Mechanics **6** (1991), 5–11.

[EgKo7] *On the negative spectrum of an elliptic operator.* Mat. Sb. **181**, 2 (1990), 147–166. (Math. USSR-Sb. **69**, 1 (1991), 155–177.)

[EgKo8] *Estimates of the negative spectrum of an elliptic operator.* Advances in Soviet Mathemat. **6** (1991) (AMS).

[EgKo9] *On the traces of negative eigen-values of an elliptic operator.* Proceedings of the Conference in Holzhau, 1994.

Egorov, Yu. V., and Schulze, B.-W.

[EgSch] *Pseudo-differential Operators, Singularities, Applications.* Birkhäuser, Basel, 1996.

Egorov, Yu. V., and Shubin, M. A.

[EgShu] *Partial Differential Equations I, II.* Encyclopaedia of Mathematical Sciences, **30** (1991), **31**, (1992). Springer Verlag, Berlin and Heidelberg.

Eidus, D. M.

[Ei] *Some inequalities for eigenfunctions*, Doklady AN SSSR **103** (1959), 1, 19–22.

Eskin, G. I.

[Es] *Boundary value problems for elliptic pseudodifferential equations.* Nauka, Moscow, 1973. (Engl. transl. Transl. Math. Monogr. **52**, Providence, RI, 1981).

Fedosov, B. V.

[Fed] *Analytic formulas for the index of elliptic operators.* Trudy Mosk. Mat. Ob-va **30** (1974), 159–241. (Trans. Moscow Math. Soc. **30** (1974), 159–240).

Fefferman, C.

[Fef] *The uncertainty principle.* Bull. Amer. Math. Soc. **9** (1983), 129–206.

Hadwiger, H.

[Ha] *Vorlesungen über Inhalt, Oberfläche und Isoperimetrie.* Springer Verlag, Berlin, Heidelberg, and New York, 1957.

Hayman, W. K.

[Hay] *Some bounds for principal frequency*, Applicable Analysis **7** (1978), 247–254.

Hörmander, L.

[Hö1] *Linear Partial Differential Operators.* Springer Verlag, Berlin, Heidelberg, and New York, 1963.

[Hö2] *The Analysis of Linear Partial Differential Operators* I, II, III, IV. Springer Verlag, Berlin, Heidelberg, and New York, 1983–1985.

Il'in, V. P.

[Il] *On the embedding theorem with a limit exponent.* Soviet Math. Doklady **96** (1954), 905–908.

Ivrii, V. Ya.

[Iv] *On asymptotics of the discrete spectrum for some operators in* $\mathbf{R}^d$. Functional Anal. Appl. **19**, 1 (1985), 73–74.

Kantorovich, L. V., and Akilov G. P.

[KaAk] *Functional Analysis.* Nauka, Moscow, 1967.

Karaa, S.

[Ka] *Valeurs propres extremales dans des problèmes de Sturm-Liouville*, Comptes Rendus Acad. Sci. Paris, Série 1, **321** (1995), 265–270.

Keller, J. B.

[Ke] *The shape of the strongest column.* Arch. Rational Mech. Anal. **5** (1960), 275–285.

Keller, J. B., and Tadjbakhsh, I.

[KeTa] *Strongest columns and isoperimetric inequalities for eigenvalues.* Journal of Applied Mech. **29** (1962), 159–164.

Kerman, R., and Sawyer, G.

[KerSa] *Weighted norm inequalities for potentials with applications to Schrödinger operators.* Bull. Amer. Math. Soc. **12**, no. 1 (1985), 112–116.

Kolmogorov, A. N., and Fomin, S. V.

[KolFo] *Functional Analysis.* Nauka, Moscow, 1989.

Kondratiev, V. A.

[Ko1] *Boundary value problems for elliptic equations in domains with conical or singular points.* Trudy Mosk. Mat. Ob-va. **16** (1967), 209–292. (Trans. Moscow Math. Soc. **16** (1967), 277–313.)

Kondratiev, V. A.

[Ko2] *On solubility of the first boundary value problem for strongly elliptic equations*, Trudy of Moscow Math. Soc. **16** (1967), 293–318.

Lagrange J. L.

[La] Sur la figure des colonnes. *Oeuvres de Lagrange*, Gauthier-Villars, Paris, 1867.

Li, P., and Yau, S.-T.

[LiYau] *On the Schrödinger equation and the eigenvalue problem.* Comm. Math. Phys. **88** (1983), 309–318.

Lieb, E.

[L1] *Bounds on the eigenvalues of the Laplace and Schroedinger operators.* Bull. Amer. Math. Soc. **82** (1976), 751–753.

[L2] *The number of bound states of one-body Schroedinger operators and the Weyl problem.* Proc. Symp. Pure Math. **36** (1980), 241–252.

[L3] *Sharp constants in the Hardy-Littlewood-Sobolev and related inequalities.* Ann. Math. **118** (1983), 349–374.

[L4] *On characteristic exponents in turbulence.* Comm. Math. Phys. **92** (1984), 473–480.

Lieb, E., and Thirring, W.

[LTh] *Inequalities for the moments of the Schroedinger Hamiltonians and their relation to Sobolev inequalities.* Studies in Math. Physics: Essays in honor of V. Bargmann. Princeton Univ. Press, Princeton, NJ, 1976.

Lions, J. L., and Magenes, E.

[LioMag] *Problèmes aux limites non-homogènes et applications.* 1. Dunod, Paris, 1968.

Lopatinsky, Ya. B.

[Lop] *On a method of reducing of boundary problem for a system of differential equations of elliptic type to regular integral equations.* Ukrain. Math. Zh. **5** (1953), 2, 123–151.

Mazya, V. G.

[Maz1] *Sobolev Spaces*, Nauka, Moscow, 1985.

[Maz2] *On the negative spectrum of the Schrödinger operator.* Izvestiya Akademii Nauk SSSR **144**, 4 (1964), 721–722.

[Maz3] *To the theory of the many-dimensional Schrödinger operator.* Doklady Akademii Nauk SSSR **28**, no.1, (1964), 1145–1148.

Mihlin, S. G.

[Mi] *Partial Differential Equations.* Nauka, Moscow, 1985.

Miranda, C.

[Mir] *Equazioni ellíptichi*, Springer, Berlin, 1965.

Morrey, C.

[Mo] *Multiple Integrals in the Calculus of Variations.* Springer Verlag, Berlin and Heidelberg, 1966.

Nečas, J.

[Ne] *Les Méthodes Directes en Théorie des Équations Elliptiques.* Masson, Paris, 1967.

Olhoff, N., and Rasmussen, S.

[OlRa] *On single and bimodal optimum buckling loads of clamped columns.* Ing. Journal Solids Struct. **13** (1977), 605–614.

Osserman, R.

[Os] *Isoperimetric inequalities.* Bull. Amer. Math. Soc. **84** (1978), 1182–1238.

Peetre, J.

[Pee] *Another approach to elliptic boundary problems.* Comm. Pure Appl. Math., **14**, no. 4 (1961), 711–731.

Pòlya, G., and Szegö, G.

[PoSz] *Isoperimetric Inequalities in Mathematical Physics.* Princeton University Press, Princeton, NJ, 1951.

Reed, M., and Simon, B.

[ReSi] *Analysis of Operators,* vol. IV. Academic Press, New York, 1978.

Rempel, S., and Schulze, B.-W.

[ReSch] *Index Theory of Elliptic Boundary Problems.* Akademie-Verlag, Berlin, 1982 (North Oxford Acad. Publ. Comp., Oxford, 1985).

Rozenblyum, G. V.

[Ro1] *Distribution of the discrete spectrum of singular differential operators.* Doklady Akademii Nauk USSR **202**, 5 (1972), 1012–1015.

[Ro2] *Distribution of the discrete spectrum of singular differential operators.* Izvestiya Vysshih Uchebnyh Zavedenij, Mathematics **1** (1976), 75–86.

Ruelle, V.

[Ru] *Large volume limit of the distribution of characteristic exponents in turbulence.* Comm. Math. Phys. **87**, 5 (1982), 287–302.

Schechter, M.

[Sche1] *Various types of boundary conditions for elliptic equations.* Comm. Pure Appl. Math. **13** (1960), 407–425.

[Sche2] *The spectrum of the Schrödinger operator.* Trans. Amer. Math. Soc. **312**, no. 1 (1991), 115–128.

Schulze, B.-W.

[Sch] *Pseudo-differential operators on manifolds with singularities.* North Holland, Amsterdam, 1991.

Seeley, R. T.

[See] *Complex powers of an elliptic operator.* Proc. Symp. in Pure Math. **10** (1967), 288–307.

Seiranian, A. P.

[Se] *On a problem of Lagrange.* Inzh. Zh., Mehanika Tverdogo Tela **19** (1984), 101–111.

Shapiro, Z. Ya.

[Sha] *On general boundary value problems for elliptic equations.* Izv. AN SSSR **17** (1953), 539–562.

Shishmarev, I. A.

[Shi] *Introduction to the Theory of Elliptic Equations.* Moscow State Univ., 1979.

Shubin, M. A.

[Sh] *Pseudo-differential operators and spectral theory.* Nauka, Moscow 1978. (Springer Verlag, Berlin and Heidelberg, 1987).

Slobodezky, L. N.

[Sl] *Generalized Sobolev spaces and their applications*, Ann. Scuola Norm. Sup. Pisa (4) **3** (1976), 697–718.

Sobolev, S. L.

[So] *Applications of Functional Analysis in Mathematical Physics*, Leningrad University, 1950 [English transl.: Amer. Math. Soc., Transl., Math. Mono. **7** (1963).]

Talenti, G.

[Ta] *Elliptic equations and rearrangements.* Ann. Scuola Norm. Sup. Pisa (4) **3** (1976), 697–718.

Taylor, M.

[Tay] *On the principal eigenvalue in a multi-connected domain*, Duke Math. J. **46**, 2 (1979), 209–292.

Vainberg, B. R.

[Va] *Asymptotic Methods in Mathematical Physics.* Moscow State Univ., 1982.

Vishik, M. I.

[Vi] *On general boundary problems for elliptic differential equations.* Trudy Mosk. Mat. Ob-va **1** (1952), 187–246. (Amer. Math. Soc. Transl. **24** (1963), 107–172).

Weidl, T.

[W] *On the Lieb-Thirring constants.* Doctoral thesis, Royal Institute of Technology, Stockholm (1995), 1–15.

Weyl, H.

[We] *Das asymptotische Verteilungsgesetz der Eigenwerte linearer partieller Differentialgleichungen.* Math. Ann. **71** (1912), 441–479.

Index

averaging, 45
averaging kernel, 45

boundary, 25
bounded functional, 10

Cauchy sequence, 2
characteristic function, 26
characteristic polynomial, 115
closed system, 6
closure, 25
compact embedding, 25
complement, 25
concave function, 181
continuation of a function, 25
continuous functional, 9
continuous spectrum, 17, 141
convergence in measure, 41
convex function, 181
convolution, 50
Courant principle, 137
critical set, 29

direct sum, 8
Dirichlet problem, 109
distance between two sets, 26
domain in $\mathbf{R}^n$, 25
domain of a linear operator, 15

eigenfunction, 153, 233
eigenvalue, 17, 153, 233
  of a pencil, 246
elliptic operator, 115
ellipticity constant, 116
essential spectrum, 141

Fourier coefficients, 6
Fourier transform, 47

Fredholm alternative, 110
Friedrichs extension, 143

Gårding inequality, 110
generalized derivative, 51
generalized solution, 110

Hölder condition, 25
Hausdorff measure, 26

image of a linear operator, 15
inequality
  Bessel, 6
  Cauchy-Bounyakovsky, 1
  Gårding, 110
  Hölder, 38
  Hardy, 89
  of triangle, 1
  Poincaré, 61
inner product, 1
inner radius, 236
isoperimetric inequality, 27

kernel of a linear operator, 15

Lagrange problem, 217
level set, 29
linear
  functional, 9
  functional, in a Banach space, 10
  operator, 15
Lipschitz condition, 25
Lopatinsky condition, 124

measurable function, 26
multi-index, 26

nondegenerate point, 248
norm, 2, 10
  of an operator, 15

operator
- adjoint, 16, 140
- bounded, 15
- closed, 143
- compact, 18
- continuous, 15
- elliptic, 115
- essentially self-adjoint, 140
- inverse, 16
- invertible, 16
- linear, 15
- maximal, 145
- minimal, 145
- positive, 143
- Schrödinger, 149
- self-adjoint, 17, 139
- semi-bounded from below, 143
- strictly positive, 143
- strongly elliptic, 109
- symmetric, 17, 140

orthogonality, 5
orthonormal base, 5

Parseval equality, 49
point of $\mathbf{R}^n$, 25
point spectrum, 17, 141
potential, 276

regular point, 17
resolvent, 17
resolvent set, 141
restriction of a function, 25

simple function, 27
space
- adjoint, 10
- Banach, 2
- complete, 2
- finite-dimensional, 3
- Hilbert, 3
- linear, 1
- normed linear, 2
- pre-Hilbert, 1
- reflexive, 10
- Schwartz, 47
- separable, 3
- Sobolev, 52

spectrum, 17, 141
stationary phase method, 248
stationary point, 248
Sturm-Liouville problem, 153
subspace, 8
summable function, 27
support of a function, 25
surface, 26
symmetrization, 32

theorem
- Banach-Saks, 13
- Clarkson, 39
- Courant, 137, 162
- Fatou, 28
- Fefferman, 277
- Friedrichs, 54
- Fubini, 28
- Gronwall, 167
- Hilbert-Schmidt, 20
- Lax-Milgram, 23
- Lebesgue, 28
- Levi, 28
- Liouville, 156
- Minkovsky, 39
- Neumann, 143
- Riesz, 10
- Riesz-Fischer, 6
- Sarde, 29
- Sturm, 163
- Valle-Poussin, 208

traces of functions, 83

variational principle, 133
Vitali covering, 29

weak convergence, 11, 14

Titles previously published in the series

OPERATOR THEORY: ADVANCES AND APPLICATIONS

BIRKHÄUSER VERLAG

Edited by **I. Gohberg,**

School of Mathematical Sciences, Tel-Aviv University, Ramat Aviv, Israel

78. **M. Demuth, B.-W. Schulze** (Eds): Partial Differential Operators and Mathematical Physics: International Conference in Holzhau (Germany), July 3-9, 1994, 1995 (ISBN 3-7643-5208-6)

79. **I. Gohberg, M.A. Kaashoek, F. van Schagen**: Partially Specified Matrices and Operators: Classification, Completion, Applications, 1995 (ISBN 3-7643-5259-0)

80. **I. Gohberg, H. Langer** (Eds): Operator Theory and Boundary Eigenvalue Problems International Workshop in Vienna, July 27-30, 1993, 1995 (ISBN 3-7643-5259-0)

81. **H. Upmeier**: Toeplitz Operators and Index Theory in Several Complex Variables, 1995 (ISBN 3-7643-5280-5)

82. **T. Constantinescu**: Schur Parameters, Factorization and Dilation Problems, 1996 (ISBN 3-7643-5285-X)

83. **A.B. Antonevich**: Linear Functional Equations. Operator Approach, 1995 (ISBN 3-7643-2931-9)

84. **L.A. Sakhnovich**: Integral Equations with Difference Kernels on Finite Intervals, 1996 (ISBN 3-7643-5267-1)

85. **Y.M. Berezansky, G.F. Us, Z.G. Sheftel**: Functional Analysis, **Vol. I**, 1996 (ISBN 3-7643-5344-9)

86. **Y.M. Berezansky, G.F. Us, Z.G. Sheftel**: Functional Analysis, **Vol. II**, 1996 (ISBN 4-7643-5345-7)

87. **I. Gohberg / P. Lancaster / P.N. Shivakumar** (Eds): Recent Developments in Operator Theory and Its Applications, International Conference in Winnipeg, October 2–6, 1994, 1996 (ISBN 3-7643-5413-5)

88. **J. van Neerven** (Ed.): The Asymptotic Behaviour of Semigroups of Linear Operators 1996 (ISBN 3-7643-5455-0)

PA - Probability and its Applications

Edited by
Th. M. Liggett / Ch. Newman / L. Pitt

Probability and its Applications *publishes research-level monographs and advanced graduate texts dealing with all aspects of probability theory and stochastic processes, as well as their connections with and applications to other areas such as mathematical statistics and statistical physics.*

R. Carmona/J. LaCroix
Spectral Theory of Random Schrödinger Operators
1990. ISBN 3-7643-3486-X

R.K. Getoor
Excessive Measures
1990. ISBN 3-7643-3492-4

K.L. Chung/R.J. Williams
Introduction to Stochastic Integration
1990. ISBN 3-7643-3386-3

G.F. Lawler
Intersections of Random Walks
1991. ISBN 3-7643-3557-2

H. Linhart/W. Zucchini
Statistik Eins
1991. ISBN 3-7643-2586-0

R.M. Blumenthal
Excursions of Markov Processes
1992. ISBN 3-7643-3575-0

S. Kwapien/W. Woyczynski
Random Series and Stochastic Integrals: Single and Multiple
1992. ISBN 3-7643-3572-6

N. Madras/G. Slade
The Self-Avoiding Walk
1992. ISBN 3-7643-3589-0

R. Aebi
Schrödinger Diffusion Processes
1996. ISBN 3-7643-5386-4
H. Holden/B. Oksendal/J. Uboe/T. Zhang
Stochastic Partial Differential Equations: A Modeling, White Noise Functional Analysis Approach
1996. ISBN 3-7643-3928-4

G.F. Lawler
Intersections of Random Walks
1996. *Softcover.* ISBN 3-7643-3892-X

N. Madras/G. Slade
The Self-Avoiding Walk
1996. *Softcover.* ISBN 3-7643-3891-1

MIX
Papier aus verantwortungsvollen Quellen
Paper from responsible sources
FSC® C105338
FSC
www.fsc.org

If you have any concerns about our products,
you can contact us on
ProductSafety@springernature.com

In case Publisher is established outside the EU,
the EU authorized representative is:
Springer Nature Customer Service Center GmbH
Europaplatz 3, 69115 Heidelberg, Germany

Printed by Libri Plureos GmbH
in Hamburg, Germany